UNITEXT - La Matematica per il 3+2

Volume 106

More information about this series at http://www.springer.com/series/5418

Giampiero Esposito

From Ordinary to Partial Differential Equations

 Springer

Giampiero Esposito
INFN Sezione di Napoli, Complesso di
    Monte Sant'Angelo
Università degli Studi di Napoli Federico II
Naples, Napoli
Italy

ISSN 2038-5722   ISSN 2038-5757   (electronic)
UNITEXT - La Matematica per il 3+2
ISBN 978-3-319-57543-8   ISBN 978-3-319-57544-5   (eBook)
DOI 10.1007/978-3-319-57544-5

Library of Congress Control Number: 2017939919

Mathematical Subject Classification (2010): 11S40, 34-01, 35-01, 35A08, 35A17, 35K05, 35L10, 35L70, 35S05, 35S30

Printed on acid-free paper

This Springer imprint is published by Springer Nature
The registered company is Springer International Publishing AG
The registered company address is: Gewerbestrasse 11, 6330 Cham, Switzerland

a Margherita

Preface

Functional equations express many profound properties in mathematics, classical and quantum physics, and the other sciences, and their unknowns are functions. Thus, the ordinary and partial differential equations (both linear and nonlinear), the pseudo-differential equations, and also the functional equations not involving any derivative at all [e.g., $f(x + y) = f(x) + f(y)$] are all branches of the huge tree of functional equations. A key question is, therefore: How are we going to study the above topics in our monograph?

It has always been my dream to write a book that enables the reader to elevate himself/herself and reach a stage where he/she clearly understands how to achieve a proof or how to set up a conceptual or algorithmic framework. The historical path remains, in my opinion, of high pedagogic value because it shows the sequence of efforts which led to the modern formulations and illustrates what has been gained by the use of modern notations and techniques. If we ignore this path, we run the risk of forming students who think that, since a few brilliant minds have already taken the trouble to understand how the physical and mathematical world can be described, they only have to learn a number of recipes to enable them to apply the well-established general principles. Such an attitude would be disastrous for science because it would fail to provide the fuel necessary to give rise to new ideas and new scientific revolutions.

Now, we can recall that several areas of modern mathematics, with all their physical implications, can indeed be seen as branches of this theory: algebraic functions, Abelian integrals, analytic functions of several complex variables, contact of curves and surfaces, envelopes of curves and surfaces, skew curves, linear and nonlinear ordinary differential equations, calculus of variations, mathematical theory of surfaces and absolute differential calculus, minimal surfaces, and linear and nonlinear partial differential equations, just to mention some of the themes frequently faced by mathematicians and physicists.

It is clear from the above that, if we were to discuss all conceivable topics, we would end up by writing a multivolume work which would hardly be readable, even if feasible. Thus, I instead conceived the following plan. The book consists of eight parts, of unequal length. Part I is devoted to linear and nonlinear ordinary differential equations. Part II studies linear elliptic equations, leading the reader from harmonic to polyharmonic functions and showing the role of the Laplace-Beltrami operator in the theory of surfaces, developing the theory of Sobolev spaces, and presenting the original and the modern derivations of the Caccioppoli–Leray inequality. Spectral theory, mixed boundary-value problems, Morrey and Campanato spaces, and pseudo-holomorphic functions are further topics studied in some detail. Part III introduces the reader to a rigorous formulation of the Euler integral and differential conditions and to variational problems with constraints; thereafter, isoperimetric problems, minimal surfaces, and the Dirichlet boundary-value problem are studied. Part IV is devoted to linear and nonlinear hyperbolic equations. Several basic concepts are defined and discussed: characteristic manifolds for first- and second-order systems, wavelike propagation, hyperbolic equations, Riemann kernel for a hyperbolic equation in two variables, Cauchy's method for integrating a first-order equation, bicharacteristics, the relation between fundamental solution and Riemann's kernel, the characteristic conoid, the Hamiltonian form of geodesic equations, fundamental solution with odd and even number of variables, examples of fundamental solution, and parametrix of the scalar wave equation in curved space-time. Part IV ends with a detailed presentation of the Fourès-Bruhat use of integral equations for solution of linear and nonlinear systems of equations in normal hyperbolic form, with application to the Cauchy problem for general relativity with nonanalytic Cauchy data. Causal structure of space-time and global hyperbolicity are eventually introduced. Part V begins with the elementary theory of parabolic equations, but the detailed evaluation of the fundamental solution of the heat equation prepares the ground for the advanced use of the fundamental solution that Nash made in his proof of continuity of solutions of parabolic and elliptic equations. The moment and G bounds, the overlap estimate, and time continuity are studied following Nash's seminal paper. Part VI studies the celebrated Poincaré work on Fuchsian functions and its application to the solution of a famous nonlinear elliptic equation, which consists in studying the kernel of Laplacian plus exponential. Thus, the reader can see in action the profound link between the theory of functions and the investigation of nonlinear elliptic equations. Part VII studies the Riemann ζ-function, with emphasis on the functional equation for the ξ-function built out of the Γ- and ζ-functions, and on the integral representation for the ξ-function. Eventually, the Riemann hypothesis on nontrivial zeros of the ζ-function is discussed. Last, but not least, Part VIII opens a window on modern theory, defining the geometric setting for pseudo-differential operators on manifolds and discussing the issue of local solvability of partial differential and pseudo-differential equations.

Although I have relied heavily upon my sources in the various chapters, I believe that my navigation path is original, and I hope it will be helpful for young researchers and teachers. I have tried to make it clear to the general reader that an interdisciplinary view of mathematics is essential to make further progress: we need the abstract concepts of algebra, the majorizations and rigor of analysis, the global and local geometric descriptions, and the stunning properties of number theory. This holds true in both mathematical and physical sciences. I have covered a selection of topics and references, frequently writing the chapters as single lectures for a strongly motivated class of graduate students in order to enhance their pedagogic value. The book is aimed at developing the sensibility and skills of readers so that they can begin advanced research after reading it. What will be the basic structures for studying functional equations in the centuries to come? When Forsyth completed his monumental treatise on the theory of differential equations [58–63] at the beginning of the nineteenth century, he was lacking many fundamental tools, e.g., the theory of distributions, weak solutions, and the space-time view provided by special and general relativity. Which tools are we still lacking? Which ones will remain useful?

I leave the reader with these thoughts, and hope that the answers to the above questions will help us, mankind, in our efforts to understand the physical world and its beautiful mathematical language.

Naples, Italy   Giampiero Esposito
March 2017

Acknowledgements

I am grateful to the students Florio Ciaglia, Fabio Di Cosmo and Francesco Pappalardo, who attended my graduate lectures on elliptic and hyperbolic equations and worked very hard on the topics I covered. Special thanks are due to Yvonne Choquet-Bruhat, for giving me permission to use, in Chaps. 19–21, part of my English translation of her masterpiece on the Cauchy problem in general relativity, and to Luigi Ambrosio for granting permission to rely upon his lecture notes in my Sects. 7.2 and 10.1. The Naples students, and in particular Paolo D'Isanto and Lorenzo Iacobacci, provided further motivation for my efforts towards completion of the manuscript. I also thank Carlo Sbordone for his kindness and for useful bibliographic informations, and Pietro Santorelli who created for me the Figure used in Chap. 23. Last but not least, Francesca Bonadei and Angela Vanegas from Springer has assisted me, making sure that my dream would come true.

Contents

Introduction

Chapter 1 begins by proving the fundamental theorem on first-order ordinary differential equations in normal form, and by studying the differentiability of the solution. As a next step, it derives the canonical form of linear second-order ordinary differential equations with variable coefficients, which is very useful for qualitative investigations. Such equations are then extended to the complex domain, proving analyticity of the solution. The Sturm-Liouville problem is then studied with homogeneous boundary conditions in a closed interval, with emphasis on Green functions and the associated integral equations. Chapter 1 ends with a brief outline of the Heun equation.

Chapter 2, after some basic examples of nonlinear equations in the real domain, studies singular points in the complex domain for nonlinear equations.

Part II begins with a review of basic properties of harmonic functions, whose consideration is suggested by several branches of physics (Newtonian gravity, electrostatics) and mathematics (e.g. complex analysis and the theory of surfaces). When the theory of surfaces was developed, the Laplace-Beltrami operator was viewed as a differential parameter of order 2, and this is shown in detail.

The reader is then introduced to the theory of distributions and Sobolev spaces, while regularity theory is introduced through the Caccioppoli derivation of integral bounds on solutions of linear elliptic equations. The concept of ellipticity is then defined in various cases of interest. After an outline of spectral theory, the De Giorgi example of Laplace equation with mixed boundary conditions is studied. As a next step, Morrey and Campanato spaces, and functions of bounded mean oscillation are studied. Part II ends by focusing on pseudo-analytic functions, with the associated generalized form of Cauchy–Riemann systems, and some remarkable properties of biharmonic and polyharmonic functions.

In Part III, a presentation is given of the basic mathematical concepts in the calculus of variations, with emphasis, in Chap. 12, on methods and rigor: statement of the problem, the Euler integral condition, the Euler differential condition, and variational problems with constraints.

Chapter 13 is devoted to some classical topics of fundamental importance, i.e., isoperimetric problems, minimal surfaces (including a diversion on the Weierstrass equations for minimal surfaces), and the variational formulation of the Dirichlet boundary-value problem.

After a brief review of properties of systems of partial differential equations in Part IV, some key concepts and techniques are introduced in detail, i.e., characteristic manifold for first- and second-order systems, wavelike propagation, hyperbolic equation, Riemann kernel for a hyperbolic equation in two variables, Cauchy's method for integration of a first-order equation, the bicharacteristics. With hindsight, the characteristics and bicharacteristics are, in space-time geometry, the null hypersurfaces and the null geodesics, respectively, but since the mathematical theory of hyperbolic equations was initiated well before the advent of special and general relativity, the nomenclature of nineteenth's century has not been completely replaced, so that the general reader should be prepared to use both nomenclatures, depending on his particular aims.

Yet other key concepts are the definition of characteristic conoid, fundamental solution and world function. We follow Hadamard in our effort of describing how to build the fundamental solution and providing examples of its evaluation. We also evaluate the parametrix for the scalar wave equation in a curved space-time model. Part IV ends with the detailed analysis by Fourès-Bruhat of linear and non-linear systems of hyperbolic equations, with application to Einstein's general relativity when the Cauchy data are not analytic.

Part V begins with a pedagogical review of linear equations in two variables, and a detailed evaluation of fundamental solution for the heat equation in one space dimension. This prepares the ground for the Nash proof of continuity of parabolic equations in the form

$$\mathrm{div}\,(C\,\mathrm{grad}T) = \frac{\partial T}{\partial t},$$

which relies also on the basic properties of fundamental solutions. After proving the so-called moment bound with a brilliant use of Fourier analysis and majorizations, Nash obtained the Hölder continuity of parabolic equations in the above form, when C fulfills the uniform ellipticity condition.

Chapter 25 is devoted to the important concepts of fundamental circle, hyperbolic group, discontinuous group, Fuchsian group, Fuchsian function, Θ-Fuchsian function, system of ζ-Fuchsian functions. There exist therefore a wide family of functions, of which the elliptic functions are only a particular case, which make it possible to integrate a large number of differential equations. Chapter 26 describes in detail how the theory of Fuchsian (or automorphic) functions was applied by Poincaré to solve a non-linear partial differential equation, i.e., the task of finding the kernel of Laplacian plus exponential. The importance of an interdisciplinary view of mathematics is therefore stressed once again, since group theory, geometry and analysis are needed to study these issues.

In Part VII, first, a detailed proof is given of the Euler theorem on the equality of an infinite product built from all prime numbers and a complex variable, with the ζ-function. We then follow Riemann in his derivation of the integral formula that relates the analytic continuation of Γ- and ζ-function with the Jacobi function. The next step is the introduction of the ξ-function with the associated functional equation. The logarithm of the ζ-function is then studied, and its link with the counting function of prime numbers. The chapter ends by presenting the Riemann conjecture on non-trivial zeros of the analytically extended ζ-function, and some remarks on more research work to be done are made.

In Part VIII, the transition from differential to pseudo-differential operators is first described, starting from the action of the Laplacian on functions admitting Fourier transform. When the pseudo-differential operators act on smooth functions (or on sections of vector bundles) over compact Riemannian manifolds, a more advanced geometric setting is necessary, which is here defined. Both the symbol and the principal (or leading) symbol turn out to be equivalence classes, here defined as well. The chapter ends with a discussion of local solvability of partial differential (Lewy) and principal-type pseudo-differential (Dencker) operators.

Part I
Ordinary Differential Equations

Chapter 1
Linear Differential Equations

1.1 Fundamental Theorem for First-Order Equations

Let us consider the first-order equation of normal form (i.e. solved with respect to the derivative)

$$y' = f(x, y), \tag{1.1.1}$$

where f is a function of class C^1 with respect to y in a field Γ. Let the point $(\bar{x}, \bar{y})$ and its rectangular neighbourhood

$$|x - \bar{x}| \leq a, \quad |y - \bar{y}| \leq b$$

belong to the interior of Γ. An integral of Eq. (1.1.1) which takes, for $x = \bar{x}$, the value $\bar{y}$, verifies the integral equation

$$y(x) = \bar{y} + \int_{\bar{x}}^{x} f(x, y(x))\mathrm{d}x. \tag{1.1.2}$$

In order to prove the existence of $y(x)$ we here use the method of successive approximations.

Thus, we begin by replacing $y(x)$ with the constant function $y_1(x) = \bar{y}$ on the right-hand side of Eq. (1.1.2). Hence we obtain a function $y_2(x)$. We then replace $y(x)$ with $y_2(x)$ and obtain a function $y_3(x)$ and so on, and we continue indefinitely, setting in general

$$y_m(x) = \bar{y} + \int_{\bar{x}}^{x} f(x, y_{m-1}(x))\mathrm{d}x. \tag{1.1.3}$$

We therefore build a sequence of functions $y_m(x)$ for which we now prove convergence to the desired function $y(x)$, for which the various $y_l(x)$ play precisely the role of successive approximations.

© Springer International Publishing AG 2017
G. Esposito, *From Ordinary to Partial Differential Equations*, UNITEXT 106,
DOI 10.1007/978-3-319-57544-5_1

For Eq. (1.1.3) (which holds for $m = 2$) to be valid for any value of m, it is sufficient that the point (x, y_m) never leaves the rectangular neighbourhood I defined above. This will be the case for $|x - \bar{x}| \leq a$ if $Ma < b$, where M denotes the maximum of $f(x, y)$ within I. Indeed, under this assumption, one finds from (1.1.3) that

$$|y_m(x) - \bar{y}| \leq M \ |x - \bar{x}| \leq Ma < b. \qquad (1.1.4)$$

We thus assume, if needed, that a has been reduced to a value less than $\frac{b}{M}$.

Now if K is the maximum of $\left|\frac{\partial f}{\partial y}\right|$ in I, the Lagrange theorem implies that, on denoting by η_1 and η_2 two values of y in the interval $(\bar{y} - b, \bar{y} + b)$, one has

$$|f(x, \eta_1) - f(x, \eta_2)| \leq K \ |\eta_1 - \eta_2| . \qquad (1.1.5)$$

From Eq. (1.1.4) and the inequality (1.1.5) one obtains, for $m = 2, 3, \ldots,$

$$|y_{m+1}(x) - y_m(x)| \leq \int_{\bar{x}}^{x} |f(x, y_m(x)) - f(x, y_{m-1}(x))| \, \mathrm{d}x$$
$$\leq K \int_{\bar{x}}^{x} |y(x) - y_{m-1}(x)| \, \mathrm{d}x, \qquad (1.1.6)$$

and hence, in light of the inequality

$$|y_2(x) - y_1(x)| = |y_2(x) - \bar{y}| \leq M \ |\bar{x} - x| , \qquad (1.1.7)$$

the successive inequalities are obtained

$$|y_3(x) - y_2(x)| \leq K \int_{\bar{x}}^{x} |y_2(x) - y_1(x)| \, \mathrm{d}x \leq MK \frac{|x - \bar{x}|^2}{2}, \qquad (1.1.8)$$

$$|y_4(x) - y_3(x)| \leq K \int_{\bar{x}}^{x} |y_3(x) - y_2(x)| \, \mathrm{d}x \leq MK^2 \frac{|x - \bar{x}|^3}{3!}, \qquad (1.1.9)$$

and therefore, in general,

$$|y_{m+1}(x) - y_m(x)| \leq MK^{m-1} \frac{|x - \bar{x}|^m}{m!}, \quad m = 1, 2, 3, \ldots. \qquad (1.1.10)$$

Thus, the series

$$y_1(x) + \left(y_2(x) - y_1(x)\right) + \left(y_3(x) - y_2(x)\right) + \cdots, \qquad (1.1.11)$$

whose partial sums are $y_1, y_2, y_3, \ldots$, is majorized by

$$S(x) \equiv |\bar{y}| + M \sum_{m=1}^{\infty} K^{m-1} \frac{|x - \bar{x}|}{m!}, \tag{1.1.12}$$

which is uniformly convergent. On defining

$$y(x) \equiv \lim_{m \to \infty} y_m(x), \tag{1.1.13}$$

the uniform convergence makes it possible the passage to the limit for $m \to \infty$ in Eq. (1.1.3), because also the integrand on the right-hand side is uniformly convergent. But, as m approaches ∞, y_m and y_{m-1} are replaced by y, and Eq. (1.1.3) reduces to Eq. (1.1.2), hence the function $y : x \to y(x)$ is the desired integral.

We can now prove uniqueness of the solution. Indeed, if η were another function satisfying Eq. (1.1.2), one would find

$$y(x) - \eta(x) = \int_{\bar{x}}^{x} \Big[f(x, y(x)) - f(x, \eta(x)) \Big] dx, \tag{1.1.14}$$

and if we exploit the inequality (1.1.5) we find therefore

$$|y(x) - \eta(x)| \le K \int_{\bar{x}}^{x} |y(x) - \eta(x)| \, dx \le K a \mu, \tag{1.1.15}$$

where μ is the maximum of the left-hand side in the interval $(\bar{x} - a, \bar{x} + a)$. In particular, at a point ξ of maximum one has

$$|y(\xi) - \eta(\xi)| = \mu \le K a \mu, \tag{1.1.16}$$

which implies that $\mu = 0$ and hence $y = \eta$ as soon as $K a < 1$. Thus, on reducing the half-amplitude a of the interval where the variable x is considered below $\frac{1}{K}$, the desired uniqueness of the solution is proved.

It should be noticed that, so far, the assumed continuity of $\frac{\partial f}{\partial y}$ has been used only to prove the condition (1.1.5). This is a Lipschitz condition, and can be stated by saying that the function $f(x, y)$ must have, with respect to y, bounded increment rations (this, strictly, does not even imply the existence without exceptions of the partial derivative). We are now in a position to state the fundamental theorem as follows [15].

Theorem 1.1 *If the function $f(x, y)$, in the neighbourhood of the point $(\bar{x}, \bar{y})$, is continuous and verifies, with respect to y, a Lipschitz condition, there exists a unique integral of Eq. (1.1.1) defined about $\bar{x}$ and taking the value $\bar{y}$ for $x = \bar{x}$.*

This theorem implies that Eq. (1.1.1) admits infinitely many integrals, each of which is characterized by the value taken at $x = \bar{x}$, which is freely specifiable in a suitably restricted neighbourhood of $\bar{y}$. Indeed, the previous considerations can be repeated by replacing $\bar{y}$ with a sufficiently close number y_0, e.g. such that $y_0 - \bar{y} < \frac{b}{2}$, and then choosing $a < \frac{b}{2M}$, so that the rectangular neighbourhood I gets replaced by another rectangular neighbourhood contained within it. Moreover, also $\bar{x}$ can be replaced by an arbitrary value x_0, provided such a value is sufficiently close. Thus, if we assign arbitrarily a point (x_0, y_0) in a suitably restricted neighbourhood of $(\bar{x}, \bar{y})$, there exists a unique integral of Eq. (1.1.1) defined in a neighbourhood of $\bar{x}$ and verifying the condition $y(x_0) = y_0$. In geometric language, the neighbourhood of the point is covered by a simple infinity of integral curves, having equation $y = y(x)$, such that, through every given point, there passes one and only one curve of such a family.

Note also that, if an integral $y(x)$ is defined in an interval (α, β) and the points $(\alpha, y(\alpha))$ and $(\beta, y(\beta))$ lie within the field Γ where the function $f(x, y)$ is defined, the fundamental theorem can be applied to these points, so that the function $y(x)$ is extended beyond the interval (α, β), with a corresponding extension of the integral curve. The extension is possible until the point (x, y) reaches the frontier of the field Γ. Such a field is therefore furrowed by integral curves, and through every internal point there passes only one of them.

The integration of Eq. (1.1.1) admits a simple but profound geometric interpretation: an integral curve is a curve of the field Γ which has, at any of its points (x, y), a tangent by means of the angular coefficient $f(x, y)$, i.e. it is the trajectory of a point which, starting from an arbitrary initial position, moves in such a way that, in every new position it reaches, its motion continues along a direction here prefixed.

1.2   Differentiability of the Solution

Let us consider the integral of Eq. (1.1.1) determined by the initial condition $y(\bar{x}) = y_0$. This is a function $y(x, y_0)$ of the two variables x and y_0, defined in two neighbourhoods of $\bar{x}$ and $\bar{y}$, respectively. The successive approximations defined by the formulae

$$y_1(x, y_0) = y_0, \quad y(x, y_0) = y_0 + \int_{\bar{x}}^{x} f(x, y_0) \mathrm{d}x, \qquad (1.2.1)$$

give rise to continuous functions of x and y_0, and one has

$$y(x, y_0) = \lim_{m \to \infty} y_m(x, y_0) = y_1(x, y_0) + \left(y_2(x, y_0) - y_1(x, y_0)\right) + \cdots, \quad (1.2.2)$$

uniformly with respect to the two variables, as is shown by the use of the majorizing series (1.1.12). This proves in turn that $y(x, y_0)$ is a continuous function of both x and y_0. This result is extended immediately to the normal system

$$\frac{\mathrm{d}y_1}{\mathrm{d}x} = f_1(x, y_1, \ldots, y_n), \ldots$$

$$\frac{\mathrm{d}y_n}{\mathrm{d}x} = f_n(x, y_1, \ldots, y_n), \qquad (1.2.3)$$

whose integral consists of continuous functions of x and of the initial values $y_1^0, \ldots, y_n^0$.

Furthermore, one proves that, if the functions $f_1, \ldots, f_n$ depend continuously on $x, y_1, \ldots, y_n$ and a finite number of parameters $\lambda, \mu, \ldots$, the integral

$$y_1\left(x, y_1^0, \ldots, y_n^0, \lambda, \mu, \ldots\right), \ldots, y_n\left(x, y_1^0, \ldots, y_n^0, \lambda, \mu, \ldots\right) \tag{1.2.4}$$

consists of continuous functions of all their arguments. For this purpose, it is sufficient to regard $\lambda, \mu, \ldots$ as new unknown functions and supplement the system (1.2.3) with the equations

$$\frac{d\lambda}{dx} = 0, \ldots, \frac{d\mu}{dx} = 0, \tag{1.2.5}$$

because then every set of values $\lambda_0, \mu_0, \ldots$ of the parameters can be viewed as a set of initial values of the functions $\lambda, \mu, \ldots$, which are constant by virtue of the added equations (1.2.5).

Let us now study the differentiability of the integral $y(x, y_0)$ with respect to its arguments. The derivative with respect to x is indeed continuous, since it coincides with $f(x, y)$. As far as y_0 is concerned, let us set

$$\delta y \equiv y(x, y_0 + \delta y_0) - y(x, y_0), \quad \frac{\delta y}{\delta y_0} \equiv z(x, \delta y), \tag{1.2.6}$$

keeping y_0 fixed and letting δy_0 vary in a neighbourhood of 0. With the understanding that a prime denotes derivative with respect to x, we find

$$\frac{\delta y}{\delta y_0} = z'(x, y_0) = \frac{f(x, y(x, y_0 + \delta y_0)) - f(x, y(x, y_0))}{\delta y_0} \varphi(x, \delta y_0) z, \tag{1.2.7}$$

where, by virtue of the Lagrange theorem,

$$\varphi(x, \delta y_0) \equiv \frac{f(x, y + \delta y) - f(x, y)}{\delta y} - \frac{\partial}{\partial y} f(x, y + \theta \delta y), \tag{1.2.8}$$

θ (that depends on x and δy_0) lying in between 0 and 1.

Let us now consider the functional equation

$$z' = \varphi(x, \delta y_0) z, \tag{1.2.9}$$

where $\varphi(x, \delta y_0)$ is a continuous function in light of the assumed continuity of $\frac{\partial f}{\partial y}$. We can therefore apply to it the fundamental theorem and the theorem on the continuous dependence of the integral on a parameter. In this case the parameter is δy_0, which varies in a neighbourhood of 0. One obtains in such a way

$$\lim_{\delta y_0 \to 0} \frac{\delta y}{\delta y_0} = \lim_{\delta y_0 \to 0} \Big(z(x, \delta y_0) - z(x, 0) \Big), \tag{1.2.10}$$

z being the integral of the previous equation that equals 1 at $x = \bar{x}$. This proves not only that the derivative

$$\frac{\partial}{\partial y_0} y(x, y)$$

exists and is continuous, but also that it is obtained by integrating the linear homogeneous equation

$$z' = \varphi(x, 0)z = \frac{\partial f}{\partial y} z, \tag{1.2.11}$$

supplemented by the initial condition $z(\bar{x}) = 1$.

The result is extended to the system (1.2.3): if z_{ij} denotes the derivative of the function $y_i(x, y_1^0, \ldots, y_n^0)$ with respect to the initial value y_{i0} of y_i, the functions z_{ij} satisfy the linear homogeneous equations

$$z_{ij} = \frac{\partial f_i}{\partial y_1} z_{1j} + \cdots + \frac{\partial f_i}{\partial y_n} z_{nj} \ (i, j = 1, 2, \ldots, n), \tag{1.2.12}$$

with the initial conditions

$$z_{ii}(\bar{x}) = 1, \ z_{ij}(\bar{x}) = 0 \ (i \neq j). \tag{1.2.13}$$

The differentiability with respect to one or more parameters $\lambda, \mu, \ldots$ is established as was done before for the continuity, assuming that $f_1, \ldots, f_n$ have continuous derivatives also with respect to $\lambda, \mu, \ldots$; the derivatives $z_1, \ldots, z_n$ of $y_1, \ldots, y_n$ with respect to λ are obtained by integrating the linear inhomogeneous system

$$z_i' = \frac{\partial f_i}{\partial y_1} z_1 + \cdots + \frac{\partial f_i}{\partial y_n} z_n + \frac{\partial f_i}{\partial \lambda} \ (i = 1, 2, \ldots, n), \tag{1.2.14}$$

with the initial conditions

$$z_k(\bar{x}) = 0, \ \forall k = 1, \ldots, n. \tag{1.2.15}$$

Thus, the following result holds [15]:

Theorem 1.2 *If the functions $f_1, \ldots, f_n$ contain, besides $x, y_1, \ldots, y_n$, one or more parameters $\lambda, \mu, \ldots$ and have, with respect to all these variables, continuous first derivatives, the integral $(y_1, \ldots, y_n)$ consists of functions possessing continuous derivatives with respect to the initial values $y_1^0, \ldots, y_n^0$ and to the parameters. On denoting by ω any whatsoever of these quantities, the derivatives*

$$\frac{\partial y_i}{\partial \omega} = z_i$$

are obtained by integrating the linear system

$$z_1' = \frac{\partial f_1}{\partial y_1} z_1 + \cdots + \frac{\partial f_1}{\partial y_n} z_n + \frac{\partial f_1}{\partial w}, \tag{1.2.16}$$

$$\cdots\cdots\cdots\cdots\cdots\cdots\cdots\cdots\cdots\cdots\cdots\cdots\cdots$$

$$z_n' = \frac{\partial f_n}{\partial y_1} z_1 + \cdots + \frac{\partial f_n}{\partial y_n} z_n + \frac{\partial f_n}{\partial w}, \tag{1.2.17}$$

with the initial conditions $z_i(w) = 0$ or $z_i(w) = 1$ depending on whether w is not or is the initial value y_i^0.

This system, which turns out to be homogeneous, i.e. with last terms identically vanishing when w is an initial value, is said to be the system of variational equations, because it is satisfied by the derivatives and hence from the partial differentials or variations of the y_i with respect to w.

In case of a single equation of order n, i.e.

$$y^{(n)} = f(x, y, y', \ldots, y^{(n-1)}), \tag{1.2.18}$$

the following theorem holds [15]:

Theorem 1.3 *If the right-hand side of Eq. (1.2.18) contains certain parameters $\lambda, \mu, \ldots$, and admits continuous first derivatives with respect to*

$$y, \ldots, y^{(n-1)}, \lambda, \mu, \ldots,$$

the integral y, possessing continuous derivatives up to the n-th order with respect to z, admits, equally well as $y', \ldots, y^{(n-1)}$, continuous first derivatives with respect to the initial values and the parameters, and, denoting by w any whatsoever of these quantities, one obtains $\frac{\partial y}{\partial w} = 1$ by integrating the linear equation

$$z^{(n)} = \frac{\partial f}{\partial y} z + \frac{\partial f}{\partial y'} z' + \cdots + \frac{\partial f}{\partial y^{(n-1)}} z^{(n-1)} + \frac{\partial f}{\partial w}, \tag{1.2.19}$$

with the initial conditions $z^i(\bar{x}) = 1$ depending on whether w is not or is the initial value of $y^{(i)}$, for all $i = 0, 1, \ldots, n - 1$.

1.3 Linear Differential Equations of Second Order

We now focus on second-order equations. Within the framework of ordinary differential equations, every second-order linear differential equation can be written in the form

$$\left[\frac{d^2}{dx^2} + p(x)\frac{d}{dx} + q(x) \right] u(x) = 0, \tag{1.3.1}$$

where x is taken to lie in the closed interval $[a, b]$, while p and q are suitably smooth functions. This equation can be brought into the Liouville form, where the coefficient of the first-order derivative vanishes. For this purpose, one sets $u(x) = \varphi(x)\psi(x)$, so that Eq. (1.3.1) reads as

$$\varphi\psi'' + (2\varphi' + p\varphi)\psi' + (\varphi'' + p\varphi' + q\varphi)\psi = 0. \tag{1.3.2}$$

Our task is achieved if the function φ solves the first-order equation

$$\frac{\varphi'}{\varphi} = -\frac{p}{2} \implies \varphi(x) = \exp\left(-\frac{1}{2}\int p(x)dx \right). \tag{1.3.3}$$

At this stage, since we can divide by φ, we can re-express Eq. (1.3.2) in the canonical form

$$\psi'' + J\psi = 0, \tag{1.3.4}$$

where the potential term turns out to be

$$J = \frac{\varphi''}{\varphi} + p\frac{\varphi'}{\varphi} + q = -\frac{1}{2}p' - \frac{p^2}{4} + q. \tag{1.3.5}$$

Once we have reduced ourselves to studying Eq. (1.3.4), one can deduce important qualitative properties. For example, if the function J is continuous for $x \in [a, b]$, and if there exist real constants ω and Ω such that

$$0 < \omega^2 < J(x) < \Omega^2, \tag{1.3.6}$$

one can compare the zeros of solutions of Eq. (1.3.4) with the zeros of solutions of the equations

$$\psi'' + \alpha^2\psi = 0, \quad \alpha = \omega \text{ or } \Omega. \tag{1.3.7}$$

Equation (1.3.7) are solved by periodic functions $\sin\alpha(x - x_0)$ which have zeros at $x_0 + \frac{k\pi}{\alpha}$, k being an integer and α taking one of the two values in (1.3.7). One can then prove that the difference δ between two adjacent zeros of a solution of Eq. (1.3.4) satifies the conditions [134]

$$\frac{\pi}{\Omega} \leq \delta \leq \frac{\pi}{\omega}. \tag{1.3.8}$$

In several problems of pure mathematics and mathematical physics one is interested in studying Eq. (1.3.1) in the complex domain, when the independent variable x is replaced by a complex variable z. One can assume that a domain S exists where both $p(z)$ and $q(z)$ are analytic, except at a finite number of poles. All points of S where both p and q are analytic are said to be ordinary points, otherwise they are called singular points. The procedure leading to the canonical form (1.3.4) and (1.3.5) remains valid in the complex domain, and an ordinary point of Eq. (1.3.1) is also an ordinary point of Eq. (1.3.4).

Let us now consider, following [141], an ordinary point b, and the domain S_b formed by a circle of radius r_b centred at b, the radius r_b being such that every point of S_b is a point of S, and is an ordinary point of the equation. We also consider a sequence of functions $v_n(z)$, analytic in S_b and defined by the recursive equations

$$v_0(z) = a_0 + a_1(z - b),\tag{1.3.9}$$

$$v_n(z) = \int_b^z (\zeta - z)J(\zeta)v_{n-1}(\zeta)d\zeta, \quad n = 1, 2, 3 \ldots \tag{1.3.10}$$

where a_0, a_1 are arbitrary constants.

Let M and μ be the upper bounds of the absolute values $|J(z)|$ and $|v_0(z)|$ in the domain S_b. It is then possible to majorize the absolute values of $v_n(z)$ in the form

$$|v_n(z)| \leq \mu M^n \frac{|z - b|^{2n}}{n!}.\tag{1.3.11}$$

One can prove (1.3.11) by induction. It is indeed true when $n = 0$, and, if it holds for all n ranging from 0 through $(m - a)$, one finds, by integration along a straight line,

$$\begin{aligned}
|v_m(z)| &= \left| \int_b^z (\zeta - z)J(\zeta)v_{m-1}(\zeta)d\zeta \right| \\
&\leq \frac{1}{(m-1)!} \int_b^z |\zeta - z| \, |J(\zeta)| \mu M^{m-1} |\zeta - b|^{2m-2} d\zeta \\
&\leq \frac{1}{(m-1)!} \mu M^m |z - b| \int_0^{|z-b|} t^{2m-2} dt \\
&< \frac{1}{m!} \mu M^m |z - b|^{2m}.
\end{aligned}\tag{1.3.12}$$

Thus, by induction, the inequality (1.3.11) holds for all values of n.

Moreover, since $|v_n(z)| \leq \frac{\mu M^n (r_b)^{2n}}{n!}$ when z lies in S_b, and

$$\sum_{n=0}^{\infty} \mu M^n \frac{(r_b)^{2n}}{n!}$$

converges, one finds that

$$v(z) = \sum_{n=0}^{\infty} v_n(z) \tag{1.3.13}$$

is a series of analytic functions uniformly convergent in S_b. On the other hand, from the definition (1.3.10) of $v_n(z)$, one can write

$$\frac{d}{dz} v_n(z) = -\int_b^z J(\zeta) v_{n-1}(\zeta) d\zeta, \tag{1.3.14}$$

$$\frac{d^2}{dz^2} v_z = -J(z) v_{n-1}(z), \tag{1.3.15}$$

hence it follows the equation

$$\frac{d^2 v}{dz^2} = \frac{d^2 v_0(z)}{dz^2} + \sum_{n=1}^{\infty} \frac{d^2 v_n(z)}{dz^2} = -J(z) v(z). \tag{1.3.15a}$$

In other words, v is a function of the complex variable z, analytic in S_b, which satisfies the differential equation

$$\left[\frac{d^2}{dz^2} + J(z) \right] v(z) = 0, \tag{1.3.15b}$$

and the initial conditions

$$v(b) = a_0, \quad v'(b) = \left\{ \frac{d}{dz} v(z) \right\}_{z=b} = a_1, \tag{1.3.16}$$

where the coefficients a_0, a_1 are arbitrary.

1.4　Sturm-Liouville Problems

Let us consider a linear differential equation of second order with coefficients real and continuous in the closed interval $[a, b]$:

$$z'' + p(t) z' + q(t) z = g(t). \tag{1.4.1}$$

This is the inhomogeneous version of Eq. (1.3.1). We aim at finding an integral $z(t)$ of Eq. (1.4.1) defined in the given closed interval and satisfying the boundary conditions

$$z(a) = \alpha, \quad z(b) = \beta. \tag{1.4.2}$$

Inspired by what we found in Sect. 1.3 we remark that, upon multiplying (1.4.1) by

$$\theta(t) = e^{\int_a^t p(s)ds},$$
(1.4.3)

one can always write it in the form

$$\frac{d}{dt}\left(\theta\frac{dz}{dt}\right) + B(t)z = -f(t), \quad B \equiv q\theta, \ f \equiv -g\theta.$$
(1.4.4)

If $u(t)$ and $v(t)$ are the integrals of the homogeneous equation associated to Eq. (1.4.1) which satisfy the initial conditions

$$u(a) = 1, \ u'(a) = 0, \quad v(a) = 0, \ v'(a) = 1,$$
(1.4.5)

their Wronskian, that fulfills $w(a) = 1$, is given by

$$w(t) = u(t)v'(t) - v(t)u'(t) = \frac{1}{\theta(t)},$$
(1.4.6)

and hence the general solution of Eq. (1.4.1) reads as

$$z = c_1 u(t) + c_2 v(t) - \int_a^t [u(s)v(t) - u(t)v(s)]f(s)ds.$$
(1.4.7)

By imposing the boundary conditions (1.4.2) the coefficients c_1 and c_2 can be found, obtaining therefore for z the expression

$$z = \alpha\left(u(t) - \frac{u(b)v(t)}{v(b)}\right) + \beta\frac{v(t)}{v(b)}$$

$$+ \frac{1}{v(b)}\int_a^b v(t)[u(s)v(b) - u(b)v(s)]f(s)ds$$

$$- \frac{1}{v(b)}\int_a^t v(b)[u(s)v(t) - u(t)v(s)]f(s)ds.$$
(1.4.8)

This formula suggests defining

$$G(t, s) = \frac{v(s)}{v(b)}[u(t)v(b) - u(b)v(t)] \text{ if } a \le s \le t \le b,$$
(1.4.9a)

$$G(t, s) = \frac{v(t)}{v(b)}[u(s)v(b) - u(b)v(s)] \text{ if } a \le t \le s \le b,$$
(1.4.9b)

so that the solution (1.4.8) is re-expressed in the form

$$z = \alpha G_s(t, a) - \beta G_s(t, b)\theta(b) + \int_a^b g(t, s)f(s)ds.$$
(1.4.10)

Our $G(t, s)$ is, by construction, a continuous and symmetric kernel function, known under the name of Burkhardt (or Green) function of the boundary-value problem [101].

The validity of (1.4.10) is subject to the restriction

$$v(b) \neq 0. \tag{1.4.11}$$

We note however that, if $B(t) \leq 0$, the validity of (1.4.11) is guaranteed. Indeed, if $v(b)$ were vanishing, since $v(a)$ has to vanish by hypothesis, multiplication by $v(t)$ of the homogeneous equation associated with (1.4.4), and integration from a to b would yield

$$-\int_a^b \theta \left(\frac{dv}{dt} \right)^2 dt + \int_a^b B v^2 dt = 0, \tag{1.4.12}$$

which would then imply, if $B \leq 0$, that $v = \text{const.} = 0$.

Let us now consider, assuming again that $B \leq 0$, the more general equation

$$\frac{d}{dt} \left(\theta \frac{dz}{dt} \right) + (\lambda A(t) + B(t)) z = -f(t) \tag{1.4.13}$$

and the associated Sturm-Liouville problem, which consists of looking for a solution satisfying the homogeneous boundary conditions

$$z(a) = z(b) = 0. \tag{1.4.14}$$

From what we said before, such a problem is equivalent to the integral equation

$$z(t) - \lambda \int_a^b G(t, s) A(s) z(s) ds = \int_a^b G(t, s) f(s) ds, \tag{1.4.15}$$

from which one can infer an existence theorem with alternative. If $A > 0$, such a theorem takes a particularly simple form. Indeed, upon defining

$$x(t) = \sqrt{A(t)}\, z(t), \quad K(t, s) = \sqrt{A(t)}\, \sqrt{A(s)}\, G(t, s), \tag{1.4.16}$$

Eq. (1.4.15) takes the form

$$x(t) - \lambda \int_a^b K(t, s) x(s) ds = \int_a^b K(t, s) \frac{f(s)}{\sqrt{A(s)}} ds, \tag{1.4.17}$$

and this is an integral equation with symmetric kernel. From the general theorems in the theory of integral equations, one derives the following result [101].

Theorem 1.4 *If $B \leq 0$ and $A > 0$, the Sturm-Liouville problem admits always one and only one solution, with the only exception of a sequence $\{\lambda_k\}$ of real values of λ, for which there exists a proper solution of the associated problem.*

Let us now point out that, to a given λ, there cannot correspond two linearly independent solutions of the homogeneous Sturm-Liouville problem, because two solutions of the homogeneous equation associated to (1.4.13), vanishing at the same point, are certainly linearly dependent. Moreover, the λ_k are all positive, as can be seen from the relation

$$\lambda_k \int_a^b A(t)(z_k)^2 dt = \int_a^b \left[\left(\frac{dz_k}{dt} \right)^2 - B(z_k)^2 \right] dt, \qquad (1.4.18)$$

fulfilled by the eigensolution z_k corresponding to λ_k. We can therefore order the λ_k in a monotonically increasing sequence, and the relative eigensolutions $z_k(t)$ can be also chosen so as to verify the orthonormality conditions

$$\int_a^b z_l(t) z_k(t) A(t) dt = \delta_{lk}. \qquad (1.4.19)$$

It is enough for this purpose to bear in mind that the eigensolutions z_k are linked to the eigensolutions x_k of the kernel $K(t, s)$ by the relations $x_k = \sqrt{A} z_k$.

By virtue of a theorem of Mercer, one has

$$\sum_{k=1}^{\infty} \frac{1}{\lambda_k} < +\infty. \qquad (1.4.20)$$

Last, the system $\{z_k\}$ is complete because, if

$$u(t) = \int_a^b K(t, s) f(s) ds = 0, \qquad (1.4.21)$$

one then has

$$-\sqrt{A(t)} f(t) = \frac{d}{dt} \left(\theta \frac{d}{dt} \frac{u}{\sqrt{A}} \right) + \frac{Bu}{\sqrt{A}} = 0. \qquad (1.4.22)$$

Note also that, if $\lambda = \lambda_k$, the solvability of Eq. (1.4.17), and hence of the Sturm-Liouville problem, is subject to the fulfillment of the compatibility condition

$$\int_a^b x_k(t) dt \int_a^b K(t, s) \frac{f(s)}{\sqrt{A(s)}} ds = 0, \qquad (1.4.23)$$

which is turned easily into

$$\int_a^b z_k(s) f(s) ds = 0. \qquad (1.4.24)$$

Analogous results and considerations hold even if we replace the homogeneous boundary conditions by more general equations, e.g.

$$a_{11}z(a) + a_{12}z'(a) = 0, \quad a_{21}z(b) + a_{22}z'(b) = 0. \tag{1.4.25}$$

If then we replace (1.4.14) and (1.4.25) by inhomogeneous conditions, the analysis is not substantially altered but for the modification of the right-hand side in Eq. (1.4.15).

1.5   Singular Points of Linear Differential Equations

Suppose that a point c is such that, although the functions p or q in Eq. (1.3.1) (either of them or both) have poles at c, the poles are of such orders that $(z - c)p(z)$, $(z - c)^2q(z)$ are analytic at c. Such a point is called a *regular (or Fuchsian) point* [132, 141] for the differential equation (1.3.1) in the complex domain. Any poles of $p(z)$ or of $q(z)$ which are not of this nature are called, by contrast, *irregular points*.

If c is a regular point, Eq. (1.3.1) may be written in the Frobenius normal form

$$\left[(z - c)^2 \frac{d^2}{dz^2} + (z - c)P\frac{d}{dz} + Q \right] u = 0, \tag{1.5.1}$$

where P and Q are analytic at c. Thus, by virtue of the Taylor theorem for functions of complex variable,

$$P = P(z - c) = p_0 + p_1(z - c) + p_2(z - c)^2 + \cdots, \tag{1.5.2}$$

$$Q = Q(z - c) = q_0 + q_1(z - c) + q_2(z - c)^2 + \cdots, \tag{1.5.3}$$

where p_k, q_k are constants. These series converge in the domain S_c consisting of a circle of radius r centred at c and of its interior, where r is so small that c is the only singular point of the equation which lies in S_c.

Let us now assume as a formal solution of Eq. (1.5.1) the function

$$u = (z - c)^\alpha \left[1 + \sum_{n=1}^{\infty} a_n(z - c)^n \right], \tag{1.5.4}$$

where $\alpha, a_1, \ldots, a_n, \ldots$ are constants to be determined as we are going to see. Such an ansatz is suggested by the considerations developed (for example) by Poincaré in his beautiful work on the irregular integrals of linear equations [114]. If the ansatz (1.5.4) is inserted into Eq. (1.5.1), one finds, by assuming that term-by-term differentiation and multiplication of series are legitimate, the fundamental equation

$$(z - c)^{\alpha} \left[\alpha(\alpha - 1) + \sum_{n=1}^{\infty} a_n(\alpha + n)(\alpha + n - 1)(z - c)^n \right]$$

$$+ (z - c)^{\alpha} P(z - c) \left[\alpha + \sum_{n=1}^{\infty} a_n(\alpha + n)(z - c)^n \right]$$

$$+ (z - c)^{\alpha} Q(z - c) \left[1 + \sum_{n=1}^{\infty} a_n(z - c)^n \right] = 0. \tag{1.5.5}$$

At this stage, we exploit the Taylor expansions (1.5.2) and (1.5.3), perform the multiplications and set to zero the coefficients of all powers of $(z - c)$. This method leads to the equations

$$F(\alpha) \equiv \alpha^2 + (p_0 - 1)\alpha + q_0 = 0, \tag{1.5.6}$$

$$a_1 \left[(\alpha + 1)^2 + (p_0 - 1)(\alpha + 1) + q_0 \right] + \alpha p_1 + q_1 = 0, \tag{1.5.7}$$

$$a_2 \left[(\alpha + 2)^2 + (p_0 - 1)(\alpha + 2) + q_0 \right] + a_1 \left[(\alpha + 1)p_1 + q_1 \right] + \alpha p_2 + q_2 = 0, \tag{1.5.8}$$

$$a_n \left[(\alpha + n)^2 + (p_0 - 1)(\alpha + n) + q_0 \right]$$

$$+ \sum_{m=1}^{n-1} a_{n-m} \left[(\alpha + n - m)p_m + q_m \right] + \alpha p_n + q_n = 0. \tag{1.5.9}$$

Equation (1.5.6) is the *indicial equation*, and determines two values for α (such roots may be coincident). If c is an irregular point, the indicial equation is at most of first degree.

On denoting by ρ_1 and ρ_2 the two roots α of Eq. (1.5.6), which are said to be the *exponents* of Eq. (1.5.1) at the point c, the subsequent equations, for chosen value of α, determine uniquely $a_1, a_2, \ldots$ provided that $F(\alpha + n)$ does not vanish when $n = 1, 2, 3, \ldots$. In other words, if $\alpha = \rho_1$, the resulting ρ_2 is not one of the numbers $\rho_1 + 1, \rho_1 + 2, \ldots$. If instead $\alpha = \rho_2$, the resulting ρ_1 is not one of the numbers $\rho_2 + 1, \rho_2 + 2, \ldots$.

Thus, if the difference of the exponents is neither vanishing nor an integer, one can always obtain two distinct series which satisfy formally (i.e., up to convergence issues) Eq. (1.5.1). A separate method is instead necessary if $\rho_1 - \rho_2 = s$ is a positive integer or 0. In such a case, we consider

$$w_1(z) \equiv (z - c)^{\rho_1} \left[1 + \sum_{n=1}^{\infty} a_n(z - c)^n \right], \tag{1.5.10}$$

and then look for the second solution in the factorized form

$$u(z) = w_1(z)\zeta(z), \tag{1.5.11}$$

where, for consistency, the function ζ has to solve the differential equation

$$(z - c)^2 \frac{d^2\zeta}{dz^2} + \left[2(z - c)^2 \frac{w_1'(z)}{w_1(z)} + (z - c)P(z - c) \right] \frac{d\zeta}{dz} = 0. \tag{1.5.12}$$

This can be viewed as a first-order equation for the function $G \equiv \frac{d\zeta}{dz}$, and hence two consecutive integrations yield

$$
\begin{aligned}
\zeta &= A + B \int \frac{dz}{[w_1(z)]^2} \exp\left[-\int \frac{P(z - c)}{(z - c)} dz \right] \\
&= A + B \int \frac{(z - c)^{-p_0}}{[w_1(z)]^2} \exp\left[-p_1(z - c) - \frac{1}{2} p_2(z - c)^2 - \cdots \right] dz \\
&= A + B \int (z - c)^{-p_0 - 2p_1} g(z) dz,
\end{aligned}
\tag{1.5.13}
$$

where A, B are arbitrary constants and $g(z)$ is analytic throughout the interior of any circle centred at c, which does not contain any singularities of $P(z - c)$ or singularities or zeros of $(z - c)^{-p_1} w_1$, and such that $g(c) = 1$. Now we write the absolutely convergent Taylor series for $g(z)$, i.e.

$$g(z) = 1 + \sum_{n=1}^{\infty} g_n (z - c)^n. \tag{1.5.14}$$

If $s \neq 0$, one can write [141]

$$
\begin{aligned}
\zeta &= A + B \int \left[1 + \sum_{n=1}^{\infty} g_n(z - c)^n \right] (z - c)^{-s-1} dz \\
&= A + B \left[-\frac{1}{s}(z - c)^{-s} - \sum_{n=1}^{s-1} \frac{g_n}{(s - n)} (z - c)^{n-s} + g_s \log(z - c) \right. \\
&\quad \left. + \sum_{n=s+1}^{\infty} \frac{g_n}{(n - s)} (z - c)^{n-s} \right].
\end{aligned}
\tag{1.5.15}
$$

Thus, the general solution of the differential equation that is analytic at all points of C but c reads as

$$Aw_1(z) + B\left[g_s w_1(z) \log(z - c) + \overline{w}(z) \right],$$

where the function $\overline{w}$ has been defined by

$$\overline{w}(z) = (z - c)^{\rho_2} \left[-\frac{1}{s} + \sum_{n=1}^{\infty} h_n (z - c)^n \right], \tag{1.5.16}$$

the h_n denoting constant coefficients.

If instead s vanishes, the resulting solution reads as

$$u(z) = A w_1(z) + B \left[w_1(z) \log(z - c) + (z - c)^{\rho_2} \sum_{n=1}^{\infty} h_n (z - c)^n \right]. \tag{1.5.17}$$

What we have obtained are examples of *regular integrals*, i.e. solutions which hold in the neighbourhood of a regular point c. A general recipe to obtain them is as follows. First we obtain $w_1(z)$, and then we determine the coefficients in a function

$$\overline{w}_1(z) = \sum_{n=0}^{\infty} b_n (z - c)^{\rho_2 + n}, \tag{1.5.18}$$

by inserting $w_1(z) \log(z - c) + \overline{w}_1(z)$ in the left-hand side of Eq. (1.5.1) and setting to zero the coefficients of all powers of $(z - c)$ in the expression so obtained.

In the neighbourhood of a point which is not a regular point, an equation of second order cannot have two regular integrals, because the indicial equation is at most of first degree. Regular integrals may be lacking, or there may be no more than one. Interestingly, a linear differential equation may be derived from another differential equation by making two or more singularities of the latter tend to coincide. If this occurs, the former equation is said to be a *confluent form* of the latter, and the limiting process alluded to is said to provide a *confluence*. The limiting form of the equation may have singularities of a simpler nature.

As an example, let us consider here the most general form of linear equation of second order having singularities only at a_1, a_2, a_3, a_4 and ∞, these five points being regular singular points with exponents α_r, β_r at a_r for all $r = 1, 2, 3, 4$, and exponents μ_1, μ_2 at the point at infinity. We can therefore write such a linear equation in the form

$$\left[\frac{d^2}{dz^2} + \left(\sum_{r=1}^{4} \frac{1 - \alpha_r - \beta_r}{(z - a_r)} \right) \frac{d}{dz} + \left(\sum_{r=1}^{4} \frac{\alpha_r \beta_r}{(z - a_r)^2} + \frac{A z^2 + 2Bz + C}{\prod_{r=1}^{4}(z - a_r)} \right) \right] u = 0, \tag{1.5.19}$$

where A occurs in the algebraic equation of second degree providing the exponents at ∞, i.e.

$$\mu^2 + \mu \left(\sum_{r=1}^{4} (\alpha_r + \beta_r) - 3 \right) + \sum_{r=1}^{4} \alpha_r \beta_r + A = 0, \tag{1.5.20}$$

while B, C are unconstrained constants. Within this framework, a particular case of special interest occurs when the difference of the two exponents at each singularity is $\frac{1}{2}$. Under such assumptions, one can write $\beta_r = \alpha_r + \frac{1}{2}$, for all $r = 1, 2, 3, 4$ and write ζ in place of z, so that Eq. (1.5.19) becomes

$$\left[\frac{d^2}{d\zeta^2} + \left(\sum_{r=1}^{4} \frac{\frac{1}{2} - 2\alpha_r}{(\zeta - a_r)} \right) \frac{d}{d\zeta} + \left(\sum_{r=1}^{4} \frac{\alpha_r \left(\alpha_r + \frac{1}{2} \right)}{(\zeta - a_r)^2} + \frac{A\zeta^2 + 2B\zeta + C}{\prod_{r=1}^{4}(\zeta - a_r)} \right) \right] u = 0,$$

(1.5.21)

where, by virtue of Eq. (1.5.20) and of the condition $\mu_2 - \mu_1 = \frac{1}{2}$, one has

$$A = \left(\sum_{r=1}^{4} \alpha_r \right)^2 - \sum_{r=1}^{4} (\alpha_r)^2 - \frac{3}{2} \sum_{r=1}^{4} \alpha_r + \frac{3}{16}.$$

(1.5.22)

Equation (1.5.21) is said to be the generalised Lamé equation. If one sets $a_1 = a_2$ in this equation, the confluence of the two singularities a_1, a_2 yields a singularity at which the exponents α, β are given by the equations

$$\alpha + \beta = 2(\alpha_1 + \alpha_2), \quad \alpha\beta = \alpha_1 \left(\alpha_1 + \frac{1}{2} \right) + \alpha_2 \left(\alpha_2 + \frac{1}{2} \right) + D, \quad (1.5.23)$$

having defined

$$D \equiv \frac{\left(A(a_1)^2 + 2Ba_1 + C \right)}{(a_1 - a_3)(a_1 - a_4)}.$$

(1.5.24)

Interestingly, the exponent-difference at the confluent singularity is not $\frac{1}{2}$, unlike the original equation, but it may take any assigned value by suitable choice of B and C. In a similar way, by exploiting the confluence of three or more singularities, one can obtain an irregular singularity.

By means of suitable confluences of the five singularities of the original Eq. (1.5.19), one can obtain six types of equations, which may be classified according to the following parameters [141]:

(i) The number of their singularities with exponent-difference $\frac{1}{2}$;
(ii) The number of their other regular singularities;
(iii) The number of their irregular singularities.

1.6   Fundamental Properties of the Heun Equation

In the modern literature on applied mathematics and theoretical physics there is a Fuchsian equation of second order of particular importance. This is known as the Heun equation [46, 78, 124], and it reads as

$$\left[\frac{d^2}{dz^2} + \left(\frac{\gamma}{z} + \frac{\delta}{(z-1)} + \frac{\epsilon}{(z-a)}\right)\frac{d}{dz} + \frac{\alpha\beta z - q}{z(z-1)(z-a)}\right]y = 0. \qquad (1.6.1)$$

With this notation, the independent variable z and the unknown function y take complex values, and $\alpha, \beta, \gamma, \delta, \epsilon, q, a$ are parameters, generally complex and arbitrary, except that a cannot take the values 0 and 1. The first five parameters are related by the linear equation

$$\gamma + \delta + \epsilon = \alpha + \beta + 1. \qquad (1.6.2)$$

Equation (1.6.1) is of Fuchsian type, with regular singularities at $z = 0, 1, a, \infty$, the exponents at these singularities being

$$(0, 1, -\gamma), \quad (0, 1, -\delta), \quad (0, 1, -\epsilon), \quad (\alpha, \beta).$$

The sum of these exponents must take the value 2, in light of the general theory of Fuchsian equations; it is precisely this property that gives rise to Eq. (1.6.2).

The seven parameters play different roles [124]. The parameter a locates the third finite singularity and can be therefore called the *singularity parameter*, while $\alpha, \beta, \gamma, \delta, \epsilon$ determine, subject to the restriction (1.6.2), the exponents at the four singularities and can be called the *exponent parameters*. Last, q is said to be the *auxiliary parameter*; there exist applications in which q is an eigenparameter. An important theorem ensures that any Fuchsian second-order differential equation with four singularities can be reduced by elementary transformation to the form (1.6.1), which is therefore the most general form of such equation.

The Heun equation was originally conceived as a generalization of the hypergeometric equation, and we here describe three ways in which the Heun reduces to the hypergeometric form [124].

(i) First, if we write (1.6.1) as

$$z(z-1)(z-a)y''(z) + \left[\gamma(z-1)(z-a) + \delta z(z-a) + \epsilon z(z-1)\right]y'(z)$$
$$+ (\alpha\beta z - q)y(z) = 0, \qquad (1.6.3)$$

and set

$$a = 1, \quad q = \alpha\beta, \qquad (1.6.4)$$

then a common factor $(z-1)$ emerges and can be taken out, leaving us with the hypergeometric equation in its standard form

$$z(1-z)y''(z) + \left[\gamma - (\alpha + \beta + 1)z\right]y'(z) - \alpha\beta y(z) = 0, \qquad (1.6.5)$$

having exploited Eq. (1.6.2).

(ii) Second, upon choosing

$$a = q = 0 \qquad (1.6.6)$$

in Eq. (1.6.3) and removing the common factor z gives the equation

$$z(1 - z)y''(z) + \left[\alpha + \beta - \delta + 1 - (\alpha + \beta + 1)z\right]y'(z) - \alpha\beta y(z) = 0, \quad (1.6.7)$$

which is again of hypergeometric form.
   (iii) Third, upon taking

$$\epsilon = 0, \quad q = a\alpha\beta, \tag{1.6.8}$$

a common factor $(z - a)$ can be taken out of the equation, which therefore becomes

$$z(1 - z)y''(z) + \left[\gamma - (\alpha + \beta + 1)z\right]y'(z) - \alpha\beta y(z) = 0. \tag{1.6.9}$$

If we assume that none of the exponent parameters γ, δ, ϵ is an integer, nor is the difference $(\alpha - \beta)$, no logarithmic terms can occur in the solution of Eq. (1.6.1), and four sets of solutions are found to exist [124]. They are as follows.

(I) **Local solutions**. Each singular point is regular and hence, in the neighbourhood of any one singularity there exist two linearly independent solutions, one corresponding to each of the exponents therein. These solutions are valid only in a circle that excludes the nearest other singularity, and may be found, at least in principle, as a power series. There will be two such local solutions about each singularity, giving eight local solutions in all.

(II) **Heun functions**. A local solution about a singularity s_1 can be continued analytically to a neighbourhood of an adjacent singularity s_2, but does not coincide in general with one of the local solutions therein. Those solutions that are local solutions *both* about $z = s_1$ and $z = s_2$, so that they are analytic in some domain including both these singularities, are said to be Heun functions. A key problem is under which conditions on the parameters in Eq. (1.6.1) a Heun function can exist. This leads eventually to an eigenvalue problem for the accessory parameter q.

(III) **Heun polynomials**. There may exist solutions which are simultaneously local at three singular points, and hence analytic in a domain containing these singularities. A necessary (but not sufficient) condition for the existence of such solutions is that one of the quantities

$$\alpha, \gamma - \alpha, \delta - \alpha, \epsilon - \alpha,$$

or $\beta, \gamma - \beta$, and so on, should be an integer, while the parameter q should take one of a finite number of characteristic values. If α is a negative integer, i.e. $\alpha = -n$, the solutions are indeed polynomials of degree n. Otherwise, powers of $z, z - 1, z - a$ occur. Interestingly, it has been proved that if a solution exists that is simultaneously a local solution about three singular points, then it must also be a local solution about the fourth singular point, and is therefore analytic in the whole finite z-plane, with suitable cuts.

(IV) **Path-multiplicative solutions**. If Γ is a closed path surrounding two, and only two, singular points, a path-multiplicative solution with respect to Γ is a solution which, when continued analytically around Γ, returns to its starting point merely multiplied by a constant σ. The multiplier σ is fixed by the parameters of the equation, and an eigenvalue problem for the accessory parameter q must be solved in order to obtain a preassigned value for σ.

Chapter 2
Non-linear Equations

2.1 First Examples of Non-linear Ordinary Differential Equations

Before describing some general properties of non-linear equations, we find it useful to present some well-known examples that remain of high pedagogical value, following [72].

(i) **Circles** In analytic geometry, the circles in a plane

$$x^2 + y^2 + 2Ax + 2By + C = 0 \qquad (2.1.1)$$

for a three-parameter family of plane curves; the corresponding differential equation is therefore of the third order. By differentiating Eq. (2.1.1) three times, one finds

$$x + y\frac{dy}{dx} + A + B\frac{dy}{dx} = 0,$$

$$1 + \left(\frac{dy}{dx}\right)^2 + (y+B)\frac{d^2y}{dx^2} = 0,$$

$$3\frac{dy}{dx}\frac{d^2y}{dx^2} + (y+B)\frac{d^3y}{dx^3} = 0. \qquad (2.1.2)$$

The parameter B can be eliminated between the last two equations, and this leads to the non-linear functional equation

$$\frac{d^3y}{dx^3}\left[1 + \left(\frac{dy}{dx}\right)^2\right] - 3\frac{dy}{dx}\left(\frac{d^2y}{dx^2}\right)^2 = 0. \qquad (2.1.3)$$

The only plane curves that satisfy this equation are circles and straight lines. Of course, any straight line is an integral curve, because Eq. (2.1.3) is satisfied if we have vanishing second derivative of y, and hence also its third derivative vanishes.

© Springer International Publishing AG 2017
G. Esposito, *From Ordinary to Partial Differential Equations*, UNITEXT 106,
DOI 10.1007/978-3-319-57544-5_2

If instead the second derivative of y does not vanish, one can write Eq. (2.1.3) in the form

$$\frac{\frac{d^3 y}{dx^3}}{\frac{d^2 y}{dx^2}} = \frac{3 \frac{dy}{dx} \frac{d^2 y}{dx^2}}{\left[1 + \left(\frac{dy}{dx}\right)^2\right]}, \tag{2.1.4}$$

from which one obtains

$$\log\left(\frac{d^2 y}{dx^2}\right) = \frac{3}{2} \log\left[1 + \left(\frac{dy}{dx}\right)^2\right] + \log(C_1), \tag{2.1.5}$$

where C_1 is a non-vanishing constant. This formula may be written in the equivalent form

$$\frac{\frac{d^2 y}{dx^2}}{\left[1 + \left(\frac{dy}{dx}\right)^2\right]^{\frac{3}{2}}} = C_1. \tag{2.1.6}$$

Now one can integrate again to find

$$\frac{\frac{dy}{dx}}{\sqrt{1 + \left(\frac{dy}{dx}\right)^2}} = C_1 x + C_2, \tag{2.1.7}$$

i.e.

$$\frac{dy}{dx} = \frac{C_1 x + C_2}{\sqrt{1 - (C_1 x + C_2)^2}}. \tag{2.1.8}$$

A third integration yields eventually

$$C_1 y + C_3 = -\sqrt{1 - (C_1 x + C_2)^2}, \tag{2.1.9}$$

and the equation of a circle is recovered.

(ii) **Conics** If a conic has no asymptote parallel to the y-axis, its equation can be written in the form

$$y = mx + n + \sqrt{Ax^2 + 2Bx + C}. \tag{2.1.10}$$

Upon differentiating twice, one finds

$$\frac{d^2 y}{dx^2} = \frac{(AC - B^2)}{(Ax^2 + 2Bx + C)^{\frac{3}{2}}}, \tag{2.1.11}$$

or the equivalent form

$$\left(\frac{d^2 y}{dx^2}\right)^{-\frac{2}{3}} = (AC - B^2)^{-\frac{1}{3}}(Ax^2 + 2Bx + C), \qquad (2.1.12)$$

which implies that the left-hand side is a trinomial of second degree in x. Thus, in order to eliminate the three constants A, B, C three differentiations are sufficient, and the desired differential equation can be written as

$$\frac{d^3}{dx^3}\left[\left(\frac{d^2 y}{dx^2}\right)^{-\frac{2}{3}}\right] = 0. \qquad (2.1.13)$$

Upon performing the three derivatives, one obtains the Halphen non-linear equation

$$40\left(\frac{d^3 y}{dx^3}\right)^3 - 45\frac{d^2 y}{dx^2}\frac{d^3 y}{dx^3}\frac{d^4 y}{dx^4} + 9\left(\frac{d^2 y}{dx^2}\right)^2\frac{d^5 y}{dx^5} = 0. \qquad (2.1.14)$$

(iii) **Parabolas** For a parabola, the coefficient A vanishes in the previous formulae, and the fractional power $-\frac{2}{3}$ of the second derivative of y is therefore a binomial of first degree. The differential equation (2.1.13) is then replaced by

$$\frac{d^2}{dx^2}\left[\left(\frac{d^2 y}{dx^2}\right)^{-\frac{2}{3}}\right] = 0, \qquad (2.1.15)$$

which leads to

$$5\left(\frac{d^3 y}{dx^3}\right)^2 - 3\frac{d^2 y}{dx^2}\frac{d^4 y}{dx^4} = 0. \qquad (2.1.16)$$

(iv) **Bernoulli's equation** This is a non-linear equation reading as

$$\frac{dy}{dx} + u_0 y + u_1 y^n = 0, \qquad (2.1.17)$$

where the exponent n is any real number different from 0 and 1. Interestingly, this non-linear equation can be reduced to a linear equation by introducing the new dependent variable $z \equiv y^{1-n}$, because

$$z \equiv y^{1-n} \implies \frac{1}{(1-n)}\frac{dz}{dx} + u_0 z + u_1 = 0. \qquad (2.1.18)$$

In general, one can reduce to the previous type any equation of the form

$$\phi\left(\frac{y}{x}\right)dx + \psi\left(\frac{y}{x}\right)dy + kx^m(x\,dy - y\,dx) = 0, \qquad (2.1.19)$$

where k and m are arbitrary numbers. Indeed, upon defining $y \equiv wx$, such an equation can be written in the form

$$[\phi(w) + w\psi(w)]\frac{dx}{dw} + x\psi(w) + kx^{m+2} = 0. \qquad (2.1.20)$$

At this stage, by considering $z \equiv x^{-(m+1)}$, one obtains a linear equation.

(v) **Riccati equation** This equation occurs frequently in the classical differential geometry of curves and surfaces [11, 30], and provides another enlightening example of a non-linear equation that can be re-expressed exactly in linear form under certain assumptions. Nowadays it is customary to write it as

$$\frac{dy}{dx} + u_2 y^2 + u_1 y + u_0 = 0, \qquad (2.1.21)$$

although Riccati limited himself to studying the equation [30]

$$\frac{dy}{dx} = ay^2 + bx^m, \qquad (2.1.22)$$

which is a particular case of Eq. (2.1.21), with

$$u_2 = -a, \ u_1 = 0, \ u_0 = -b, \ m = 0. \qquad (2.1.23)$$

In general, Eq. (2.1.21) cannot be integrated by quadratures, but, *if a particular integral is known, the general integral can be found by two quadratures*. Indeed, let y_1 be a particular integral. Upon considering the split $y = y_1 + z$, the equation for z has the same form but does not contain any term independent of z, since $z = 0$ must be an integral. The non-linear equation for z reads as

$$\frac{dz}{dx} + (u_1 + 2u_2 y_1)z + u_2 z^2 = 0, \qquad (2.1.24)$$

and this equation suggests defining

$$u \equiv \frac{1}{z} \qquad (2.1.25)$$

in order to achieve a linear equation. This proves what we anticipated.

This simple result has several important consequences. The general integral of the linear equation for u reads as

$$u = Cf(x) + \phi(x). \qquad (2.1.26)$$

By virtue of (2.1.25) and (2.1.26), the general integral of the Riccati equation is

$$y = y_1 + \frac{1}{Cf(x) + \phi(x)} = \frac{Cf_1(x) + \phi_1(x)}{Cf(x) + \phi(x)}. \tag{2.1.27}$$

In other words, *once a particular solution is known, the general integral of the Riccati equation is a rational function of first degree in the integration constant.*

Conversely, every differential equation of first order having this property is a Riccati equation. Indeed, if f, ϕ, f_1, ϕ_1 are any four functions of x, all functions y represented by Eq. (2.1.27), where C is an arbitrary constant, are integrals of an equation of first order, which is obtained by solving Eq. (2.1.27) for C and then deriving with respect to x. This method leads to

$$C = \frac{\phi_1 - y\phi}{yf - f_1}, \tag{2.1.28}$$

and the corresponding differential equation is

$$(yf - f_1)\left(\frac{d\phi_1}{dx} - \phi\frac{dy}{dx} - y\frac{d\phi}{dx}\right)$$
$$- (\phi_1 - y\phi)\left(\frac{dy}{dx}f + y\frac{df}{dx} - \frac{df_1}{dx}\right) = 0, \tag{2.1.29}$$

which is of the form (2.1.21).

Let us now consider four particular integrals y_1, y_2, y_3, y_4 that correspond to the values C_1, C_2, C_3 and C_4, respectively, of the constant C. By virtue of the theory of the anharmonic ratio, one can write the relation

$$\frac{(y_4 - y_1)}{(y_4 - y_2)} + \frac{(y_3 - y_1)}{(y_3 - y_2)} = \frac{(C_4 - C_1)}{(C_4 - C_2)} + \frac{(C_3 - C_1)}{(C_3 - C_2)}. \tag{2.1.30}$$

This can be verified also by a patient calculation, and proves that *the anharmonic ratio of any four particular integrals of the Riccati equation is constant.*

This theorem makes it possible to find without quadratures the general integral of a Riccati equation, *provided that one knows three of its particular integrals y_1, y_2, y_3.* Every other integral must be such that the anharmonic ratio

$$\frac{(y - y_1)}{(y - y_2)} + \frac{(y_3 - y_1)}{(y_3 - y_2)}$$

is constant. The general integral is then obtained by equating this ratio to an arbitrary constant. By construction, y is a rational function of first degree in this constant, which proves that the previous property holds only for the Riccati equation.

If instead we knew only two particular integrals, y_1 and y_2, we might complete the integration by performing one quadrature; as a matter of fact, after the split $y = y_1 + z$, the equation obtained for the unknown function z has the integral $y_2 - y_1$.

The linear equation in u has therefore the particular integral $\frac{1}{(y_2-y_1)}$. The general integral of the differential equation in u is then found by a single quadrature.

This paragraph is only an elementary introduction to the Riccati equation. For a thorough investigation from the point of view of differential geometry, we refer the reader to Sect. 7.2.2 of Carinena et al. [22].

2.2 Non-linear Differential Equations in the Complex Domain

The study of solutions of non-linear equations exhibits difficulties resulting from the existence of singularities which are not fixed, but movable. In this section we introduce the reader to the results of Briot and Bouquet, relying upon the monograph by Valiron [134].

If a first-order differential equation is given, here written in the form

$$\frac{dy}{dx} = f(x, y), \tag{2.2.1}$$

we know that it admits a unique solution taking for $x = x_0$ the value y_0, when the function $f(x, y)$ is analytic at the point (x_0, y_0). This solution is holomorphic within a circle centred at x_0, and the coefficients of its expansion can be evaluated one after the other.

Suppose now that, for $x = x_0$, $y = y_0$, the function $f(x, y)$ is not analytic, whereas $\frac{1}{f(x,y)}$ is analytic. Let us set

$$g(x, y) \equiv \frac{1}{f(x, y)}. \tag{2.2.2}$$

One has $g(x_0, y_0) = 0$, otherwise $f(x, y)$ would be analytic. One can thus say that $f(x, y)$ is infinite at the point (x_0, y_0). In order to study the solutions of Eq. (2.2.1) in the neighbourhood of the point (x_0, y_0), one can reduce this point to the origin, by considering the equation

$$\frac{dx}{dy} = g(x, y) = a_{1,0}x + a_{0,1}y + \cdots . \tag{2.2.3}$$

The existence theorem holds for this equation. It admits a unique solution that vanishes at $y = 0$, and which is holomorphic for $|y|$ sufficiently small. Since its derivative vanishes at $y = 0$, this solution is of the form

$$x = c_p y^p + \cdots , \tag{2.2.4}$$

unless it is identically vanishing. In the general case where the solution has the form (2.2.4), one obtains, by performing the inversion,

$$y = \gamma_p x^{\frac{1}{p}} + \cdots . \tag{2.2.5}$$

Thus, *there exists a unique solution taking at $x = 0$ the value 0, it is an analytic function admitting the origin as an algebraic critical point.*

For the Eq. (2.2.3) to admit for solution 0, it is necessary that the right-hand side vanishes identically at $x = 0$, hence it contains as a factor a power of x, and this is sufficient. In this case, there exists no solution of Eq. (2.2.1) taking at $x = x_0$ the value y_0. As examples, the equations

$$\frac{dy}{dx} = \frac{1}{2y}, \quad \frac{dy}{dx} = \frac{1}{2xy} \tag{2.2.6}$$

are solved, respectively, by

$$y = \sqrt{x + C}, \quad y = \sqrt{\log(Cx)}. \tag{2.2.7}$$

If $C = 0$, the first curve in Eq. (2.2.7) passes through the origin, whereas the second curve in (2.2.7) never meets the origin.

Briot and Bouquet have studied the case in which the function on the right-hand side of Eq. (2.2.1) is indeterminate at the point (x_0, y_0) and analytic about this point. We shall assume that this point is the origin. Let us consider the equation

$$x \frac{dy}{dx} = f(x, y), \tag{2.2.8}$$

where $f(x, y)$ is analytic about the origin and vanishes at this point. One has

$$f(x, y) = a_{1,0} x + a_{0,1} y + \cdots , \tag{2.2.9}$$

and one can write Eq. (2.2.8), by changing the notation:

$$xy' - \lambda y = ax + \varphi(x, y), \tag{2.2.10}$$

where the series $\varphi(x, y)$ is an entire function that begins with terms of second degree. Let us look for a holomorphic solution, necessarily vanishing at the origin, given by

$$y = c_1 x + c_2 x^2 + \cdots . \tag{2.2.11}$$

The formal evaluation of coefficients of this series is possible if λ is not a positive integer. One has

$$(-\lambda + 1)c_1 = a, \ldots, (-\lambda + n)c_n = P_n(c_1, c_2, \ldots, c_{n-1}). \tag{2.2.12}$$

With this notation, P_n is a polynomial in $c_1, c_2, \ldots, c_{n-1}$, and with respect to the coefficients of the function $\varphi(x, y)$; the coefficients of this polynomial are positive. One can prove convergence of the series (2.2.11) so obtained by using the method of majorant functions. Let us denote by H the minimum of the numbers $|1 - \lambda|$, $|2 - \lambda|, \ldots$. This minimum exists because these numbers are not vanishing and have as the only accumulation point the value $+\infty$, and is a positive number by construction. Let

$$F(X, Y) = \frac{M}{\left(1 - \frac{X}{r}\right)\left(1 - \frac{Y}{R}\right)} - M - M\frac{Y}{R} \equiv AX + \Phi(X, Y) \qquad (2.2.13)$$

be a majorant of the right-hand side of Eq. (2.2.10). Let us consider the implicit equation

$$HY - F(X, Y) = 0. \qquad (2.2.14)$$

If one looks for a solution in the form of an entire series $Y = C_1 X + C_2 X^2 + \cdots$, the coefficients will be given by the equalities

$$HC_1 = A, \ldots, \quad HC_n = P_n(C_1, C_2, \ldots, C_{n-1}), \qquad (2.2.15)$$

where the polynomial P_n, in $C_1, \ldots, C_{n-1}$ and with respect to the coefficients of $\Phi(X, Y)$, is the same as in Eq. (2.2.12). The series Y majorizes therefore the series (2.2.11) because $H \leq |n - \lambda|$. Or, by virtue of the existence theorem of implicit functions in the analytic case, Eq. (2.2.14) admits a holomorphic solution vanishing at $X = 0$ because the derivative of the left-hand side with respect to Y, evaluated at $X = 0$, is $H \neq 0$. The series $Y = C_1 X + C_2 X^2 + \cdots$ converges in a certain circle, hence also the series (2.2.11). We have therefore proved the following result of Briot and Bouquet:

Theorem 2.1 *If λ is not a positive integer, the differential equation (2.1.10), where $\varphi(x, y)$ is analytic about the origin and has neither terms independent of x and y nor terms of first degree in x and y, admits a unique holomorphic solution that vanishes at $x = 0$.*

The assessment of the simplest cases shows that this holomorphic solution is not necessarily the only solution vanishing at the origin. As a first example, let us consider the equation

$$xy' - \lambda y = ax, \qquad (2.2.16)$$

where a and λ are some constants. This is homogeneous and can be integrated straight away. One has, if $\lambda - 1 \neq 0$, the general integral

$$y = \frac{a}{(1 - \lambda)}x + Cx^\lambda, \qquad (2.2.17)$$

where C is an arbitrary constant. If we assume that λ is a non-integer positive number, the solution holomorphic at the origin corresponds to $C = 0$. If λ has a positive real part, all solutions approach 0 if x approaches 0 by following a suitable path. Indeed, if r is the modulus and ψ is the argument of x, and if $\lambda = \alpha + i\beta$, one has

$$\left|x^\lambda\right| = r^\alpha e^{-\psi\beta}. \tag{2.2.18}$$

Since α is positive, it suffices that r approaches 0 and ψ remains fixed for y to approach 0.

If $\alpha \le 0$ and $\beta \ne 0$, it is sufficient to assume that ψ is a function $\psi(r)$ such that

$$\lim_{r \to 0}\left(\alpha \log(r) - \beta\psi(r)\right) = -\infty, \tag{2.2.19}$$

for y to approach 0. One can assume that x approaches 0 on a logarithmic spiral $\psi(r) = k \log(r)$, k being a constant such that $\alpha - k\beta > 0$. Thus, the only case in which the holomorphic integral is the only that approaches 0 when x approaches 0 is that one where $\lambda \le 0$.

2.3 Integrals Not Holomorphic at the Origin

In the general case of Eq. (2.2.10), λ not being a positive integer, Briot and Bouquet obtained integrals that can approach 0 when $x \to 0$ following a suitable path. It is necessary to assume that λ is not a real number ≤ 0.

Let us denote by z the solution holomorphic at $x = 0$ and let us set $y = z + u$. Equation (2.2.10) is then transformed into the equation

$$x\frac{du}{dx} = u\left(\lambda + \alpha_{1,0}x + \alpha_{0,1}u + \cdots\right). \tag{2.3.1}$$

On bearing in mind the investigation of the previous example, let us set

$$u = x^\lambda v, \tag{2.3.2}$$

v being the new unknown function, which will solve the equation

$$x\frac{dv}{dx} = v\left[\alpha_{1,0}x + \alpha_{0,1}x^\lambda v + \cdots\right]. \tag{2.3.3}$$

One can look for a solution having the form of double series in the variables x and x^λ, i.e.

$$v = \sum_{m,n=0}^{\infty} c_{m,n}x^{m+\lambda n}. \tag{2.3.4}$$

The subsequent coefficients will be evaluated one after the other. One has indeed

$$(m + \lambda n)c_{m,n} = Q_{m,n}(\alpha_{p,q}, c_{\mu,\nu}), \ m + n > 0, \tag{2.3.5}$$

$Q_{m,n}$ being a polynomial in $(\alpha_{p,q}, c_{\mu,\nu})$, with $\mu + \nu < m + n$, and having positive coefficients. The coefficient $c_{0,0}$ is arbitrary. One will replace $c_{m,n}$ by a positive number $C_{m,n} > |c_{m,n}|$ if one replaces in $Q_{m,n}$ the $\alpha_{p,q}$ by some positive numbers $A_{p,q} > |\alpha_{p,q}|$, similarly $c_{\mu,\nu}$ by $C_{\mu,\nu} \geq |c_{\mu,\nu}|$, and if $m + \lambda n$ is replaced by a positive number $H < |m + \lambda n|$. If λ is not negative, there exists a positive number H less than all numbers $|m + \lambda n|$, $m \geq 0$, $n \geq 0$. One can take for the $A_{p,q}$ the coefficients of a majorant of the crossed terms of the right-hand of Eq. (2.3.3), expressed in the form

$$\sum_{p,q} A_{p,q} X^{p + \Lambda q V q} = \frac{M}{\left(1 - \frac{X}{r}\right)\left(1 - \frac{V X^{\Lambda}}{R}\right)} - M, \ \Lambda \geq |\lambda|. \tag{2.3.6}$$

The coefficients of the expansion of the solution of the equation

$$H(V - C_{0,0}) = \frac{MV}{\left(1 - \frac{X}{r}\right)\left(1 - \frac{V X^{\Lambda}}{R}\right)} - MV, \ \Lambda \geq |\lambda| \tag{2.3.7}$$

which takes the value $C_{0,0}$ at $X = 0$, are given by the equalities

$$HC_{m,n} = Q_{m,n}(A_{p,q}, C_{\mu,\nu}), \ m + n > 0. \tag{2.3.8}$$

This solution is a majorant of the series (2.3.4). It is legitimate to assume that Λ is an integer; one then says that Eq. (2.3.7) admits a holomorphic solution equal to $C_{0,0}$ at $X = 0$, provided that $C_{0,0}$ is sufficiently small. The series (2.3.4) converges when $|x^{\lambda}|$ and $|x|$ are small enough for the series

$$\sum_{m,n} |c_{m,n}| X^{m + \Lambda n}$$

to converge for $X < X_0$. One obtains in such a way the second theorem of Briot and Bouquet relative to Eq. (2.2.10).

Theorem 2.2 *When λ is neither a positive integer nor a real number ≤ 0, every solution of Eq. (2.2.10), which takes a value sufficiently close to the one of the solution holomorphic at the origin for a value of x sufficiently small, approaches 0 when x approaches 0 along some paths suitably chosen.*

Case where λ is a Positive Integer. If one reverts to Eq. (2.2.10) and one assumes at the beginning $\lambda = 1$, the equation that provides the coefficient c_1 of the series (2.2.11) reduces to $a = 0$; there is no solution holomorphic at the origin if $\lambda = 1$, $a \neq 0$.

Suppose that $\lambda = 1$, $a = 0$. If one defines $y \equiv xz$, one obtains the equation

$$\frac{dz}{dx} = \frac{\varphi(x, zx)}{x^2}, \tag{2.3.9}$$

whose right-hand side is analytic at the origin. There exists a solution holomorphic at the origin, $z = \psi(x)$, taking at $x = 0$ an arbitrary value z_0 provided that $|z_0|$ is small enough. *Equation (2.2.10) admits an infinity of solutions holomorphic at the origin and vanishing at this point.*

If $\lambda = 1, a \neq 0$, the circumstances are different. Let us consider the particular case of the equation

$$xy' - y = ax, \quad a = \text{const}, \tag{2.3.10}$$

which is the Euler homogeneous equation whose general integral is

$$y = x[C + a \log x], \tag{2.3.11}$$

C being the arbitrary constant. There is no solution holomorphic at the origin, but all solutions approach 0 when x approaches 0 in a suitable way (if r is the modulus of x and θ its argument, it suffices that θr approaches 0).

This feature is general. One shows that, when $\lambda = 1, a \neq 0$, one obtains solutions depending on an arbitrary constant and which are given by an entire series in the variables x and $x \log x$; they approach 0 with x under the conditions which have been specified.

Case $\lambda = n$, n being a positive integer. One performs the transformation

$$y = \frac{ax}{(1 - \lambda)} + xz, \tag{2.3.12}$$

which leads to the equation

$$xz' - (\lambda - 1)z = a_1 x + \varphi_1(x, z), \tag{2.3.13}$$

which has indeed the same form as (2.2.10), but with λ replaced by $\lambda - 1$. A finite number of such transformations brings back to the case $\lambda = 1$ that we have already considered. We can therefore state what follows:

In all cases where λ is a positive integer, two circumstances may occur: either there is no solution holomorphic at the origin, or there are infinitely many such solutions. There always exist infinitely many solutions, depending on a parameter, that approach 0 when x approaches 0 along some suitably chosen paths.

General Case If one assumes of dealing with an equation of the form

$$\frac{dy}{dx} = \frac{Y}{X}, \tag{2.3.14}$$

where X and Y are entire series in x and y, convergent in the neighbourhood of the origin and vanishing at the origin:

$$Y = ax + by + \cdots, \quad X = a'x + b'y + \cdots, \tag{2.3.15}$$

one can set (as if X and Y were being reduced to their terms of first degree)

$$y = xz. \tag{2.3.16}$$

One obtains therefore the equation

$$x\frac{dz}{dx} = \frac{a + (b - a')z - b'z^2 + x\psi(x, z)}{a' + b'z + x\varphi(x, z)}, \tag{2.3.17}$$

φ and ψ being entire series. If $|a| + |a'| \neq 0$, one is brought back to the case studied previously: if $a \neq 0$, the differential coefficient is infinite at the origin; if $a = 0$, it is not determined. When one passes to the variable $y = xz$, a solution z admitting a simple pole at the origin will provide a solution holomorphic at the origin; a solution z holomorphic and not vanishing at the origin will provide a solution y of Eq. (2.3.14), holomorphic and vanishing at the origin.

In order to obtain the solutions of Eq. (2.3.17) which are holomorphic at the origin, one can set $z = z_0 + Z$, z_0 being the value of this solution at the origin. Hence Eq. (2.3.17) becomes

$$x\frac{dZ}{dx} = \frac{a + (b - a')z_0 - b(z_0)^2 + \cdots}{a' + b'z_0 + \cdots}. \tag{2.3.18}$$

Since the left-hand side vanishes at $x = 0$ if Z is holomorphic, one has the condition

$$a + (b - a')z_0 - b(z_0)^2 = 0, \tag{2.3.19}$$

and if $a' + b'z_0 \neq 0$, one is led to an equation of the type considered when we studied the case when $f(x, y)$ is not determined at the point (x_0, y_0), which admits, in general, a holomorphic solution. Equation (2.3.18) will therefore have, in general, two solutions holomorphic at the origin, corresponding to the two roots of Eq. (2.3.19).

If a and a' vanish, but b and b' are not both 0, one exchanges the roles of x and y, which can provide solutions $y(x)$ approaching 0 when x approaches 0.

If a, a', b, b' are all vanishing, the method used above shows that there exist, in general, three holomorphic integrals of Eq. (2.3.17), and hence three integrals of Eq. (2.3.14) holomorphic and vanishing at the origin.

Note that, when a root of Eq. (2.3.19) is also a root of $a' + b'z_0$, Eq. (2.3.18) for Z can take the new form

$$x^2\frac{dZ}{dx} = \alpha x + \beta Z + \cdots . \tag{2.3.20}$$

There may exist no solution holomorphic at the origin. This will indeed be the case for the equation

$$x^2\frac{dZ}{dx} = \beta Z + \alpha x, \tag{2.3.21}$$

whose general integral

$$Z = e^{-\frac{\beta}{x}} \left[C + \alpha \int e^{\frac{\beta}{x}} \frac{dx}{x} \right], \quad C = \text{const}, \qquad (2.3.22)$$

admits, no matter what value is taken by C, the origin as an isolated singularity.

Part II
Linear Elliptic Equations

Chapter 3
Harmonic Functions

3.1 Motivations for the Laplace Equation

Our analysis of linear elliptic equations begins with the Laplace-Beltrami operator, whose consideration is suggested by several branches of physics and mathematics. Since we aim at encouraging an interdisciplinary view of mathematics, we list (and anticipate) them here already. The minimal set of examples is as follows.

(i) Newtonian potentials are solutions of Laplace's equation in the absence of masses. Conversely, solutions of Laplace's equation are always Newtonian potentials.

(ii) In electrostatics, the electric field $\vec{E}$ satisfies the equation $\operatorname{div}\vec{E} = 0$, and hence can be derived from a scalar potential according to $\vec{E} = -\operatorname{grad}(\varphi)$, provided that every closed curve in the region under consideration can be contracted to a point. But then we look for a scalar function φ of class C^2 satisfying the equation

$$- \operatorname{divgrad}(\varphi) = -\triangle \varphi = 0. \tag{3.1.1}$$

(iii) In complex analysis, holomorphic functions f of a complex variable $z = x + iy$ are continuous functions of z for which the first derivative $f'(z)$ exists. This suffices to prove continuity of f' itself [71], without having to assume that f is of class C^1, and hence the Cauchy-Riemann conditions are found to hold among the real and imaginary parts of $f(z) = u(x, y) + iv(x, y)$, i.e.

$$\frac{\partial u}{\partial x} = \frac{\partial v}{\partial y}, \ \frac{\partial u}{\partial y} = -\frac{\partial v}{\partial x}. \tag{3.1.2}$$

Equation (3.1.2) imply that both u and v satisfy Eq. (3.1.1), i.e.

$$\left(\frac{\partial^2}{\partial x^2} + \frac{\partial^2}{\partial y^2}\right) u = 0, \ \left(\frac{\partial^2}{\partial x^2} + \frac{\partial^2}{\partial y^2}\right) v = 0. \tag{3.1.3}$$

© Springer International Publishing AG 2017
G. Esposito, *From Ordinary to Partial Differential Equations*, UNITEXT 106,
DOI 10.1007/978-3-319-57544-5_3

(iv) Once Gauss proved that the theory of surfaces (i.e. smooth two-dimensional Riemannian manifolds, in modern language) can be based entirely on a pair of symmetric differential forms, called first and second fundamental form, the resulting systematic investigation of differential invariants and differential parameters by Lamé and Beltrami led to an expression of the Laplace-Beltrami operator in geometric language. The beautiful lectures of Darboux, Bianchi, and the development of the absolute differential calculus by Ricci-Curbastro and Levi-Civita (with contributions by Bianchi himself), led to the appropriate invariant setting for surfaces (see Chap. 4).

We might add to such a list the study of the functional determinant of the Laplacian in quantum field theory, but one does not really need quantum physics to discover the Laplacian, hence we are happy with the main motivations brought to the attention of our reader.

3.2 Geometry of the Second Derivatives in the Laplacian

The operator divgrad in Eq. (3.1.1) takes the familiar form in Cartesian coordinates on $\mathbf{R}^3$

$$\triangle \equiv \frac{\partial^2}{\partial x^2} + \frac{\partial^2}{\partial y^2} + \frac{\partial^2}{\partial z^2}, \tag{3.2.1}$$

and, on $\mathbf{R}^n$,

$$\triangle = \sum_{k=1}^{n} \frac{\partial^2}{\partial (x^k)^2}. \tag{3.2.2}$$

The second partial derivatives occurring in (3.2.2) result actually from taking the vector (or linear) space $\mathbf{R}^n$ endowed with the Euclidean metric

$$\delta_E = \sum_{i,j=1}^{n} \delta_{ij} \mathrm{d}x^i \otimes \mathrm{d}x^j, \tag{3.2.3}$$

where $\delta_{ij} = 1$ if $i = j$, 0 otherwise. Such a metric has the inverse δ^{-1} defined by the familiar condition

$$\sum_{l=1}^{n} \delta_{il}(\delta^{-1})^{lj} = \delta_i{}^j, \tag{3.2.4}$$

and hence the form (3.2.2) of the operator $\triangle$ results from the geometric definition

$$\triangle \equiv \sum_{i,j=1}^{n} (\delta^{-1})^{ij} \frac{\partial}{\partial x^j} \frac{\partial}{\partial x^i}. \tag{3.2.5}$$

In other words, we deal here with the Riemannian manifold $(\mathbf{R}^n, \delta_E)$. The Laplacian is actually

$$P \equiv -\triangle . \tag{3.2.6}$$

Thus, the preservation of the minus sign in Eq. (3.1.1) was intentional. By virtue of the minus sign, the leading symbol of the Laplacian is a positive-definite quadratic form, and its spectrum is bounded from below. On generic Riemannian manifolds, we shall learn soon how to write in a more general form the operator (3.2.6).

3.3 The Three Green Identities

A function $U(x, y, z)$ is said to be *harmonic* at a point $P(x, y, z)$ if its second derivatives exist and are continuous and satisfy the Laplace equation (3.1.1) throughout some neighbourhood of that point. U is said to be *harmonic in a domain*, if it is harmonic at all points of that domain. It is said to be harmonic in a closed region if it is continuous in the region, and harmonic at all interior points of the region. We begin by studying bounded regions, but in the case of infinite domains or regions a supplementary condition will be imposed later on.

Let R be a closed regular region of three-dimensional Euclidean space, and let U and W be two functions defined in R, and continuous in R together with their partial derivatives of first order. Moreover, we assume that U is of class C^2 in R. The divergence theorem holds in R with the vector having components

$$X = W\frac{\partial U}{\partial x}, \ Y = W\frac{\partial U}{\partial y}, \ Z = W\frac{\partial U}{\partial z}, \tag{3.3.1}$$

and, since $\operatorname{div}(W\operatorname{grad}U) = W \triangle U + \operatorname{grad}W \cdot \operatorname{grad}U$, it reads as

$$\int\int_R\int \left(W \triangle U + \operatorname{grad}U \cdot \operatorname{grad}W\right)dV = \int\int_S W\frac{\partial U}{\partial n}dS, \tag{3.3.2}$$

where n denotes the outwardly directed normal to the surface S bounding R. Equation (3.3.2) is the *first Green identity*. If U is harmonic and of class C^2 over the whole of R including S, and if W is set to 1, we find a simple but non-trivial theorem [88]:

Theorem 3.1 *The integral of the normal derivative of a function vanishes:*

$$\int\int_S \frac{\partial U}{\partial n}dS = 0, \tag{3.3.3}$$

when extended over the boundary of any closed regular region in which the function is harmonic and continuously differentiable. Conversely, if the integral of the normal derivative vanishes when extended over the boundary of any closed regular region in a domain, the function is harmonic in such a domain.

In particular, if $W = U$, and if U is harmonic in R, the first Green identity yields

$$\int\int_R\int \text{grad}\,U \cdot \text{grad}\,U\,dV = \int\int_S U\frac{\partial U}{\partial n}\,dS. \tag{3.3.4}$$

If U is taken to vanish on S, we then find that the volume integral of $\text{grad}\,U \cdot \text{grad}\,U$ vanishes. But this function is continuous and never negative, hence the gradient of U has to vanish at all points of R, which implies that U is constant in R. By hypothesis U vanishes on the boundary S of R, and being continuous in the closed region U must vanish throughout the whole of R. Hence we have proved

Theorem 3.2 *If the function U is harmonic and of class C^2 in a closed regular region R, and vanishes at all points of the boundary S of R, it vanishes at all points of R.*

Thus, if some functions U_1 and U_2 are both harmonic in R, and take the same boundary values, their difference $U_1 - U_2$ is harmonic in R and vanishes on the boundary. Hence it vanishes throughout R by virtue of Theorem 3.2, and one arrives at

Theorem 3.3 *A function, harmonic and continuously differentiable in a closed regular region R, is uniquely determined by its values on the boundary of R.*

Moreover, by inspection of Eq. (3.3.4), we arrive at once at another result:

Theorem 3.4 *If the function U is single-valued, continuously differentiable and harmonic in the closed regular region R, and if its normal derivative vanishes at every point of the boundary of R, then U is constant in R. Moreover, a function, single-valued and harmonic in R, is determined, up to an additive constant, by the values of its normal derivative on the boundary.*

Let us now assume that both U and W are continuously differentiable in R and have continuous partial derivatives of the second order in R. One can then write the first Green identity (3.3.2) and, in addition, the identity obtained by interchanging U and W. If such an equation is subtracted from Eq. (3.3.2), one obtains the *second Green identity*

$$\int\int_R\int \left(U\,\triangle\,W - W\,\triangle\,U\right)dV = \int\int_S \left(U\frac{\partial W}{\partial n} - W\frac{\partial U}{\partial n}\right)dS. \tag{3.3.5}$$

By inspection, the second Green identity leads to

Theorem 3.5 *If U and W are harmonic and continuously differentiable in the closed regular region R, then*

$$\int\int_S \left(U\frac{\partial W}{\partial n} - W\frac{\partial U}{\partial n}\right)dS = 0, \tag{3.3.6}$$

where S is the boundary of R.

In the case of infinite regions, however, the divergence theorem used so far does not hold without further assumptions on the functions under examination. For example, one may be interested in functions harmonic outside a given bounded surface, motivated by the physics of conductors in electrostatics. Following again Kellogg, we shall mean by an infinite regular region a region bounded by a regular surface, and containing all sufficiently distant points. For the functions U and W considered so far, one imposes, in the presence of infinite regions, the additional requirement that, on denoting by ρ the distance from any fixed point, the functions

$$\rho U, \ \rho^2 \frac{\partial U}{\partial x}, \ \rho^2 \frac{\partial U}{\partial y}, \ \rho^2 \frac{\partial U}{\partial z}; \ \rho W, \ \rho^2 \frac{\partial W}{\partial x}, \ \rho^2 \frac{\partial W}{\partial y}, \ \rho^2 \frac{\partial W}{\partial z}$$

must be bounded in absolute value for all sufficiently large ρ. If this requirement is fulfilled, the functions U and W are said to be *regular at infinity*. This is indeed the case of Newtonian potentials of bounded distributions. From now on, a function *harmonic in an infinite domain or region* is meant to be harmonic and regular at infinity. With this nomenclature, the function $W = 1$ used before is not regular at infinity, so that Theorem 3.1 cannot be derived as we did for bounded regions. But if Theorem 3.1 is applied to the portion R' of R included within and on a regular surface Σ enclosing the whole boundary S of R, we obtain

Theorem 3.1' *If R is a regular infinite region, and U is harmonic and continuously differentiable in R, the integral*

$$\int\int_\Sigma \frac{\partial U}{\partial n} dS$$

has one and the same value when extended over the boundary of any finite regular region containing the whole boundary of R in its interior.

We are now going to derive the third Green identity. For this purpose, let R be any regular region, bounded or infinite, and let $P(x, y, z)$ be any interior point. We take for W, in the second Green identity (3.3.5), the function

$$W = \frac{1}{r}, \tag{3.3.7}$$

where r is the Euclidean distance from P to $Q(\xi, \eta, \zeta), \xi, \eta, \zeta$ being now the variables of integration in that identity, in place of x, y, z. Since P is interior to R, the identity cannot be applied to the whole region R, hence we surround P with a small sphere σ centred at P, and remove from R the interior of such a sphere. On denoting by R' the resulting region one can write, since $\frac{1}{r}$ is harmonic in R',

$$-\int\int_{R'}\int \frac{1}{r} \triangle U \, dV = \int\int_S \left(U \frac{\partial}{\partial \nu} \frac{1}{r} - \frac{1}{r} \frac{\partial U}{\partial \nu} \right) dS$$
$$+ \int\int_\sigma \left(U \frac{\partial}{\partial \nu} \frac{1}{r} - \frac{1}{r} \frac{\partial U}{\partial \nu} \right) dS. \tag{3.3.8}$$

With this notation, ν is the normal to the boundary of R, pointing outwards from R, so that on the sphere σ it has the direction opposite to the radius r. Hence the last integral in (3.3.8) may be written as

$$\int\int_\Omega \left(U\frac{1}{r^2} + \frac{1}{r}\frac{\partial U}{\partial r}\right)r^2 d\Omega = \overline{U}4\pi + \int\int_\Omega r\frac{\partial U}{\partial r}d\Omega, \qquad (3.3.9)$$

where $\overline{U}$ is the value of U at some point of σ, and the integration is performed with respect to the solid angle subtended at P by the element of σ. The limit of the integral over σ in (3.3.8), as the radius of the small sphere σ approaches 0, is therefore $4\pi U(P)$, while the volume integral on the left-hand side converges to the integral over R. In this way one arrives eventually at the *third Green identity*

$$U(P) = -\frac{1}{4\pi}\int\int_R\int \frac{(\Delta U)}{r}dV + \frac{1}{4\pi}\int\int_S \frac{\partial U}{\partial \nu}\frac{1}{r}dS - \frac{1}{4\pi}\int\int_S U\frac{\partial}{\partial \nu}\frac{1}{r}dS.$$
$$(3.3.10)$$

This holds under the assumptions that U and its partial derivatives of the first order are continuous in R, and that its partial derivatives of the second order are continuous in the interior of R, and that the volume integral is convergent if R is infinite. In this case, also regulatity at infinity of U is assumed.

This formula admits of a remarkable physical interpretation, because the first term on the right-hand side is the potential of a volume distribution of density $-\frac{(\Delta U)}{4\pi}$, the second is the potential of a distribution on the boundary S of R, of density $\frac{1}{4\pi}\frac{\partial U}{\partial \nu}$, while the third is the potential of a double distribution on S of moment $-\frac{U}{4\pi}$. Thus, not only do harmonic functions behave as Newtonian potentials, but so also do any functions with sufficient differentiability. In particular, the third Green identity (3.3.10) implies the following result:

Theorem 3.6 *A function, harmonic and continuously differentiable in a closed regular region R may be represented as the sum of the potentials of a simple and of a double distribution on the boundary of R.*

Let now T be any domain, regular or not, and let U be harmonic in T. Then U is harmonic in any sphere lying entirely in T, and is thus, in that sphere, the potential of the so-called Newtonian spreads on the surface. But such spreads are known to be real-analytic at the points of free space, and hence in the interior of the sphere. Since such a sphere can be described about any point whatsoever of T, one obtains

Theorem 3.7 *If U is harmonic in a domain, it is real-analytic at all points of that domain.*

The property of having only real-analytic solutions is shared by the Laplace equation and by a class of partial differential equations. Conversely, from the theory of complex-analytic functions, one can re-obtain all known properties of harmonic functions (see, e.g., [2], but we do not follow this path, since the theory of harmonic functions can be built without abandoning real analysis and real differential geometry.

3.4 Mean-Value Theorem

The third Green identity (3.3.10) leads to a general proof of the Gauss theorem on the arithmetic mean. For this purpose, let U be harmonic and continuously differentiable in R. Then (3.3.10) yields

$$U(P) = \frac{1}{4\pi} \int \int_S \frac{\partial U}{\partial \nu} \frac{1}{r} dS - \frac{1}{4\pi} \int \int_S U \frac{\partial}{\partial \nu} \frac{1}{r} dS. \tag{3.4.1}$$

If R is bounded by a sphere S, and P is the centre of the sphere, the first integral vanishes, by Theorem 3.1, because r is constant on S. Hence Eq. (3.4.1) reduces to

$$U(P) = \frac{1}{4\pi r^2} \int \int_S U \, dS. \tag{3.4.2}$$

The result (3.4.2) has been obtained by assuming that U and its derivatives of the first order are continuous in the closed sphere, but the derivatives of U do not occur in the integral formula, and no assumption on their behaviour at the boundary is needed. The integral formula holds for any interior concentric sphere. We can state the result as follows.

Theorem 3.8 Gauss' Theorem of the Arithmetic Mean. *If the function U is harmonic in a sphere, the value of U at the centre of the sphere is the arithmetic mean of its values on the surface.*

Conversely, if R is a closed region, and W a function which is continuous in R, and whose value at any interior point of the region is the arithmetic mean of its values on the surface of any sphere with that point as centre, which lies entirely in R, one has

$$W(P) = \frac{1}{4\pi r^2} \int_0^\pi \int_0^{2\pi} W(Q) r^2 \sin \theta \, d\varphi \, d\theta, \quad r \le a, \tag{3.4.3}$$

where Q has the spherical coordinates (r, φ, θ) with P as origin, and a is the distance from P to the nearest boundary point of R.

Remarkably, the function W is also its own arithmetic mean over the volumes of spheres. This property is proved by multiplying both sides of Eq. (3.4.3) by r^2 and then integrating from 0 to r. One thus obtains, since $W(P)$ is independent of r,

$$W(P) \frac{r^3}{3} = \frac{1}{4\pi} \int_0^r \int_0^\pi \int_0^{2\pi} W(Q) \rho^2 \sin \theta \, d\varphi \, d\theta \, d\rho, \quad r \le a, \tag{3.4.4}$$

i.e.,

$$W(P) = \frac{3}{4\pi r^3} \int \int_{\Sigma} \int W(Q)\mathrm{d}V, \tag{3.4.5}$$

where Σ is the sphere of radius r about P. With suitable cares, following Kellogg [88], one can prove that $W(P)$ has continuous derivatives at the interior points of R, and that the derivatives of W of first order are also their own arithmetic means over the volumes of spheres in any region in the interior of R:

$$\frac{\partial W}{\partial z} = \frac{3}{4\pi a^3} \int \int \int \frac{\partial W}{\partial z} \mathrm{d}V, \tag{3.4.6}$$

the integral being taken over the region bounded by Σ. The process can be repeated as many times as is necessary, and this shows that the partial derivatives of any given order of W exist and are continuous in any region interior to R. In particular,

$$\frac{\partial^2 W}{\partial z^2} = \frac{3}{4\pi a^3} \int \int_{\Sigma} \frac{\partial W}{\partial z} \cos(n, z)\mathrm{d}S \implies \Delta W = \frac{3}{4\pi a^3} \int \int_{\Sigma} \frac{\partial W}{\partial n} \mathrm{d}S, \tag{3.4.7}$$

at all points a distance more than a from the boundary. The last integral in (3.4.7) vanishes: if in (3.4.3) one cancels the constant factor r^2 inside and outside the integral, and differentiates the resulting equation with respect to r, which is possible because the derivatives of W are continuous, one finds

$$\int_0^\pi \int_0^{2\pi} \frac{\partial W}{\partial r} \sin\theta \, \mathrm{d}\varphi \, \mathrm{d}\theta = \int \int_{\Sigma} \frac{\partial W}{\partial n} \mathrm{d}S = 0. \tag{3.4.8}$$

At any interior point of R the function W is therefore harmonic, and we obtain the converse of Gauss' theorem, stated as follows.

Theorem 3.9 *If the function W is continuous in the closed region R, and at every interior point of R has as value the arithmetic mean of its values on any sphere with centre at that point and lying in R, then W is harmonic in R.*

This theorem may even be used to define harmonic functions in the first place. Another result about the converse of a property is as follows.

Theorem 3.10 *If the function U is continuous in a region R, and has continuous derivatives of the first order in the interior of R, and if the integral*

$$\int \int_S \frac{\partial U}{\partial n} \mathrm{d}S$$

vanishes when extended over the boundary of all regular regions interior to R, or even if only over all spheres, then U is harmonic in the interior of R.

In order to prove the last theorem stated in this section, let P be an interior point of R, and Σ a sphere, of radius r, about P, and lying in the interior of R. By virtue of the hypothesis, one can write

$$0 = \int \int_{\Sigma} \frac{\partial U}{\partial n} dS = \int_0^{\pi} \int_0^{2\pi} \frac{\partial U}{\partial r} r^2 \sin \theta \, d\varphi \, d\theta, \tag{3.4.9}$$

and hence, since r is constant on Σ,

$$\int_0^{\pi} \int_0^{2\pi} \frac{\partial U}{\partial r} \sin \theta \, d\varphi \, d\theta = 0. \tag{3.4.10}$$

Integration of this equation from 0 to r yields

$$\int_0^{\pi} \int_0^{2\pi} U(Q) \sin \theta \, d\varphi \, d\theta - 4\pi U(P) = 0, \tag{3.4.11}$$

where the integral is taken over the sphere of radius r about P. This formula is equivalent to

$$U(P) = \frac{1}{4\pi r^2} \int \int_{\Sigma} U(Q) dS. \tag{3.4.12}$$

This holds at first for spheres interior to R, and by continuity for spheres in R. By virtue of (3.4.12) the function U is its own arithmetic mean on the surfaces of spheres in R, and hence it is harmonic in R by Theorem 3.9. The desired proof is therefore obtained.

So far we have been dealing mainly with $\mathbf{R}^3$, but our presentation can be extended to $\mathbf{R}^n$ endowed with the Euclidean distance of vectors

$$d_E(x, y) \equiv \sqrt{\sum_{k=1}^{n} (x^k - y^k)^2}. \tag{3.4.13}$$

Two basic concepts are the *open ball* $B_r(x)$ centred at x with radius r, defined by

$$B_r(x) \equiv \{ y \in \mathbf{R}^n : d_E(x, y) < r \}, \tag{3.4.14}$$

and the corresponding *sphere*, which is the boundary of B_r, i.e.

$$\partial B_r(x) \equiv \{ y \in \mathbf{R}^n : d_E(x, y) = r \}. \tag{3.4.15}$$

In particular, the volume of the unit ball in $\mathbf{R}^n$ is given by

$$\omega_n = \frac{\pi^{\frac{n}{2}}}{\Gamma \left(\frac{n}{2} + 1 \right)}, \tag{3.4.16}$$

where Γ is the Gamma function.

3.5   The Weak Maximum Principle

The weak maximum principle for harmonic functions states that a function which is harmonic is bounded inside a domain by its values on the boundary. This does not rule out the possibility that a non-constant function attains an interior maximum. More precisely, the property that we are going to prove can be stated as follows.

Theorem 3.11 *If Ω is a bounded, connected open set in $\mathbf{R}^n$ and the function $U \in C^2(\Omega) \cap C(\overline{\Omega})$ is harmonic in Ω, then*

$$\max_{\overline{\Omega}} U = \max_{\partial\Omega} U, \ \min_{\overline{\Omega}} U = \min_{\partial\Omega} U. \tag{3.5.1}$$

In order to prove it, one can use an argument based on the non-positivity of the second derivative at an interior maximum. Indeed, on denoting by ε a positive parameter, one can consider the 1-parameter family of functions

$$U^\varepsilon(x) = U(x) + \varepsilon \sum_{k=1}^{n} (x^k)^2. \tag{3.5.2}$$

Since U is taken to be harmonic in Ω, one finds [87]

$$\triangle U^\varepsilon = \varepsilon \sum_{l=1}^{n} \frac{\partial}{\partial x^l} \sum_{k=1}^{n} 2x^k \delta_l^k = 2n\varepsilon > 0. \tag{3.5.3}$$

If U^ε has a local maximum at an interior point, then $\triangle U^\varepsilon \le 0$ by the second derivative test, which contradicts Eq. (3.5.3). Thus, U^ε has no interior maximum, and it attains its maximum on the boundary. If $|x| \le R$ for all $x \in \Omega$, one finds therefore the inequalities

$$\sup_\Omega U \le \sup_\Omega U^\varepsilon \le \sup_{\partial\Omega} U^\varepsilon \le \sup_{\partial\Omega} U + \varepsilon R^2. \tag{3.5.4}$$

Upon letting ε approach 0 from the right, one gets

$$\sup_\Omega U \le \sup_{\partial\Omega} U. \tag{3.5.5}$$

If the same argument is applied to $-U$, one finds the minorization

$$\inf_\Omega U \ge \inf_{\partial\Omega} U. \tag{3.5.6}$$

By virtue of (3.5.5) and (3.5.6), the theorem expressed by (3.5.1) holds.

3.6 Derivative Estimates

An important feature of the Laplace equation obeyed by harmonic functions is that one can estimate the derivatives of a solution in a ball by means of the solution on a larger ball. To state it properly, we introduce some standard notation in the modern literature. The set A is said to be compactly contained in the set B if the closure $\overline{A}$ of A is contained in B, and if such a closure is a compact set. Moreover, for functions $f \in L^1_{loc}(\Omega)$ over a ball $B_r(x)$ compactly contained in Ω, or the corresponding sphere $\partial B_r(x)$, one defines the averages

$$\fint_{B_r(x)} f \, dx \equiv \frac{1}{\omega_n r^n} \int_{B_r(x)} f \, dx, \tag{3.6.1}$$

$$\fint_{\partial B_r(x)} f \, dS \equiv \frac{1}{n \omega_n r^{n-1}} \int_{\partial B_r(x)} f \, dS. \tag{3.6.2}$$

Now we can state and prove what follows.

Theorem 3.12 *If the function* $U \in C^2(\Omega)$ *is harmonic in the open set* Ω, *within which the open ball* $B_r(x)$ *is compactly contained, then for any integer i such that $1 \le i \le n$ one has the inequality*

$$|\partial_i U(x)| \le \frac{n}{r} \max_{\overline{B}_r(x)} |U|. \tag{3.6.3}$$

In the proof, since the function U is taken to be smooth, differentiation of the Laplace equation for U with respect to x^i shows that also $\partial_i U$ is harmonic, hence by the mean value property for balls, jointly with the divergence theorem,

$$\partial_i U = \fint_{B_r(x)} \partial_i U \, dx = \frac{1}{\omega_n r^n} \int_{\partial B_r(x)} U \nu_i \, dS. \tag{3.6.4}$$

On taking the absolute value of this equation, and using the estimate

$$\left| \int_{\partial B_r(x)} U \nu_i \, dS \right| \le n \omega_n r^{n-1} \max_{\overline{B}_r(x)} |U|, \tag{3.6.5}$$

one obtains the desired result. A corollary of this theorem is that a bounded harmonic function on $\mathbf{R}^n$ is constant, because all components of its gradient are majorized by functions that approach 0.

Appendix 3.a: Frontier of a Set and Manifolds with Boundary

Since the concept of boundary plays an important role in this section, we consider it in greater detail in this appendix.

Given a subset E of $\mathbf{R}^n$, it closure is denoted by $\overline{E}$, the interior by E° and the *frontier* by

$$\partial E \equiv \frac{\overline{E}}{E^\circ}. \tag{3.a1}$$

A richer concept is the one of manifold with boundary, a prototype of which is the open ball of Euclidean space, whose boundary is the corresponding sphere as we have seen in (3.4.15). Following Schwartz [125], a manifold with boundary, of class C^m and dimension n, is a part V of a manifold $\widetilde{V}$ of class C^m and dimension n, bounded, identical with the closure of its interior:

$$V = \overline{V^\circ}, \tag{3.a2}$$

whose frontier $\partial V = \Sigma$ is a hypersurface of $\widetilde{V}$, sub-manifold of class C^m and dimension $n - 1$. This frontier is then said to be the boundary of V, also denoted by bV.

Some of the topological properties which hold when V is a ball of an Euclidean space, and bV the corresponding sphere, can be extended to the general case. If Σ is empty, then V is simply an ordinary manifold (also called without boundary or closed). Let us therefore consider the case where Σ is not empty. Its complementary set $C\Sigma$ within $\widetilde{V}$ is the union of two disjoint open sets, i.e. the interior V° of V, and the complementary set CV, exterior to V. None of these two open sets can be empty. Indeed, if V° were empty, then by virtue of (3.a2) V would be necessarily empty as well, being the closure of an empty set. On the other hand, if CV were empty, then V would be identical with $\widetilde{V}$, and hence its frontier Σ would be empty. It then follows that $C\Sigma$, being the union of two disjoint and non-empty open sets, is certainly not connected. It contains at least two regions. If we denote by $(\Omega_i)_{i \in I}$ the connected components or regions of $C\Sigma$, each of them, if it has at the same time points in common with V° and with CV, must have points in common with Σ, which is instead impossible because it lies in the complementary set of Σ; thus, each of the regions Ω_i lies entirely in V° or entirely in CV. In other words, one can obtain a partition of the set I in the form of union of two complementary sets J and K, and one has

$$V^\circ = \cup_{i \in J}(\Omega_i), \quad CV = \cup_{i \in K}\Omega_i. \tag{3.a3}$$

The interior and the exterior of V are unions of regions of the complementary set of Σ. If one finds that Σ splits V exactly into two regions, then these two regions are necessarily V° and CV.

Chapter 4
Mathematical Theory of Surfaces

4.1 Quadratic Differential Forms

Assuming that the reader is familiar with the algebraic theory of quadratic forms, we now move on to study quadratic differential forms, bearing in mind the Gauss findings, according to which two-dimensional surfaces can be completely characterized by their first and second fundamental forms. We shall arrive eventually at the most general form of the Laplacian in metric differential geometry, without assuming any prior knowledge of tensor calculus or relativity.

If $x^1, x^2, \ldots, x^n$ are n independent variables and $dx^1, dx^2, \ldots, dx^n$ their differentials, let us consider the *quadratic differential form* [11]

$$f = \sum_{r,s=1}^{n} a_{rs} dx^r \otimes dx^s, \tag{4.1.1}$$

where the coefficients a_{rs} are given functions of the x variables. In general we shall regard the $x^1, x^2, \ldots, x^n$ as n real variables and shall assume that the $a_{rs}(x)$ are real- and single-valued functions, finite and continuous, with partial derivatives of first and second order finite and continuous as well. Strictly, their order of differentiability will be as large as necessary for the applications of interest. The determinant of the $n \times n$ matrix of functions a_{rs} is nowhere vanishing by hypothesis.

If the n variables x are expressed by means of n new variables x' through the formulae

$$x'^i = f^i(x^1, x^2, \ldots, x^n), \quad \forall i = 1, 2, \ldots, n, \tag{4.1.2}$$

where the f^i are taken to satisfy the same differentiability conditions of the a_{rs}, the differentials of the x^i undergo the linear transformation

© Springer International Publishing AG 2017
G. Esposito, *From Ordinary to Partial Differential Equations*, UNITEXT 106,
DOI 10.1007/978-3-319-57544-5_4

$$\mathrm{d}x^r = \sum_{i=1}^{n} p_i^r \mathrm{d}x'^i, \quad p_i^r \equiv \frac{\partial x^r}{\partial x'^i}, \tag{4.1.3}$$

and hence f is changed into a new quadratic differential form

$$f' = \sum_{r,s=1}^{n} b_{rs} \mathrm{d}x'^r \otimes \mathrm{d}x'^s, \tag{4.1.4}$$

where

$$b_{rs} = \sum_{i,k=1}^{n} a_{ik} \frac{\partial x^i}{\partial x'^r} \frac{\partial x^k}{\partial x'^s}. \tag{4.1.5}$$

If P denotes the functional determinant of the linear substitution (4.1.3), i.e.

$$P \equiv \frac{\partial(x^1, x^2, \ldots, x^n)}{\partial(x'^1, x'^2, \ldots, x'^n)} = \det \frac{\partial x^i}{\partial x'^j}, \tag{4.1.6}$$

one has

$$b = \det b_{rs} = a P^2. \tag{4.1.7}$$

Moreover, the inverse A^{ij} of a_{rs}, defined by the condition

$$\sum_{i=1}^{n} a_{li} A^{ik} = \delta_l^k, \tag{4.1.8}$$

and the inverse B^{ij} of b_{rs}, are related by

$$A^{rs} = \sum_{i,k=1}^{n} B^{ik} \frac{\partial x^r}{\partial x'^i} \frac{\partial x^s}{\partial x'^k}. \tag{4.1.9}$$

4.2   Invariants and Differential Parameters

Suppose now that we deal with an expression formed with the coefficients a_{rs} of the form f and their derivatives of first, second,...n-th order [11]

$$\varphi\left(a_{rs}, \frac{\partial a_{rs}}{\partial x^t}, \frac{\partial^2 a_{rs}}{\partial x^t \partial x^u}, \ldots\right)$$

of such a nature that, upon operating a change whatsoever of the variables x into new variables x', it gets changed into the same expression

$$\varphi\left(b_{rs}, \frac{\partial b_{rs}}{\partial x''}, \frac{\partial^2 b_{rs}}{\partial x'' \partial x''''}, \dots\right)$$

formed in the same way with the coefficients b_{rs} of the transformed differential form f' and with their derivatives; such an expression φ is then said to be a *differential invariant of the form f*.

Another case of interest is obtained when in the expression φ, besides the coefficients of the differential form f and their derivatives, there also occur a number of arbitrary functions $U, V, \dots$ jointly with their derivatives, in such a way that, under a change whatsoever of variables, one finds again

$$\varphi\left(a_{rs}, \frac{\partial a_{rs}}{\partial x^l}, \dots, U, V, \dots, \frac{\partial U}{\partial x^i}, \frac{\partial V}{\partial x^k} \dots\right)$$
$$= \varphi\left(b_{rs}, \frac{\partial b_{rs}}{\partial x''}, \dots, U', V', \dots, \frac{\partial U'}{\partial x'^i}, \frac{\partial V'}{\partial x'^k} \dots\right), \tag{4.2.1}$$

where $U' \equiv U(x(x'))$, $V' = V(x(x'))$, ... If Eq. (4.2.1) holds, the expression φ is said to be a *differential parameter*. A differential parameter is said to be of order r if the highest order derivatives of the arbitrary functions occurring in it are of order r. As can be seen, the differential parameters, relative to a quadratic form f, are expressions formed with the coefficients of f, with a certain number of arbitrary functions and with the derivatives of the coefficients a_{rs} and functions $U, V \dots$, whose form is preserved under a change whatsoever of variables. If the arbitrary functions are missing, a differential parameter reduces to a differential invariant.

The method based upon the use of differential parameters and differential forms is as follows. Suppose that a differential form ψ is known, quadratic or of higher degree, whose coefficients are formed with those of the fundamental form f and with their derivatives, jointly with a certain number of arbitrary functions and their derivatives. Let the form ψ be of such a nature that, under a change whatsoever of variables, it gets changed into the differential form ψ' formed in the same way with respect to the transformed form f' and to the arbitrary functions. In such a case the form ψ is said to be *covariant to the form f*. If f and ψ are viewed as algebraic forms (in the differentials dx^i), by building their absolute invariants, one obtains differential parameters or differential invariants of f, depending on whether or not there occur arbitrary functions in the coefficients of the differential form ψ.

4.3   Differential Parameters of First Order

If U is an arbitrary function of class C^1 of the variables $x^1, x^2, \ldots, x^n$, in the square of its first differential

$$dU \otimes dU = \sum_{r,s=1}^{n} \frac{\partial U}{\partial x^r} \frac{\partial U}{\partial x^s} dx^r \otimes dx^s \qquad (4.3.1)$$

one has a quadratic differential form manifestly covariant to the given form f. On denoting by λ an arbitrary parameter, one can consider

$$\varphi \equiv \sum_{r,s=1}^{n} \left(a_{rs} + \lambda \frac{\partial U}{\partial x^r} \frac{\partial U}{\partial x^s} \right) dx^r \otimes dx^s, \qquad (4.3.2)$$

which is a differential form covariant to f as well. The coefficients of the various powers of λ in the ratio

$$\frac{\det \varphi_{rs}}{\det a_{rs}}$$

will be therefore the first differential parameters, formed with the arbitrary function U. In particular, the differential parameter given by the coefficient of the first power of λ, denoted by $\triangle_1 U$, has manifestly the value

$$\triangle_1 U \equiv \sum_{r,s=1}^{n} A^{rs} \frac{\partial U}{\partial x^r} \frac{\partial U}{\partial x^s}. \qquad (4.3.3)$$

It is called, following Beltrami, the *first differential parameter* of the function U.

Let now V be a second arbitrary function; in the tensor product of the first differentials

$$dU \otimes dV = \sum_{r,s=1}^{n} \frac{\partial U}{\partial x^r} \frac{\partial V}{\partial x^s} dx^r \otimes dx^s \qquad (4.3.4)$$

one has again a differential form covariant to f, and the same considerations developed above prove that the expression (the notation being the one of Beltrami)

$$\nabla(U, V) \equiv \sum_{r,s=1}^{n} A^{rs} \frac{\partial U}{\partial x^r} \frac{\partial V}{\partial x^s} \qquad (4.3.5)$$

is a differential parameter of first order with the two arbitrary functions U and V. It is called the *mixed differential parameter* of U and V. Of course, if one sets $V = U$ in Eq. (4.3.5), one obtains the first differential parameter $\triangle_1 U$.

4.4 Equivalence of Quadratic Forms; Christoffel Formulae

Two quadratic differential forms

$$f = \sum_{r,s=1}^{n} a_{rs} dx^r \otimes dx^s, \quad f' = \sum_{r,s=1}^{n} b_{rs} dx'^r \otimes dx'^s \tag{4.4.1}$$

are said to be *equivalent*, the a_{rs} being given functions of the independent variables $x^1, \ldots, x^n$, and similarly for the b_{rs} and the independent variables $x'^1, \ldots, x'^n$, if it is possible to assign to the x's some values in terms of the x' variables in such a way that the form f is turned into the form f'. For the equivalence of the two differential forms it is therefore necessary and sufficient that one can determine n unknown functions

$$x^i = x^i(x'^1, \ldots, x'^n), \quad \forall i = 1, \ldots, n, \tag{4.4.2}$$

in such a way as to fulfill the $\frac{n(n+1)}{2}$ simultaneous partial differential equations of first order

$$b_{rs} = \sum_{i,k=1}^{n} a_{ik} \frac{\partial x^i}{\partial x'^r} \frac{\partial x^k}{\partial x'^s}. \tag{4.4.3}$$

Since $\frac{n(n+1)}{2} > n$, it is clear that the equivalence of the two differential forms requires particular relations among the coefficients. In order to assess the compatibility of Eqs. (4.4.3), it is convenient to study the equations resulting from differentiation of (4.4.3). We are going to obtain the Christoffel formulae, which imply that all second-order derivatives of the unknown functions x are expressed through their first derivatives with respect to the independent variables. For this purpose, let us differentiate Eq. (4.4.3) with respect to any whatsoever of the x' variables, let it be x'^t, and let us bear in mind that $b_{rs} = b_{rs}(x')$, whereas for the derivatives of a_{rs} one has to apply the rule for taking derivatives of composite functions, i.e.

$$\frac{\partial a_{ik}}{\partial x'^t} = \sum_{l=1}^{n} \frac{\partial a_{ik}}{\partial x^l} \frac{\partial x^l}{\partial x'^t}. \tag{4.4.4}$$

Hence one finds

$$\frac{\partial b_{rs}}{\partial x'^t} = \sum_{i,k,l=1}^{n} \frac{\partial a_{ik}}{\partial x^l} \frac{\partial x^i}{\partial x'^r} \frac{\partial x^k}{\partial x'^s} \frac{\partial x^l}{\partial x'^t}$$

$$+ \sum_{i,k=1}^{n} a_{ik} \left(\frac{\partial^2 x^i}{\partial x'^r \partial x'^t} \frac{\partial x^k}{\partial x'^s} + \frac{\partial x^i}{\partial x'^r} \frac{\partial^2 x^k}{\partial x'^s \partial x'^t} \right). \tag{4.4.5}$$

In this equation we interchange the index s with t, and we permute in the triple sum on the right-hand side the summation indices k, l, hence obtaining

$$\frac{\partial b_{rt}}{\partial x'^{s}} = \sum_{i,k,l=1}^{n} \frac{\partial a_{il}}{\partial x^{k}} \frac{\partial x^{i}}{\partial x'^{r}} \frac{\partial x^{k}}{\partial x'^{s}} \frac{\partial x^{l}}{\partial x'^{t}}$$
$$+ \sum_{i,k=1}^{n} a_{ik} \left(\frac{\partial^{2} x^{i}}{\partial x'^{r} \partial x'^{s}} \frac{\partial x^{k}}{\partial x'^{t}} + \frac{\partial x^{i}}{\partial x'^{r}} \frac{\partial^{2} x^{k}}{\partial x'^{s} \partial x'^{t}} \right). \qquad (4.4.6)$$

In this last equation we interchange r with s and, in the triple sum on the right-hand side, as in the second term of the double sum, we interchange i with k. This leads to

$$\frac{\partial b_{st}}{\partial x'^{r}} = \sum_{i,k,l=1}^{n} \frac{\partial a_{kl}}{\partial x^{i}} \frac{\partial x^{i}}{\partial x'^{r}} \frac{\partial x^{k}}{\partial x'^{s}} \frac{\partial x^{l}}{\partial x'^{t}}$$
$$+ \sum_{i,k=1}^{n} a_{ik} \left(\frac{\partial^{2} x^{i}}{\partial x'^{r} \partial x'^{s}} \frac{\partial x^{k}}{\partial x'^{t}} + \frac{\partial x^{k}}{\partial x'^{s}} \frac{\partial^{2} x^{i}}{\partial x'^{r} \partial x'^{t}} \right). \qquad (4.4.7)$$

At this stage, Eqs. (4.4.6) and (4.4.7) are added, subtracting eventually Eq. (4.4.5). In such a linear combination of equations, we then divide both members by 2, finding

$$\frac{1}{2} \left(\frac{\partial b_{rt}}{\partial x'^{s}} + \frac{\partial b_{st}}{\partial x'^{r}} - \frac{\partial b_{rs}}{\partial x'^{t}} \right)$$
$$= \sum_{l=1}^{n} \frac{\partial x^{l}}{\partial x'^{t}} \left\{ \sum_{i,k=1}^{n} \frac{1}{2} \left(\frac{\partial a_{il}}{\partial x^{k}} + \frac{\partial a_{kl}}{\partial x^{i}} - \frac{\partial a_{ik}}{\partial x^{l}} \right) \frac{\partial x^{i}}{\partial x'^{r}} \frac{\partial x^{k}}{\partial x'^{s}} \right.$$
$$+ \sum_{i=1}^{n} a_{il} \frac{\partial^{2} x^{i}}{\partial x'^{r} \partial x'^{s}} \Bigg\}. \qquad (4.4.8)$$

This formula suggests defining the 3-index *Christoffel symbol of first kind*, for which we here use the notation (see comments a few lines below)

$$\gamma[i, k, l] \equiv \frac{1}{2} \left(\frac{\partial a_{il}}{\partial x^{k}} + \frac{\partial a_{kl}}{\partial x^{i}} - \frac{\partial a_{ik}}{\partial x^{l}} \right). \qquad (4.4.9)$$

In analogous way, we set

$$\gamma'[i, k, l] \equiv \frac{1}{2} \left(\frac{\partial b_{il}}{\partial x'^{k}} + \frac{\partial b_{kl}}{\partial x'^{i}} - \frac{\partial b_{ik}}{\partial x'^{l}} \right). \qquad (4.4.10)$$

With this notation, Eq. (4.4.8) reads

$$\gamma'[r, s, t] = \sum_{l=1}^{n} \frac{\partial x^l}{\partial x'^t} \left\{ \sum_{i,k=1}^{n} \gamma[i, k, l] \frac{\partial x^i}{\partial x'^r} \frac{\partial x^k}{\partial x'^s} + \sum_{i=1}^{n} a_{il} \frac{\partial^2 x^i}{\partial x'^r \partial x'^s} \right\}. \quad (4.4.11)$$

We now want to solve for the n second derivatives $\frac{\partial^2 x^i}{\partial x'^r \partial x'^s}$, $\forall i = 1, \dots, n$, keeping fixed r, s.

For this purpose we multiply Eq. (4.4.11) by $B^{mt} \frac{\partial x^c}{\partial x'^m}$, summing with respect to m, t from 1 through n and bearing in mind Eq. (4.1.9). Hence one obtains

$$\sum_{m=1}^{n} \left(\sum_{t=1}^{n} B^{mt} \gamma'[r, s, t] \right) \frac{\partial x^c}{\partial x'^m} = \sum_{i,k=1}^{n} \left(\sum_{l=1}^{n} A^{cl} \gamma[i, k, l] \right) \frac{\partial x^i}{\partial x'^r} \frac{\partial x^k}{\partial x'^s} + \frac{\partial^2 x^c}{\partial x'^r \partial x'^s}. \quad (4.4.12)$$

The form of Eq. (4.4.12) suggests defining the 3-index *Christoffel symbols of second kind*, i.e.

$$\Gamma\{c, [i, k]\} \equiv \sum_{l=1}^{n} A^{cl} \gamma[i, k, l]. \quad (4.4.13)$$

These symbols are more frequently written in the modern literature in the form Γ^c_{ik}, but we find it misleading, because in tensor calculus (or absolute differential calculus) such an expression denotes the components of a tensor field of type $(1, 2)$, i.e. once contravariant and twice covariant, whereas the Christoffel symbols of second kind do not have the homogeneous transformation property of tensors (they are instead tensor affinities [138]). Thus, the old-fashioned notation of the founding fathers Ricci-Curbastro, Bianchi and Levi-Civita is actually clearer. We have decided to use γ for Christoffel symbols of first kind, and Γ for Christoffel symbols of second kind.

By virtue of (4.4.12) and (4.4.13), we can express the desired second derivatives of the unknown functions in the form

$$\frac{\partial^2 x^c}{\partial x'^r \partial x'^s} = \sum_{m=1}^{n} \Gamma'\{m, [r, s]\} \frac{\partial x^c}{\partial x'^m} - \sum_{i,k=1}^{n} \Gamma\{c, [i, k]\} \frac{\partial x^i}{\partial x'^r} \frac{\partial x^k}{\partial x'^s}. \quad (4.4.14)$$

These are the Christoffel formulae for the second derivatives of the unknown functions x^c.

4.5 Properties of Christoffel Symbols

The Christoffel symbols make it possible to obtain the general expression of the Laplacian on scalar functions in Riemannian geometry, and hence we devote a section to their properties, without assuming that the reader is already familiar with them. This may well be the case for those readers who either know only the index-free notation or have studied only flat n-dimensional Euclidean space.

The symbols of first kind are, by definition, symmetric in the first two indices, i.e.

$$\gamma[i, k, l] = \gamma[k, i, l].$$ (4.5.1)

By virtue of (4.4.13), also the symbols of second kind are symmetric in the two indices

$$\Gamma \{l, [i, h]\} = \Gamma \{l, [h, i]\}.$$ (4.5.2)

By construction, first derivatives of the coefficients of the differential form f are a linear combination of Christoffel symbols of first kind, i.e.

$$\frac{\partial a_{ik}}{\partial x^l} = \gamma[i, l, k] + \gamma[k, l, i],$$ (4.5.3)

and the definition (4.4.13) can be inverted to give

$$\gamma[i, k, l] = \sum_{c=1}^{n} a_{lc} \Gamma \{c, [i, k]\}.$$ (4.5.4)

Thus, Eq. (4.5.3) can be re-expressed in the form

$$\frac{\partial a_{ik}}{\partial x^l} = \sum_{m=1}^{n} \left[\Gamma \{m, [i, l]\} a_{km} + \Gamma \{m, [k, l]\} a_{im} \right].$$ (4.5.5)

The logarithmic derivative of $a = \det a_{rs}$ occurs frequently in the applications, and it can be re-expressed concisely by means of Christoffel symbols. Indeed, relying upon the differentiation rule for determinants, according to which

$$\frac{1}{a} \frac{\partial a}{\partial x^l} = \sum_{i,k=1}^{n} A^{ik} \frac{\partial a_{ik}}{\partial x^l},$$ (4.5.6)

one finds, by virtue of (4.5.3),

$$\frac{1}{a} \frac{\partial a}{\partial x^l} = \sum_{i,k=1}^{n} A^{ik} \left(\gamma[i, l, k] + \gamma[k, l, i] \right) = 2 \sum_{i,k=1}^{n} A^{ik} \gamma[i, l, k],$$ (4.5.7)

where in the second equality we have exploited relabelling of summation indices and the symmetry of A^{ik}. The definition (4.4.13) of symbols of second kind yields therefore

$$\frac{\partial}{\partial x^l} \log \sqrt{a} = \sum_{i=1}^{n} \Gamma\{i, [i, l]\} = \sum_{i,k=1}^{n} A^{ik} \gamma[i, l, k].$$

(4.5.8)

Now we add to both sides of Eq. (4.5.8) the term $\sum_{i,k=1}^{n} A^{ik} \gamma[i, k, l]$ and exploit Eq. (4.5.3) to write the resulting equation in the form

$$\frac{\partial}{\partial x^l} \log \sqrt{a} + \sum_{i,k=1}^{n} A^{ik} \gamma[i, k, l] = \sum_{i,k=1}^{n} A^{ik} \frac{\partial a_{kl}}{\partial x^i}.$$

(4.5.9)

The right-hand side of (4.5.9) can be re-expressed by differentiation with respect to x^i of the identity (4.1.8) for the inverse of a_{rs}, i.e.

$$\sum_{i,k=1}^{n} A^{ik} \frac{\partial a_{kl}}{\partial x^i} = -\sum_{i,k=1}^{n} a_{kl} \frac{\partial A^{ik}}{\partial x^i},$$

(4.5.10)

and hence (4.5.9) reads as

$$\sum_{i,k=1}^{n} A^{ik} \gamma[i, k, l] = -\frac{\partial}{\partial x^l} \log \sqrt{a} - \sum_{i,k=1}^{n} a_{kl} \frac{\partial A^{ik}}{\partial x^i}.$$

(4.5.11)

4.6 The Laplacian Viewed as a Differential Parameter of Order 2

By exploiting Eqs. (4.4.14) we can now build a quadratic differential form, covariant to the given form f, whose coefficients are formed with those of f and with the first and second derivatives of an arbitrary function U.

By denoting as we did before $U' \equiv U(x(x'))$, i.e. what becomes U upon expressing the x by means of the x', we have, by using the rule for derivative of composite functions,

$$\frac{\partial U'}{\partial x'^r} = \sum_{l=1}^{n} \frac{\partial U}{\partial x^l} \frac{\partial x^l}{\partial x'^r},$$

(4.6.1)

and hence

$$\frac{\partial^2 U'}{\partial x'^r \partial x'^s} = \sum_{l,m=1}^{n} \frac{\partial^2 U}{\partial x^m \partial x^l} \frac{\partial x^l}{\partial x'^r} \frac{\partial x^m}{\partial x'^s} + \sum_{l=1}^{n} \frac{\partial U}{\partial x^l} \frac{\partial^2 x^l}{\partial x'^r \partial x'^s},$$

(4.6.2)

i.e., by virtue of (4.4.14),

$$\frac{\partial^2 U'}{\partial x'^r \partial x'^s} - \sum_{m=1}^{n} \Gamma' \{m, [r, s]\} \frac{\partial U'}{\partial x'^m} = \sum_{m,l=1}^{n} \left[\frac{\partial^2 U}{\partial x^m \partial x^l} - \sum_{i=1}^{n} \Gamma \{i, [m, l]\} \frac{\partial U}{\partial x^i} \right] \frac{\partial x^m}{\partial x'^r} \frac{\partial x^l}{\partial x'^s}. \quad (4.6.3)$$

This equation suggests defining

$$U_{ml} = \frac{\partial^2 U}{\partial x^m \partial x^l} - \sum_{i=1}^{n} \Gamma \{i, [m, l]\} \frac{\partial U}{\partial x^i}, \qquad (4.6.4)$$

and denoting as before by a prime the same quantity in the x' coordinates, Eq. (4.6.3) reveals a homogeneous transformation law under coordinate change:

$$U'_{rs} = \sum_{m,l=1}^{n} U_{ml} \frac{\partial x^m}{\partial x'^r} \frac{\partial x^l}{\partial x'^s}. \qquad (4.6.5)$$

This formula proves that the combinations of first and second derivatives of the arbitrary function U, denoted in Eq. (4.6.4) with the symbol U_{ml}, are as a matter of fact the coefficients of a quadratic form *covariant* to the fundamental form f, because from Eq. (4.6.5) there follows the equality [11]

$$\sum_{r,s=1}^{n} U'_{rs} dx'^r \otimes dx'^s = \sum_{m,l=1}^{n} U_{ml} dx^m \otimes dx^l. \qquad (4.6.6)$$

Following Ricci [120], the U_{rs} are said to be the covariant second derivatives of the function U, built with respect to the fundamental form f.

By operating on the covariant form

$$f_U \equiv \sum_{r,s=1}^{n} U_{rs} dx^r \otimes dx^s, \qquad (4.6.7)$$

that was called the covariant second differential of U, as we did on the tensor product of the first differential of U with itself, we come to the conclusion that the coefficients of the various powers of λ in the ratio

$$\frac{\det(a_{rs} + \lambda U_{rs})}{\det(a_{rs})} = \lambda P_1^{(2)} + \lambda^2 P_2^{(2)} + \dots \qquad (4.6.8)$$

are differential parameters of second order of U. In particular, the coefficient of λ, denoted always by $\triangle_2 U$, is said to be the *second differential parameter* of U, and is given by

$$\triangle_2 U = \sum_{i,k=1}^{n} A^{ik} U_{ik} = \sum_{i,k=1}^{n} A^{ik} \left[\frac{\partial^2 U}{\partial x^i \partial x^k} - \sum_{l=1}^{n} \Gamma \{l, [i, k]\} \frac{\partial U}{\partial x^l} \right]. \quad (4.6.9)$$

By inserting into (4.6.9) the relation (4.4.13) among Christoffel symbols of second and first kind, and making use of (4.5.11) and (4.1.8) one finds

$$\Delta_2 U = \sum_{i,\,k=1}^{n} A^{ik} \frac{\partial^2 U}{\partial x^i \partial x^k} + \frac{1}{\sqrt{a}} \sum_{m,\,l=1}^{n} \frac{\partial \sqrt{a}}{\partial x^l} A^{ml} \frac{\partial U}{\partial x^m} + \sum_{i,\,k=1}^{n} \frac{\partial A^{ik}}{\partial x^i} \frac{\partial U}{\partial x^k}$$

$$= \frac{1}{\sqrt{a}} \sum_{i=1}^{n} \frac{\partial}{\partial x^i} \left(\sum_{k=1}^{n} A^{ik} \sqrt{a} \frac{\partial U}{\partial x^k} \right), \tag{4.6.10}$$

where in the last equality we have used the Leibniz rule for partial derivatives and a relabelling of dummy indices. This is the *Laplace-Beltrami operator* on scalar functions in Riemannian geometry.

For the differential forms occurring in the theory of surfaces, it is sufficient to consider the differential parameters already introduced, i.e.

$$\Delta_1 U, \quad \nabla(U, V), \quad \Delta_2 U,$$

because any other differential parameter, as was shown by Darboux [32] in Volume III of his Lectures on the Theory of Surfaces, can be reduced to the repeated application of the three differential parameters just written down.

4.7 Isothermal Systems

By means of the differential parameters, one can solve the problem of evaluating the coefficients of the form obtained by transformation of the first fundamental form

$$f_1 = E\,du \otimes du + F(du \otimes dv + dv \otimes du) + G\,dv \otimes dv, \tag{4.7.1}$$

when the variables u, v are replaced by new arbitrary variables φ, ψ. Let indeed

$$f_1' = E_1 d\varphi \otimes d\varphi + F_1(d\varphi \otimes d\psi + d\psi \otimes d\varphi) + G_1 d\psi \otimes d\psi \tag{4.7.2}$$

be the differential form obtained by applying the map $(u, v) \rightarrow (\varphi, \psi)$ to the form f_1. By virtue of the fundamental property of differential parameters, the values of

$$\Delta_1 \varphi, \quad \nabla(\varphi, \psi), \quad \Delta_1 \psi$$

evaluated for the original form f_1 are equal to those evaluated for the transformed differential form f_1'. But for the latter one finds [11, 32]

$$\Delta_1 \varphi = \frac{E \left(\frac{\partial \varphi}{\partial v} \right)^2 - 2F \frac{\partial \varphi}{\partial u} \frac{\partial \psi}{\partial v} + G \left(\frac{\partial \varphi}{\partial u} \right)^2}{(EG - F^2)}$$

$$= \frac{G_1}{(E_1 G_1 - (F_1)^2)}, \tag{4.7.3}$$

$$\nabla(\varphi, \psi) = \frac{E \frac{\partial \varphi}{\partial v} \frac{\partial \psi}{\partial v} - F \left(\frac{\partial \varphi}{\partial u} \frac{\partial \psi}{\partial v} + \frac{\partial \varphi}{\partial v} \frac{\partial \psi}{\partial u} \right) + G \frac{\partial \varphi}{\partial u} \frac{\partial \psi}{\partial u}}{(EG - F^2)}$$

$$= -\frac{F_1}{(E_1 G_1 - (F_1)^2)}, \tag{4.7.4}$$

$$\Delta_1 \psi = \frac{E_1}{(E_1 G_1 - (F_1)^2)}, \tag{4.7.5}$$

Thus, the first fundamental form, after passing to the new variables φ, ψ, is completely expressed in terms of differential parameters according to

$$E_1 = \frac{\Delta_1 \psi}{\left((\Delta_1 \varphi)(\Delta_1 \psi) - \nabla^2(\varphi, \psi) \right)}, \quad F_1 = -\frac{\nabla(\varphi, \psi)}{\left((\Delta_1 \varphi)(\Delta_1 \psi) - \nabla^2(\varphi, \psi) \right)}, \tag{4.7.6}$$

$$G_1 = \frac{\Delta_1 \varphi}{\left((\Delta_1 \varphi)(\Delta_1 \psi) - \nabla^2(\varphi, \psi) \right)}. \tag{4.7.7}$$

As can be seen, the new coordinate lines, for which φ and ψ are constant, are orthogonal provided that the mixed differential parameter $\nabla(\varphi, \psi)$ vanishes.

By virtue of (4.7.6) and (4.7.7) one can prove that there exist infinitely many changes of variables for which

$$E_1 = G_1, \quad F_1 = 0. \tag{4.7.8}$$

For this purpose, $\varphi(u, v)$ is taken to be a harmonic function, i.e.

$$\Delta_2 \varphi = 0 \implies \frac{\partial}{\partial u} \left(\frac{G \frac{\partial \varphi}{\partial u} - F \frac{\partial \varphi}{\partial v}}{\sqrt{EG - F^2}} \right) + \frac{\partial}{\partial v} \left(\frac{E \frac{\partial \varphi}{\partial v} - F \frac{\partial \varphi}{\partial u}}{\sqrt{EG - F^2}} \right) = 0. \tag{4.7.9}$$

This implies that there exists another function $\psi(u, v)$ for which Eq. (4.7.9) reduces to the identity

$$\frac{\partial^2 \psi}{\partial u \partial v} - \frac{\partial^2 \psi}{\partial v \partial u} = 0,$$

so that its first derivatives read as

$$\frac{\partial \psi}{\partial u} = -\frac{E\frac{\partial \varphi}{\partial v} - F\frac{\partial \varphi}{\partial u}}{\sqrt{EG - F^2}}, \quad \frac{\partial \psi}{\partial v} = \frac{G\frac{\partial \varphi}{\partial u} - F\frac{\partial \varphi}{\partial v}}{\sqrt{EG - F^2}}. \tag{4.7.10}$$

Hence the function ψ, determined up to an additive constant from Eqs. (4.7.10), is itself harmonic, and is called the *conjugate solution* of φ. Furthermore, the transformed fundamental form reads as

$$f_1' = \lambda\left(d\varphi \otimes d\varphi + d\psi \otimes d\psi\right), \quad \lambda \equiv \frac{1}{\Delta_1 \varphi} = \frac{1}{\Delta_1 \psi}. \tag{4.7.11}$$

The particular orthogonal system (φ, ψ), for which the first fundamental form takes the expression (4.7.11), is said to be an *isothermal system*. The occurrence of two harmonic functions suggests considering a holomorphic function having φ and ψ as real and imaginary parts, respectively. This point of view is described in the following section.

4.8 Holomorphic Functions Associated with Isothermal Systems

The previous result is so important that it deserves the effort of obtaining an independent derivation. For this purpose, let us regard the first fundamental form f_1 as obtainable from the tensor product of complex-valued 1-forms, i.e.

$$f_1 = \left(\sqrt{E}du + \left(F + i\sqrt{EG - F^2}\right)\frac{dv}{\sqrt{E}}\right) \otimes \left(\sqrt{E}du + \left(F - i\sqrt{EG - F^2}\right)\frac{dv}{\sqrt{E}}\right). \tag{4.8.1}$$

From the integral calculus one knows that there exist integrating factors of the first 1-form within round brackets on the right-hand side of (4.8.1). Let one of them be $\mu + i\nu$, and let φ and ψ be the real and imaginary part of the complex-valued function such that

$$(\mu + i\nu)\left[\sqrt{E}du + \left(F + i\sqrt{EG - F^2}\right)\frac{dv}{\sqrt{E}}\right] = d\varphi + id\psi, \tag{4.8.2}$$

which implies, by complex conjugation,

$$(\mu - i\nu)\left[\sqrt{E}du + (F - i\sqrt{EG - F^2})\frac{dv}{\sqrt{E}}\right] = d\varphi - id\psi. \tag{4.8.3}$$

On taking the tensor product of (4.8.2) with (4.8.3), and defining

$$\lambda \equiv (\mu^2 + \nu^2)^{-1}, \tag{4.8.4}$$

one finds the metric in the first equation of (4.7.11). The search for isothermal systems depends on the ability of integrating the equation

$$\sqrt{E}du + (F + i\sqrt{EG - F^2})\frac{dv}{\sqrt{E}} = 0,$$ (4.8.5)

for which the first fundamental form vanishes identically. The imaginary lines of the surface, determined by this equation, are therefore said to have vanishing length. The isothermal systems have a peculiar property: *they divide the surface into infinitesimal squares.* Indeed, the quadrangle enclosed on the surface by two lines φ, ψ and by the two infinitesimally close lines in the system $(\varphi + d\varphi, \psi + d\psi)$ can be viewed, up to infinitesimals of higher order, as a rectangle. Now if the system (φ, ψ) is isothermal, and if such functions are allowed to increase by constant infinitesimal increments $d\varphi$, $d\psi$, taking furthermore $d\varphi = d\psi$, the property just stated gets proved.

Last we remark, on reverting to the formulae (4.7.10), that if the original system (u, v) were already isothermal, they would become simply

$$\frac{\partial \varphi}{\partial u} = \frac{\partial \psi}{\partial v}, \frac{\partial \varphi}{\partial v} = -\frac{\partial \psi}{\partial u}.$$ (4.8.6)

These are the well-known Cauchy-Riemann equations, which express that the function $\varphi + i\psi$ is a holomorphic function of the complex variable $u + iv$. Moreover, since the replacement $\psi \to -\psi$ does not affect the characteristic form (4.7.11) of f_1, we find by inspection that, if an isothermal system (φ, ψ) is known on the surface S, any other isothermal system (φ', ψ') can be obtained by setting

$$\varphi' + i\psi' = F(\varphi \pm i\psi),$$ (4.8.7)

where F denotes an arbitrary function of complex variable. Every complex-valued function $\varphi + i\psi$ formed from a pair of harmonic functions satisfying the Cauchy-Riemann equations is said to be a *complex variable on the surface.* In geometric language, a two-dimensional, smooth and orientable Riemann surface is a one-complex-dimensional manifold [23].

4.9   Isothermal Parameters

Given an isothermal system, with corresponding metric of the surface

$$f_I = \lambda(du_1 \otimes du_1 + dv_1 \otimes dv_1),$$ (4.9.1)

if one changes the parameters that determine the coordinate lines by setting

$$u_1 \equiv \varphi(u), \ v_1 \equiv \psi(v),$$ (4.9.2)

the metric (4.9.1) takes the form

$$f_I = \lambda\Big(\varphi'^2(u)du \otimes du + \psi'^2(v)dv \otimes dv\Big), \tag{4.9.3}$$

where the quotient

$$\frac{E}{G} = \frac{\lambda\varphi'^2(u)}{\lambda\psi'^2(v)}$$

is the ratio between a function of u only and a function of v only. Conversely, if in an orthogonal system (u, v) with first fundamental form

$$f = Edu \otimes du + Gdv \otimes dv \tag{4.9.4}$$

one has

$$\frac{E}{G} = \frac{U(u)}{V(v)}, \tag{4.9.5}$$

one can write also

$$E = \lambda U, \ G = \lambda V, \tag{4.9.6}$$

so that the first fundamental form becomes

$$f = \lambda(Udu \otimes du + Vdv \otimes dv), \tag{4.9.7}$$

and upon changing the parameters (u, v) by setting

$$u_1 \equiv \int \sqrt{U}du, \ v_1 \equiv \int \sqrt{V}dv, \tag{4.9.8}$$

one recovers Eq. (4.9.1).

After these remarks, it is easy to express the condition for the lines $\varphi = $ constant to form an isothermal system, jointly with the orthogonal trajectories. For this purpose it is necessary and sufficient that, after changing suitably the parameter φ by setting

$$\varphi_1 \equiv F(\varphi), \tag{4.9.9}$$

the function φ_1 turns out to be a harmonic function:

$$\Delta_2\, \varphi_1 = 0. \tag{4.9.10}$$

Now the definition of the Laplace–Beltrami operator (or second differential parameter), jointly with the rule for derivatives of composite functions, yields

$$\Delta_2\left[F(\varphi)\right] = F'(\varphi)\,\Delta_2\,\varphi + F''(\varphi)\,\Delta_1\,\varphi, \qquad (4.9.11)$$

and hence the condition (4.9.10) is fulfilled provided that

$$\frac{\Delta_2\varphi}{\Delta_1\varphi} = -\frac{F''(\varphi)}{F'(\varphi)} = -\frac{\mathrm{d}}{\mathrm{d}\varphi}\log F'(\varphi). \qquad (4.9.12)$$

Since the right-hand side depends only on φ, the same must hold for the left-hand side. Thus, we have proved the following property:

Theorem 4.1 *The necessary and sufficient condition for the lines $\varphi = $ constant to form an isothermal system, jointly with the orthogonal trajectories, is that the ratio of first and second differential parameter of φ should depend on φ only.*

4.10 Lie Theorem on the Lines of an Isothermal System

If, in a doubly orthogonal isothermal system, the lines $\varphi = $ constant are known, the ψ lines of the other system are obtained by quadratures. Indeed, upon replacing φ with $F(\varphi)$, Eqs. (4.7.10) are turned into

$$\frac{\partial\psi}{\partial u} = -F'(\varphi)\frac{E\frac{\partial\varphi}{\partial v} - F\frac{\partial\varphi}{\partial u}}{\sqrt{EG - F^2}}, \quad \frac{\partial\psi}{\partial v} = F'(\varphi)\frac{G\frac{\partial\varphi}{\partial u} - F\frac{\partial\varphi}{\partial v}}{\sqrt{EG - F^2}}, \qquad (4.10.1)$$

while from Eq. (4.9.12) one finds [11, 120]

$$F'(\varphi) = \exp\left(-\int\frac{\Delta_2\varphi}{\Delta_1\varphi}\mathrm{d}\varphi\right). \qquad (4.10.2)$$

Such a result can be stated as follows.

Theorem 4.2 *If the lines $\varphi = $ constant belong to a doubly isothermal system, upon writing the differential equation of orthogonal trajectories in the form*

$$\frac{E\frac{\partial\varphi}{\partial v} - F\frac{\partial\varphi}{\partial u}}{\sqrt{EG - F^2}}\mathrm{d}u - \frac{G\frac{\partial\varphi}{\partial u} - F\frac{\partial\varphi}{\partial v}}{\sqrt{EG - F^2}}\mathrm{d}v = 0, \qquad (4.10.3)$$

an integrating factor is immediately given by

$$\mu = \exp\left(-\int\frac{\Delta_2\varphi}{\Delta_1\varphi}\mathrm{d}\varphi\right). \qquad (4.10.4)$$

Remarkably, Sophus Lie went on further, and he proved that if, for the lines of an isothermal system, one knows only a differential equation of first order

$$M\,du + N\,dv = 0, \tag{4.10.5}$$

of which they are the integrals, one can obtain by quadratures the equation of such lines in finite terms [11].

Chapter 5
Distributions and Sobolev Spaces

5.1 The Space $\mathcal{D}(\Omega)$ and Its Strong Dual

If E and F are two topological vector spaces, one denotes by $\mathcal{L}(E, F)$ the set of all continuous linear maps of E into F. This is, in turn, a vector space. In particular, if F coincides with the space $\mathbf{C}$ of complex numbers, $\mathcal{L}(E, \mathbf{C})$ is the vector space of all continuous linear functionals on E, and is said to be the *dual space* of E. Various topologies can be introduced in it, and we assume, following Magenes and Stampacchia [98], of having introduced the *strong topology*. This means that a fundamental set of neighbourhoods of the origin (Appendix 5.a) is obtained by taking the subsets $W(B, U)$ in $\mathcal{L}(E, \mathbf{C})$ such that:

(i) B is a bounded set (Appendix 5.a) whatsoever of E.
(ii) U is an arbitrary neighbourhood of 0 in $\mathbf{C}$.
(iii) x' belongs to $W(B, U)$ if the value $x'(x)$ that it takes, evaluated at a point whatsoever x of B, belongs to U.

We shall denote by E' the dual of E once the strong topology has been introduced; E' is also said to be the *strong dual* of E, and it is therefore a topological vector space. From now on, $\langle x', x \rangle$ denotes the value taken by x' (element of E') at the point x of E; the symbol $\langle x', x \rangle$ denotes therefore the duality among E and E'. If x is fixed in E, the complex number $\langle x', x \rangle$ obtained by varying x' in E' defines a continuous linear form on E', by the very definition of E':

$$x' \to \langle x', x \rangle.$$

Such a form is therefore an element of the strong dual of E' (which is also said to be the strong bi-dual of E, and is denoted by E'').

Let now Ω be an open set of the n-dimensional Euclidean space $\mathbf{R}^n$, with closure denoted by $\overline{\Omega}$. We consider the complex-valued functions

$$\varphi : x \in \Omega \to \varphi(x) \in \mathbf{C}$$

© Springer International Publishing AG 2017
G. Esposito, *From Ordinary to Partial Differential Equations*, UNITEXT 106,
DOI 10.1007/978-3-319-57544-5_5

of class C^∞ and compact support contained in Ω. Such a space can be viewed as a vector space with respect to the complex body, and is denoted by $\mathcal{D}(\Omega)$ or $C_c^\infty(\Omega)$.

Let $\{A\} \equiv \{A_0, A_1, \ldots, A_\nu, \ldots\}$ be a sequence of non-empty open sets contained in Ω, such that the closure of A_ν is contained in $A_{\nu+1}$ and such that every compact set of Ω is contained in A_ν for a sufficiently large value of ν. If Ω coincides with $\mathbf{R}^n$, one can take e.g.

$$A_\nu = \left\{ x : |x| = \sqrt{(x^1)^2 + \cdots + (x^n)^2} < \nu + 1 \right\}.$$

Let then $\{\varepsilon\} = \{\varepsilon_0, \varepsilon_1, \ldots, \varepsilon_\nu, \ldots\}$ a decreasing sequence of positive numbers, and $\{m\} \equiv \{m_0, m_1, \ldots, m_\nu, \ldots\}$ a sequence of non-negative integers, increasing and divergent. Following again Magenes and Stampacchia [98], we denote by U the set

$$U\left(\{m\}, \{\varepsilon\}, \{A\}\right) \equiv \left\{\varphi \in \mathcal{D}(\Omega) : \forall \nu \; |D^p \varphi(x)| \leq \varepsilon_\nu \text{ if } |p| \leq m_\nu, x \notin A_\nu \right\}, \tag{5.1.1}$$

where

$$D^p \equiv \frac{\partial^{p_1 + \cdots + p_n}}{\partial(x^1)^{p_1} \ldots \partial(x^n)^{p_n}}, \quad |p| \equiv \sum_{k=1}^{n} p_k, \tag{5.1.2}$$

the p_k being non-negative integers. As the sequences $\{m\}, \{\varepsilon\}, \{A\}$ are varying one obtains therefore a fundamental set of neighbourhoods of the origin in $\mathcal{D}(\Omega)$. Upon introducing such a topology, the space $\mathcal{D}(\Omega)$ is then a topological vector space, locally convex, Hausdorff and complete (Appendix).

The *strong dual* of $\mathcal{D}(\Omega)$ is said to be the space $\mathcal{D}'(\Omega)$ of *distributions* on Ω. The elements of such a strong dual are denoted by T. A distribution T is therefore a continuous linear form on $\mathcal{D}(\Omega)$, while $\langle T, \varphi \rangle$ denotes a continuous bilinear form on the product of spaces $\mathcal{D}'(\Omega) \times \mathcal{D}(\Omega)$.

For example, the measures μ in Ω and the functions summable in every compact subset of Ω can be seen as distributions when they are identified with the linear and continuous functionals on $\mathcal{D}(\Omega)$, respectively, i.e.

$$\langle \mu, \varphi \rangle = \int_\Omega \varphi \, d\mu, \tag{5.1.3}$$

and

$$\langle f, \varphi \rangle = \int_\Omega f(x)\varphi(x) dx. \tag{5.1.4}$$

A *space of distributions* is every topological vector sub-space E of $\mathcal{D}'(\Omega)$ such that the injection of E in $\mathcal{D}'(\Omega)$ is continuous. A space of distributions is said to be *normal* if:

(i) The following chain of inclusions holds:

$$\mathcal{D}(\Omega) \subset E \subset \mathcal{D}'(\Omega), \tag{5.1.5}$$

(ii) The injections of E in $\mathcal{D}'(\Omega)$ and of $\mathcal{D}(\Omega)$ in E are continuous.

(iii) The space $\mathcal{D}(\Omega)$ is dense in E.

Given a normal space E of distributions, its dual E' can be identified, from the algebraic point of view, with the strong dual $\mathcal{D}'(\Omega)$.

An important operation is the derivative with respect to x^i of a distribution T on Ω. This is a new distribution, denoted by D_i or $\frac{\partial T}{\partial x^i}$, defined by the equation

$$\left\langle \frac{\partial T}{\partial x^i}, \varphi \right\rangle = -\left\langle T, \frac{\partial \varphi}{\partial x^i} \right\rangle. \tag{5.1.6}$$

Every distribution admits therefore infinitely many derivatives in the sense of distributions, and one has

$$\langle D^p T, \varphi \rangle = (-1)^{|p|} \langle T, D^p \varphi \rangle. \tag{5.1.7}$$

A fundamental theorem of Schwartz states that the derivation of distributions is a continuous linear map of $\mathcal{D}'(\Omega)$ onto $\mathcal{D}'(\Omega)$.

5.2 The Space $C^{1,\alpha}(\Omega)$ and Its Abstract Completion

Let Ω be a bounded open set of $\mathbf{R}^n$, with frontier $\Gamma = \partial\Omega$. Unless otherwise stated, Ω is taken to be of class C^∞ for simplicity. This means that, to every point $x \in \Gamma$ one can associate a hypersphere $\Gamma(x)$ centred at x in such a way that the part of Γ contained in $\Gamma(x)$ can be represented, with respect to a system of axes $\xi_1, \ldots, \xi_n$ with origin at x, in the form

$$\xi_n = \psi(\xi_1, \ldots, \xi_{n-1}), \tag{5.2.1}$$

where ψ is a function defined in a suitable neighbourhood of the origin, and here indefinitely differentiable and vanishing with its first derivatives at the origin. For such a domain there exists the tangent hyperplane to Γ at every point x of Γ. In Eq. (5.2.1) the axis ξ_n is taken to be the inward-pointing normal.

Let us now consider the vector space of complex-valued functions defined on the closure of Ω and of class C^1; for all $\alpha > 1$ we denote with $C^{1,\alpha}(\Omega)$ such a space, endowed with the norm

$$\|f\| \equiv \left\{ \int_\Omega |f(x)|^\alpha dx \right\}^{\frac{1}{\alpha}} + \left\{ \int_\Omega \sum_{k=1}^n \left| \frac{\partial f}{\partial x^k} \right|^\alpha dx \right\}^{\frac{1}{\alpha}}. \tag{5.2.2}$$

Such a space is not complete. It is of the utmost importance to consider the *abstract completion* of $C^{1,\alpha}(\Omega)$. This is denoted by $\widehat{C}^{1,\alpha}(\Omega)$ and is defined as follows.

Every point of $\widehat{C}^{1,\alpha}(\Omega)$ consists of the $(n+1)$-tuple $(f, \psi_1, \ldots, \psi_n)$ of functions of $L^\alpha(\Omega)$, the space of functions of α-th power summable in Ω and endowed with the norm

$$\|h\|_{L^\alpha(\Omega)} \equiv \left(\int_\Omega |h(x)|^\alpha \mathrm{d}x \right)^{\frac{1}{\alpha}}, \tag{5.2.3}$$

the functions f and ψ_j being linked by the n relations

$$\int_\Omega f \frac{\partial \varphi}{\partial x^j} \, \mathrm{d}x = - \int_\Omega \varphi \psi_j \, \mathrm{d}x, \ \forall j = 1, \ldots, n, \tag{5.2.4}$$

for all $\varphi \in \mathcal{D}(\Omega)$. The function ψ_j is said to be the *generalized derivative* of f with respect to x^j. From Sect. 5.1 it follows that the function f is a distribution T such that

$$T \in L^\alpha(\Omega), \ D_j T \in L^\alpha(\Omega), \ D_j T = \psi_j, \tag{5.2.5}$$

the derivation being meant in the sense of distributions, according to (5.1.6) and (5.1.7). One also says that the vector with components $(\psi_1, \ldots, \psi_n)$ is the *weak gradient* of f.

5.3   The Sobolev Space $H^{1,\alpha}(\Omega)$

If $\alpha > 1$, the first Sobolev space that we consider is a space of distributions T defined as follows:

$$H^{1,\alpha}(\Omega) \equiv \Big\{ T \in \mathcal{D}'(\Omega) : T \in L^\alpha(\Omega), \ D_j T \in L^\alpha(\Omega),$$

$$\|T\| \equiv \|T\|_{L^\alpha(\Omega)} + \sum_{j=1}^n \|D_j T\|_{L^\alpha(\Omega)} \Big\}. \tag{5.3.1}$$

This turns out to be normed and complete, hence a Banach space, as can be seen upon bearing in mind the completeness of $L^\alpha(\Omega)$ and continuity of the derivation in $\mathcal{D}'(\Omega)$ (Sect. 5.1).

Under the assumptions on the bounded open set Ω made so far, a theorem ensures that

Theorem 5.1 *The abstract completion $\widehat{C}^{1,\alpha}(\Omega)$ is isomorphic to $H^{1,\alpha}(\Omega)$. But one can prove this isomorphism also under weaker assumptions on Ω* [98].

5.4 The Spaces $C^{k,\alpha}$ and $H^{k,\alpha}$

When k is a non-negative integer whatsoever, and $|p| = \sum_{k=1}^{n} p_k$, one can consider also the space

$$C^{k,\alpha}(\Omega) \equiv \left\{ f : \overline{\Omega} \to C, f \in C^p(\Omega) : \|f\| \equiv \sum_{|p| \leq k} \|D^p f\|_{L^\alpha(\Omega)} \right\}. \quad (5.4.1)$$

Such a space reduces of course to $C^{1,\alpha}(\Omega)$ if $k = 1$, and its abstract completion is denoted by $\widehat{C}^{k,\alpha}(\Omega)$.

The associated Sobolev space $H^{k,\alpha}(\Omega)$ (cf. [128–130]) is a space of distributions defined, for all $k = 0, 1, 2, \ldots$ and for all $\alpha > 1$ as

$$H^{k,\alpha}(\Omega) \equiv \left\{ T \in \mathcal{D}'(\Omega) : D^p T \in L^\alpha(\Omega), \|T\| = \sum_{|p| \leq k} \|D^p T\|_{L^\alpha(\Omega)} \right\}, \quad (5.4.2)$$

with the understanding that $D^{(0,\ldots,0)} T = T$. As in the case $k = 1$, one can prove the fundamental isomorphism

$$H^{k,\alpha}(\Omega) \cong \widehat{C}^{k,\alpha}(\Omega). \quad (5.4.3)$$

When $\alpha = 2$, the resulting Sobolev space is a Hilbert space, denoted by $H^k(\Omega)$ for brevity. If $k = 1$, the elements of $H^1(\Omega)$ are functions first studied by Beppo Levi, who, however, did not want to associate his name to this branch of mathematics.

On reverting to generic values of non-negative k and $\alpha > 1$, a fundamental theorem of Sobolev can be stated as follows.

Theorem 5.2 *If Ω is a bounded open set of class C^∞, and if $u \in H^{k,\alpha}(\Omega)$, and β is the real number*

$$\frac{1}{\beta} = \frac{1}{\alpha} - \frac{k}{n}, \quad (5.4.4)$$

then $u \in L^\beta(\Omega)$ if $\beta > 0$, whereas u is a Hölder function in the closure of Ω if $\beta < 0$. Moreover, the injection of $H^{k,\alpha}(\Omega)$ in $L^\rho(\Omega)$ for all $\rho < \beta$ is completely continuous.

Another problem, different from the one solved by Sobolev in Theorem 5.2, consists in specifying the particular nature of quasi-continuous functions for the functions belonging to the Sobolev space $H^{k,\alpha}(\Omega)$. This is part of the general problem of functional completion of a functional space.

5.5   The Trace Map for Elements of $H^{k,\alpha}(\Omega)$

For any function $u \in H^{1,\alpha}(\Omega)$, one can define its trace on $\Gamma = \partial\Omega$ in various ways. The *trace map* γu is a function with α-th power summable on Γ, i.e. $u \in L^{\alpha}(\Gamma)$, for which

$$\lim_{\substack{x \to \xi \text{ on } \nu_\xi}} u(x) = \gamma u(\xi) \tag{5.5.1}$$

for almost all points ξ of Γ, ν_ξ being the inward-pointing normal axis to Γ at the point ξ.

Next, on considering a family of open sets Ω_t within Ω, whose frontiers Γ_t are parallel to Γ, one proves that the trace $\gamma^{(t)} u$ of the restriction of u to Ω_t, when brought back upon Γ, is such that

$$\lim_{t \to 0} \int_\Gamma \left| \gamma^{(t)} u - \gamma u \right|^\alpha d\sigma = 0. \tag{5.5.2}$$

This equation expresses that $\gamma^{(t)} u$ converges on average of order α to γu on the system $\{\Gamma_t\}$.

For a function of the space $H^{k,\alpha}(\Omega)$, with $k > 1$, one can define in analogous way the traces $\gamma_s u$ of the interior normal derivatives of order s:

$$\frac{\partial^s u}{\partial \nu^s}, s = 0, 1, \ldots, k - 1.$$

The notation γu is used for the vector

$$(\gamma_0 u, \gamma_1 u, \ldots, \gamma_{k-1} u).$$

The map $u \to \gamma u$ is a continuous linear map of $H^{k,\alpha}(\Omega)$ into (but not onto)

$$\left[L^{\alpha}(\Gamma) \right]^k = L^{\alpha}(\Gamma) \times \cdots \times L^{\alpha}(\Gamma) \text{ (}k\text{ factors)}.$$

This map coincides with the ordinary trace if $u \in C^{k,\alpha}(\Omega)$. Sobolev has also proved

Theorem 5.3 *The function γu belongs to the space $L^\varsigma(\Gamma)$, where*

$$\varsigma \equiv \frac{\alpha(n-1)}{n + \alpha(s - k)}. \tag{5.5.3}$$

5.6 The Space $H_0^{k,\alpha}(\Omega)$ and Its Strong Dual

The problem naturally arises of characterizing the set of these traces, i.e. the codomain of the linear map $u \to \gamma u$. By construction, this is a vector sub-space of $[L^\alpha(\Gamma)]^k$, denoted in the literature by $S^{k,\alpha}(\Gamma)$. If $\alpha = 2$, one sets for brevity $S^{k,2}(\Gamma) = S^k(\Gamma)$. The codomain space $S^{k,\alpha}(\Gamma)$ is algebraically isomorphic to the vector space obtained by taking the quotient

$$\frac{H^{k,\alpha}(\Omega)}{H_0^{k,\alpha}(\Omega)},$$

where $H_0^{k,\alpha}(\Omega)$ is the closed vector sub-space of $H^{k,\alpha}(\Omega)$ consisting of those functions u having vanishing trace $\gamma u = 0$. The algebraic isomorphism just mentioned can be made topological as well by defining, as norm in $S^{k,\alpha}(\Gamma)$,

$$\|\gamma u\|_{S^{k,\alpha}(\Gamma)} \equiv \inf \|v\|_{H^{k,\alpha}(\Omega)}, \tag{5.6.1}$$

where v is an arbitrary element of the equivalence class $\frac{H^{k,\alpha}(\Omega)}{H_0^{k,\alpha}(\Omega)}$, to which there corresponds the image γu. The map $u \to \gamma u$ is then a continuous linear map of $H^{k,\alpha}(\Omega)$ onto $S^{k,\alpha}(\Gamma)$.

It is rather interesting, in pure mathematics, to characterize intrinsically the space $S^{k,\alpha}(\Gamma)$, by using only the values taken by γu upon Γ. In order to give an idea of this kind of characterization, one can recall for example that $S^{1,\alpha}(\Gamma)$ consists of all and only the functions $\varphi \in L^\alpha(\Gamma)$ such that the integral

$$I(\alpha, n) \equiv \int_{\Gamma \times \Gamma} \int \frac{|\varphi(x) - \varphi(y)|^\alpha}{\overline{xy}^{\alpha+n+2}} d\sigma_x \, d\sigma_y \tag{5.6.2}$$

is finite.

The sub-space $H_0^{k,\alpha}(\Omega)$ introduced earlier can be shown to coincide with the closure of $\mathcal{D}(\Omega)$ in $H^{k,\alpha}(\Omega)$. In particular, one has $H_0^{0,\alpha}(\Omega) = H^{0,\alpha}(\Omega) = L^\alpha(\Omega)$. It is useful to consider the dual space of $H_0^{k,\alpha}(\Omega)$. Since the space $\mathcal{D}(\Omega)$ is dense in $H_0^{k,\alpha}(\Omega)$, the desired dual can be identified with a sub-space of distributions upon Ω. On considering the real number β such that

$$\frac{1}{\alpha} + \frac{1}{\beta} = 1, \tag{5.6.3}$$

one agrees in the literature to denote the desired dual by

$$H^{-k,\beta}(\Omega) \equiv \left(H_0^{k,\alpha}(\Omega) \right)', \; k \geq 0, \beta > 1, \beta = \frac{\alpha}{(\alpha - 1)}. \tag{5.6.4}$$

The theorem that we are now going to prove characterizes completely the dual space of $H_0^{k,\alpha}$.

Theorem 5.4 *The space $H^{-k,\beta}(\Omega)$ consists of all and only distributions T on Ω such that*

$$T = \sum_{|p| \leq k} D^p f_p, \quad f_p \in L^\beta(\Omega). \tag{5.6.5}$$

Proof Let us indeed consider the continuous linear maps of $H_0^{k,\alpha}(\Omega)$ into $L^\alpha(\Omega)$: $u \to D^p u, |p| \leq k$. Since $L^\alpha(\Omega)$ is locally convex, the general theory of topological vector spaces guarantees that every continuous linear functional on $H_0^{k,\alpha}(\Omega)$, i.e. any element T of the strong dual $H^{-k,\beta}(\Omega)$ can be written in the form

$$\langle T, u \rangle = \sum_{|p| \leq k} (-1)^{|p|} \langle f_p, D^p u \rangle, \tag{5.6.6}$$

where $f_p \in L^\beta(\Omega)$. Note now that, since $\mathcal{D}(\Omega)$ is dense in $H_0^{k,\alpha}(\Omega)$, the left-hand side of Eq. (5.6.6) is specified by its restriction to $\mathcal{D}(\Omega)$. But if $\varphi \in \mathcal{D}(\Omega)$, then by virtue of the rule (5.1.7) for derivatives of distributions Eq. (5.6.6) becomes

$$\langle T, \varphi \rangle = \sum_{|p| \leq k} (-1)^{|p|} \langle f_p, D^p \varphi \rangle = \sum_{|p| \leq k} \langle D^p f_p, \varphi \rangle, \tag{5.6.7}$$

from which Eq. (5.6.5) follows.

Conversely, if $T \in \mathcal{D}'(\Omega)$, and Eq. (5.6.5) holds, one has

$$\langle T, \varphi \rangle = \sum_{|p| \leq k} \langle D^p f_p, \varphi \rangle = \sum_{|p| \leq k} (-1)^{|p|} \langle f_p, D^p \varphi \rangle$$

$$= \sum_{|p| \leq k} (-1)^{|p|} \int_\Omega f_p(x) D^p \varphi(x) \mathrm{d}x, \tag{5.6.8}$$

from which, since $f_p \in L^\beta(\Omega)$ and $\mathcal{D}(\Omega)$ is dense in $H_0^{k,\alpha}(\Omega)$, one finds that T is a continuous linear form on $H_0^{k,\alpha}(\Omega)$, i.e. $T \in H^{-k,\beta}(\Omega)$. $\square$

It is also useful to consider the space $H^{k,\alpha}(\mathbf{R}^n)$ of distributions T on $\mathbf{R}^n$ such that $D^p T \in L^\alpha(\mathbf{R}^n)$, with $|p| \leq k$, endowed with the norm

$$\|T\| = \sum_{|p| \leq k} \left\| D^p T \right\|_{L^\alpha(\mathbf{R}^n)}. \tag{5.6.9}$$

This is a normal space of distributions on $\mathbf{R}^n$, and its strong dual is denoted by $H^{-k,\beta}(\mathbf{R}^n)$, with β given by Eq. (5.6.4). Also for this space there holds a theorem analogous to the one just proved, according to which the distributions of $H^{-k,\beta}(\mathbf{R}^n)$ are all and only those for which one can write

$$T = \sum_{|p| \leq k} D^p f_p, \quad f_p \in L^\beta(\mathbf{R}^n). \tag{5.6.10}$$

5.7 Sub-spaces of $H^{k,\alpha}(\mathbf{R}^n)$

It may be of interest to consider the sub-space of distributions of $H^{k,\alpha}(\mathbf{R}^n)$ having support in the closure of Ω. This sub-space is denoted by $H^{k,\alpha}_{\overline{\Omega}}$. If $k = 0$, this can be identified, by construction, with the space $L^\alpha(\Omega)$. The case $k = 1$ is already much richer, and the following theorem holds [98]:

Theorem 5.5 *Every distribution of $H^{1,\alpha}_{\overline{\Omega}}$ is a function $u \in L^\alpha(\mathbf{R}^n)$, vanishing outside $\overline{\Omega}$, for which there exist n functions $u_i \in L^\alpha(\Omega)$, $i = 1, \ldots, n$, such that one has*

$$\langle D_i u, \varphi \rangle = \int_\Omega \varphi u_i \, dx + \int_\Gamma \varphi \gamma_0 u \cos(x^i, \nu) d\sigma, \tag{5.7.1}$$

for every $\varphi \in \mathcal{D}(\mathbf{R}^n)$, where $D_i u$ is the derivative of u in $\mathcal{D}'(\mathbf{R}^n)$, $\gamma_0 u$ is the trace of u on Γ and ν is the inward-pointing normal to Ω on Γ.

Proof Since $\mathcal{D}(\mathbf{R}^n)$ is dense in $H^{1,\alpha}(\mathbf{R}^n)$, there follows that, for any $u \in H^{1,\alpha}(\mathbf{R}^n)$, there exists a sequence $\{\psi_r\}$ of elements of $\mathcal{D}(\mathbf{R}^n)$ such that

$$\psi_r \to u \text{ in } L^\alpha(\mathbf{R}^n); \ \psi_r \to \gamma_0 u \text{ in } L^\alpha(\Gamma). \tag{5.7.2}$$

Moreover, there exist n functions u_i belonging to $L^\alpha(\mathbf{R}^n)$ for which $\frac{\partial \psi_r}{\partial x_r} \to u_i$ in $L^\alpha(\mathbf{R}^n)$.

As a next step, let us point out that, for ψ and $\varphi \in \mathbf{D}(\mathbf{R}^n)$, one has the Green formula

$$\int_\Omega \psi \frac{\partial \varphi}{\partial x^i} \, dx = - \int_\Omega \varphi \frac{\partial \psi}{\partial x^i} \, dx - \int_\Gamma \varphi \psi \cos(x^i, \nu) \, d\sigma. \tag{5.7.3}$$

Thus, on passing to the limit in Eq. (5.7.3) written for ψ_r one has that, for any $u \in H^{1,\alpha}(\mathbf{R}^n)$, one has the following formula:

$$\int_\Omega u \frac{\partial \varphi}{\partial x^i} \, dx = - \int_\Omega \varphi u_i \, dx - \int_\Gamma \varphi \gamma_0 u \cos(x^i, \nu) \, d\sigma. \tag{5.7.4}$$

Such an equation holds, in particular, if $u \in H^{1,\alpha}_{\overline{\Omega}}$. On the other hand, by virtue of the definition of derivative in $\mathcal{D}'(\mathbf{R}^n)$, one has the equality

$$\langle D_i u, \varphi \rangle = -\langle u, D_i \varphi \rangle, \tag{5.7.5}$$

and, since u has support contained in $\overline{\Omega}$, one has

$$\langle u, D_i \varphi \rangle = \int_\Omega u \, D_i \varphi \, dx, \tag{5.7.6}$$

and hence Eq. (5.7.1) holds by virtue of (5.7.2)–(5.7.6). $\qquad \square$

This theorem shows that, for any $u \in H_{\overline{\Omega}}^{1,\alpha}$, the derivative $D_i u$ in $\mathcal{D}'(\mathbf{R}^n)$ is the sum of a function $u_i \in L^\alpha(\Omega)$ and of a distribution on Γ; the function u_i coincides with the derivative of u in $\mathcal{D}'(\Omega)$, and the distribution on Γ is equal to $\gamma_0 u \cos(x^i, \nu)$. One can proceed in the same way to characterize the space $H_{\overline{\Omega}}^{k,\alpha}$ with $k > 1$.

5.8 Fundamental Solution and Parametrix of a Linear Equation

There is a lot more to learn about Sobolev spaces, but for the time being we take a break and define instead, in this chapter devoted to the distributional setting, a basic concept in the analysis of linear partial differential operators.

Let us consider, for simplicity, a scalar differential operator P with C^∞ coefficients

$$Pu = \sum_{\mu,\nu=1}^{n} (g^{-1})^{\mu\nu} \nabla_\nu \nabla_\mu u + \sum_{\mu=1}^{n} a^\mu \nabla_\mu u + bu, \qquad (5.8.1)$$

that is defined on a connected open set Ω of a n-dimensional geometry (M, g). A *fundamental solution* of P is a distribution denoted here by $(\mathcal{G}_q(p), \phi(p))$ which is a function of the point q such that

$$P\mathcal{G}_q - \delta_q = 0. \qquad (5.8.2a)$$

This means that

$$(P\mathcal{G}_q, \phi) = \phi(q), \quad \phi \in C_c^\infty(\Omega). \qquad (5.8.2b)$$

We stress that the fundamental solution is not uniquely defined, unlike the case of Green's functions, for each of which there is a specific boundary condition.

A C^∞ *parametrix* of P is instead a distribution π_q such that

$$P\pi_q - \delta_q = \omega \in C^\infty(\Omega). \qquad (5.8.3)$$

This concept is of interest because, although a fundamental solution is a useful tool for solving linear partial differential equations, some of the problems in which it plays a role can be handled better by means of functions possessing a singularity that is not annihilated but merely *smoothed out* by the differential operator under investigation. The smoothing might even be so weak that the singularity is actually augmented but acquires a less rapid growth than was to be anticipated from the order of the differential operator [67].

In order to understand these remarks let us consider, for $n \geq 3$ and a smooth function f with compact support, i.e. $f \in C_c^\infty(\mathbf{R}^n)$, the inhomogeneous equation $\triangle u = f$, where $\triangle$ (see Eq. (3.2.2)) is minus the Laplacian (with our convention, $\triangle$ is defined in the standard way, but the Laplacian has symbol given by $\sum_{k=1}^n (\xi_k)^2 = |\xi|^2$). Upon taking the Fourier transform of both sides, one finds

$$- |\xi|^2 \hat{u}(\xi) = \hat{f}(\xi), \qquad (5.8.4)$$

and hence, for the inverse operator Q of $\triangle$, or fundamental solution, we can write

$$
\begin{aligned}
u(x) = (Qf)(x) &= -(2\pi)^{-\frac{n}{2}} \int_{\mathbf{R}^n} e^{i\xi \cdot x} \frac{\hat{f}(\xi)}{|\xi|^2} d\xi \\
&= -\frac{\Gamma\left(\frac{n}{2} - 1\right)}{4\pi^{\frac{n}{2}}} \int_{\mathbf{R}^n} \frac{f(y)}{|x - y|^{n-2}} dy.
\end{aligned} \qquad (5.8.5)
$$

If, instead of the Laplacian on $\mathbf{R}^n$, we deal with a general partial differential operator with constant coefficients which can be written as a polynomial $P = p(D)$, where $D \equiv -i\left(\frac{\partial}{\partial x^1}, \ldots, \frac{\partial}{\partial x^n}\right)$, so that

$$p(D)u = f \in C_c^\infty(\mathbf{R}^n), \qquad (5.8.6)$$

the solution can be formally expressed as

$$u(x) = (Qf)(x) = (2\pi)^{-\frac{n}{2}} \int_{\mathbf{R}^n} e^{i\xi \cdot x} q(\xi) \hat{f}(\xi) d\xi, \qquad (5.8.7)$$

where the amplitude $q(\xi)$ is the inverse of the symbol $p(\xi)$ of $p(D)$, and the integration contour must avoid the zeros of $p(\xi)$. However, if the operator P has variable coefficients on a subset U of $\mathbf{R}^n$, i.e.

$$P = p(x, D) = \sum_{\alpha : |\alpha| \leq k} a_\alpha(x) D_x^\alpha, \ a_\alpha \in C^\infty(U), \qquad (5.8.8)$$

one can no longer solve the equation $Pu = f$ by Fourier transform. One can however *freeze* the coefficients at a point $x_0 \in U$ and consider P as a perturbation of $p(x_0, D)$, which is hence a differential operator with constant coefficients. In this way the amplitude of P reduces to $q(\xi) = \frac{1}{p_0(x_0, \xi)}$, and if we let x_0 vary in U, we obtain the *approximate solution operator*

$$(Qf)(x) = (2\pi)^{-\frac{n}{2}} \int_{\mathbf{R}^n} e^{i\xi \cdot x} q(x, \xi) \hat{f}(\xi) d\xi \qquad (5.8.9)$$

with amplitude $q(x, \xi) = p(x, \xi)^{-1}$, for $x \in U$, $f \in C_0^\infty(\mathbf{R}^n)$.

However, to solve the inhomogeneous equation $Pu = f$, we do not strictly need the full inverse operator or fundamental solution, but, as we said before, it is enough to know a parametrix, i.e. a quasi-inverse modulo a regularizing operator. One can provide a first example of parametrix by reverting to the study of constant coefficient operators. A parametrix Q can then be constructed by choosing its amplitude as

$$q(\xi) = \frac{\chi(\xi)}{p(\xi)}, \tag{5.8.10}$$

where $\chi(\xi)$ is a suitably chosen $C^\infty(\mathbf{R}^n)$ function which is identically zero in a disk about the origin and identically 1 for large ξ. In this way the integral formula

$$(Qf)(x) = (2\pi)^{-\frac{n}{2}} \int_{\mathbf{R}^n} e^{i\xi \cdot x} q(\xi) \hat{f}(\xi) d\xi, \tag{5.8.11}$$

is not affected by the convergence problems that would be met if the amplitude were taken to be just $p(\xi)^{-1}$. If Q is the integral operator above, it is no longer true that $PQ = QP = I$, but we have [12]

$$PQf = f + Rf, \ f \in C^\infty(U), \tag{5.8.12}$$

$$(Rf)(x) \equiv (2\pi)^{-\frac{n}{2}} \int_{\mathbf{R}^n} r(x - \xi) f(\xi) d\xi, \tag{5.8.13}$$

where the Fourier transform of r is $\chi - 1$. Thus, r is a smooth function and R turns out to be a smoothing operator. Such a class of smoothing operators is fully under control, and hence a parametrix Q serves just as well as a full fundamental solution, for which R vanishes identically.

Appendix 5.a: Some Basic Concepts of Topology

In Sect. 5.1, some of the concepts we rely upon are as follows.

(i) If x is a point of a topological space E, and if $B(x) \equiv \{I(x)\}$ is the set of all neighbourhoods $I(x)$ of x, the subset $G(x)$ of $B(x)$ is said to be a *fundamental system of neighbourhoods* of x if

$$\forall S \in B(x), \ \exists \sigma \in G(x) : \ \sigma \subset S. \tag{5a.1}$$

(ii) A set B of a topological vector space E is bounded if, no matter how a neighbourhood U of the origin of E is chosen, there exists a number λ such that, $\forall x \in B$, $\lambda x \in U$.

(iii) The support of a function φ is the closed set obtained by intersection of all closed sets out of which φ vanishes. With mathematical notation, one writes

$$\text{supp}(\varphi) = \cap_i C_i, \quad C_i = \text{closed set} : \varphi(x) = 0 \; \forall x \notin C_i. \tag{5a.2}$$

(iv) A topological space is Hausdorff if, for any pair of distinct points x, y of E, there exist disjoint neighbourhoods $I(x)$ and $I(y)$ of such points:

$$\forall x, y \in E : x \neq y, \; \exists I(x), I(y) : I(x) \cap I(y) = \emptyset. \tag{5a.3}$$

(v) A topological vector space E is *locally convex* if it is endowed with a fundamental system of neighbourhoods of the origin which are convex.

(vi) Given a vector space F and a vector sub-space E of F, the linear map

$$i : E \rightarrow F \; : i(x) = x, \; \forall x \in E \tag{5a.4}$$

is said to be the *canonical map* or *injection* or *immersion* of E in F.

(vii) The linear map $x \rightarrow f(x)$ of a vector space E into another vector space F is said to be an *algebraic isomorphism* of E on F if it is bijective between E and a sub-space of F (or possibly the whole of F). Moreover, if E and F are topological spaces, the algebraic isomorphism is said to be also *topological* if the map $x \rightarrow f(x)$ is also continuous with its inverse. One then says that E can be identified with F.

Chapter 6
The Caccioppoli-Leray Theorem

6.1 Second-Order Linear Elliptic Equations in n Variables

Following Caccioppoli [14], we here consider the most general elliptic equation of second order, with n independent variables $x^1, \ldots, x^n$, written in the form

$$Pu = \sum_{i,k=1}^{n} A^{ik} p_{ik} + \sum_{i=1}^{n} B^i p_i + Cu = f, \qquad (6.1.1)$$

where the A^{ik}, B^i, C, f depend on the x^i, and

$$p_i \equiv \frac{\partial u}{\partial x^i}, \quad p_{ik} \equiv \frac{\partial^2 u}{\partial x^i \partial x^k}. \qquad (6.1.2)$$

Since we are dealing with elliptic theory, the quadratic form $\sum_{i,k=1}^{n} A^{ik} \xi_i \xi_k$, which is the leading symbol of P, is taken to be positive-definite, and the discriminant $\det A^{ik}$ can be set to 1. In the early fifties it was well known that, on extending classical results of potential theory, one can assign for solutions of Eq. (6.1.1) some precise bounds, that pertain to derivatives and Hölder coefficients. In such bounds there occur both A^{ik}, B^i, C and the known term f. From such inequalities, Schauder and Caccioppoli obtained existence theorems for the Dirichlet problem. There exist however also integral bounds, i.e. majorizations of mean values pertaining to derivatives of u, by means of analogous mean values for f (e.g. the root mean square value of f). Such inequalities can be applied to the investigation of Eq. (6.1.1) under assumptions more general than usual on the known term and coefficient functions. This was the beginning of regularity theory, and in the modern view elliptic equations are characterized by studying regularity properties of their solutions.

Caccioppoli obtained general majorization formulae for root mean square values of first and second derivatives. More precisely, on considering a solution u of

© Springer International Publishing AG 2017
G. Esposito, *From Ordinary to Partial Differential Equations*, UNITEXT 106,
DOI 10.1007/978-3-319-57544-5_6

Eq. (6.1.1) in a domain D with frontier S, he did prove how the integrals (here $\triangle_1$ is the first differential parameter of Sect. 4.3)

$$J_D^{(1)} \equiv \int_D \sum_{i=1}^n (p_i)^2 d\tau = \int_D \triangle_1 u \, d\tau, \; J_D^{(2)} \equiv \int_D \sum_{i,k=1}^n (p_{ik})^2 d\tau = \sum_{i=1}^n \int_D \triangle_1 p_i \, d\tau$$

(6.1.3)

can be bounded by means of the integrals

$$\int_D f^2 \, d\tau, \; J_D \equiv \int_D u^2 \, d\tau,$$

(6.1.4)

as well as by means of the integral extended to S

$$\overline{J}_S \equiv \int_S \overline{u}^2 \, d\sigma,$$

(6.1.5)

where $\overline{u} \equiv u|_S$. On defining a norm of the solution u in the domain D according to whatever of the two prescriptions

$$\|u\|_D \equiv J_D + J_D^{(1)} + J_D^{(2)},$$

(6.1.6)

$$\|u\|_D = J_D + J_D^{(2)},$$

(6.1.7)

and a norm of u on the frontier S either as

$$\|\overline{u}\|_S \equiv \overline{J}_S + \overline{J}_S^{(2)},$$

(6.1.8)

or as

$$\|\overline{u}\|_S \equiv \overline{J}_S + \overline{J}_S^{(1)} + \overline{J}_S^{(2)},$$

(6.1.9)

Caccioppoli obtained integral bounds of the kind

$$\|u\|_D \leq K \left[\|\overline{u}\|_S + J_D + \int_D f^2 \, d\tau \right],$$

(6.1.10)

with suitable refinements in the case of a domain D' contained in D.

6.2   The Leray Lemma and Its Proof

A key technical ingredient in the Caccioppoli analysis was the following lemma.

Theorem 6.1 *If the function φ verifies, for values of the independent variable $\rho \in$]0, r], the inequality*

$$\omega(\rho)\varphi(\rho) - \lambda \int_0^\rho \varphi^2(t)dt + \delta(\rho) \geq 0, \tag{6.2.1}$$

where λ is a positive constant and ω, δ are non-negative functions of ρ, δ being also non-decreasing, one has the inequality

$$\int_0^\rho \varphi^2(t)dt \leq \frac{\delta(r)}{\lambda} + \frac{\int_0^r \omega^2(t)dt}{\lambda^2(r-\rho)^2}. \tag{6.2.2}$$

Proof As a first step, one defines the function

$$\psi : \rho \to \psi(\rho) \equiv \int_0^\rho \varphi^2(t)dt - \frac{\delta(r)}{\lambda}, \tag{6.2.3}$$

for which the first derivative exists and is equal to

$$\psi'(\rho) = \varphi^2(\rho). \tag{6.2.4}$$

Now by virtue of the assumption (6.2.1) one finds

$$\frac{\omega(\rho)\varphi(\rho)}{\lambda} \geq \int_0^\rho \varphi^2(t)dt - \frac{\delta(\rho)}{\lambda}. \tag{6.2.5a}$$

The assumption $\delta(r) - \delta(\rho) \geq 0$ and the definition (6.2.3) suggest adding and subtracting the ratio $\frac{\delta(r)}{\lambda}$ on the right-hand side of (6.2.5a). Thus, if $\psi(\rho) > 0$, one finds

$$\frac{\omega(\rho)\varphi(\rho)}{\lambda} \geq \psi(\rho) + \frac{(\delta(r) - \delta(\rho))}{\lambda} \geq \psi(\rho) \implies \frac{\omega^2\varphi^2}{\lambda^2} \geq \psi^2. \tag{6.2.5b}$$

The relations (6.2.4) and (6.2.5b) lead in turn to

$$\frac{\psi'}{\lambda^2\psi^2} = \frac{\varphi^2}{\lambda^2\psi^2} = \frac{\omega^2\varphi^2}{\lambda^2} \frac{1}{\psi^2\omega^2} \geq \frac{1}{\omega^2}, \tag{6.2.6}$$

where the left-hand side is a derivative, i.e.

$$-\frac{1}{\lambda^2}\frac{d}{d\rho}\psi^{-1}(\rho) \geq \frac{1}{\omega^2(\rho)}. \tag{6.2.7}$$

From this local condition one obtains, by integration from ρ to r,

$$\frac{1}{\lambda^2}\left[\frac{1}{\psi(\rho)} - \frac{1}{\psi(r)}\right] \geq \int_\rho^r \frac{dt}{\omega^2(t)}. \tag{6.2.8}$$

As a second step, we consider the square of the integral from ρ to r of the function equal to 1 in such an interval. Indeed, it is clear that

$$\left[\int_\rho^r dt\right]^2 = (r - \rho)^2, \tag{6.2.9}$$

whereas, upon multiplying and dividing the integrand by $\omega(t)$ and then applying the Cauchy-Schwarz inequality, one also finds

$$\left[\int_\rho^r dt\right]^2 \leq \int_\rho^r \omega^2(t)dt \int_\rho^r \frac{dt}{\omega^2(t)}. \tag{6.2.10}$$

From (6.2.9) and (6.2.10) one gets therefore the useful inequality

$$\int_\rho^r \frac{dt}{\omega^2(t)} \geq \frac{(r - \rho)^2}{\int_\rho^r \omega^2(t)dt}. \tag{6.2.11}$$

As a third step, we consider both (6.2.8) and (6.2.11) and find

$$\frac{1}{\psi(\rho)} \geq \frac{1}{\psi(r)} + \frac{\lambda^2(r - \rho)^2}{\int_\rho^r \omega^2(t)dt} \geq \frac{\lambda^2(r - \rho)^2}{\int_\rho^r \omega^2(t)dt}. \tag{6.2.12}$$

This lower bound for $\frac{1}{\psi(\rho)}$ becomes an upper bound for $\psi(\rho)$, and the definition of $\psi(\rho)$ in (6.2.3) yields then the desired inequality (6.2.2), because one has

$$\psi(\rho) \leq \frac{\int_\rho^r \omega^2(t)dt}{\lambda^2(r - \rho)^2} \leq \frac{\int_0^r \omega^2(t)dt}{\lambda^2(r - \rho)^2}. \tag{6.2.13}$$

□

The Leray lemma just proved has been applied also to the global theory of pseudo-analytic functions by Caccioppoli [18, 19], and to the solution of Hilbert's nineteenth problem by De Giorgi [35]. It is therefore of the utmost important in three branches of mathematics: regularity theory for elliptic equations, complex analysis, and calculus of variations.

6.3   Caccioppoli's Proof of Integral Bounds: Part 1

Let us consider Eq. (6.1.1) with coefficients and known term possessing continuous first derivatives, in a bounded domain D whose frontier consists of smooth manifolds on which the coordinates are functions of class C^2 of $(n - 1)$ (local) parameters; moreover, the solution u is taken to be of class C^3 also upon S.

Following Caccioppoli [14], we denote by $\mathcal{D}$ a regular domain contained in D, by Γ the frontier of $\mathcal{D}$, by Γ' a portion of Γ, and we introduce the notation

$$U[\mathcal{D}] \equiv \sqrt{\int_{\mathcal{D}} u^2 \mathrm{d}\tau}, \ \ U[\Gamma'] \equiv \sqrt{\int_{\Gamma'} u^2 \mathrm{d}\sigma}, \tag{6.3.1}$$

$$P[\mathcal{D}] \equiv \sqrt{\int_{\mathcal{D}} \sum_{i=1}^{n} (p_i)^2 \mathrm{d}\tau}, \ \ P[\Gamma'] \equiv \sqrt{\int_{\Gamma'} \sum_{i=1}^{n} (p_i)^2 \mathrm{d}\sigma}, \tag{6.3.2}$$

$$R[\mathcal{D}] \equiv \sqrt{\int_{\mathcal{D}} \sum_{i,k=1}^{n} (p_{ik})^2 \mathrm{d}\tau}, \ \ R[\Gamma'] \equiv \sqrt{\int_{\Gamma'} \sum_{i,k=1}^{n} (p_{ik})^2 \mathrm{d}\sigma}, \tag{6.3.3}$$

$$F[\mathcal{D}] \equiv \sqrt{\int_{\mathcal{D}} f^2 \mathrm{d}\tau}, \ \ F[\Gamma'] \equiv \sqrt{\int_{\Gamma'} f^2 \mathrm{d}\sigma}. \tag{6.3.4}$$

Furthermore, we set $u = \overline{u}$ on S. Such a value of the solution on the frontier S depends, in the neighbourhood of each point of S, on the local parameters $\xi_1, \ldots, \xi_{n-1}$. On denoting by S' a portion of S, one defines with respect to $\overline{u}$ the quantities analogous to $U[\mathcal{D}], P[\mathcal{D}], R[\mathcal{D}]$, hence determining a complete covering of S by means of a finite number of neighbourhoods $I_1, I_2, \ldots$, and setting

$$\overline{U}[S'] \equiv \sqrt{\int_{S'} \overline{u}^2 \mathrm{d}\sigma}, \ \ \overline{P}[S'] \equiv \sqrt{\sum_{h} \int_{I_h'} \sum_{i=1}^{n-1} \left(\frac{\partial \overline{u}}{\partial \xi_i}\right)^2 \mathrm{d}\sigma}, \tag{6.3.5}$$

$$\overline{R}[S'] \equiv \sqrt{\sum_{h} \int_{I_h'} \sum_{i,k=1}^{n-1} \left(\frac{\partial^2 \overline{u}}{\partial \xi_i \partial \xi_k}\right)^2 \mathrm{d}\sigma}, \tag{6.3.6}$$

where $I_h' \equiv S' \cap I_h$.

As a first step, Caccioppoli multiplied Eq. (6.1.1) by $u \mathrm{d}\tau$ and then integrated over $\mathcal{D}$, using the Leibniz rule to re-express the integrand, and then the Stokes' theorem. For example, by virtue of the identity

$$\frac{\partial}{\partial x^k}(A^{ik} p_i u) = u \frac{\partial A^{ik}}{\partial x^k} p_i + A^{ik} p_i p_k + A^{ik} p_{ik} u, \tag{6.3.7}$$

one finds

$$\int_{\mathcal{D}} \sum_{i,k=1}^{n} A^{ik} p_{ik} u \, d\tau$$

$$= \int_{\mathcal{D}} \sum_{i,k=1}^{n} \left[\frac{\partial}{\partial x^k} (A^{ik} p_i u) - A^{ik} p_i p_k - u \frac{\partial A^{ik}}{\partial x^k} p_i \right] d\tau$$

$$= -\int_{\Gamma} u \sum_{i,k=1}^{n} A^{ik} p_i \cos(\nu, x_k) d\sigma - \int_{\mathcal{D}} \sum_{i,k=1}^{n} A^{ik} p_i p_k \, d\tau$$

$$- \int_{\mathcal{D}} u \sum_{i,k=1}^{n} \frac{\partial A^{ik}}{\partial x^k} p_i \, d\tau,$$

(6.3.8)

and the non-local equation resulting from Eq. (6.1.1) reads as

$$-\int_{\Gamma} u \sum_{i,k=1}^{n} A^{ik} p_i \cos(\nu, x_k) d\sigma - \int_{\mathcal{D}} \sum_{i,k=1}^{n} A^{ik} p_i p_k \, d\tau - \int_{\mathcal{D}} u \sum_{i,k=1}^{n} \frac{\partial A^{ik}}{\partial x^k} p_i \, d\tau$$

$$+ \int_{\mathcal{D}} u \sum_{i=1}^{n} B^i p_i \, d\tau + \int_{\mathcal{D}} C u^2 \, d\tau - \int_{\mathcal{D}} u f \, d\tau = 0.$$

(6.3.9)

Now we use the Cauchy-Schwarz inequality, e.g.

$$\int_{\mathcal{D}} u f \, d\tau \leq \sqrt{\int_{\mathcal{D}} u^2 \, d\tau} \sqrt{\int_{\mathcal{D}} f^2 \, d\tau},$$

(6.3.10)

and point out that, as is suggested by the prototype elliptic operator of second order, i.e. the Laplacian $-\triangle$, ellipticity of Eq. (6.1.1) means that it must be possible to associate a positive-definite quadratic form to $A^{ik} p_{ik}$, and that the sup of the absolute value of the A^{ik} must be majorized by a finite positive number. Moreover, all cosines in (6.3.9) are of course majorized by 1. Hence one finds, for some $a > 0, b > 0, \alpha > 0$, and $c \equiv \max|C|$, the lower bound

$$a \sqrt{\int_{\Gamma} u^2 \, d\sigma \int_{\Gamma} \sum_{i=1}^{n} (p_i)^2 \, d\sigma} - \alpha \int_{\mathcal{D}} \sum_{i=1}^{n} (p_i)^2 \, d\tau + b \sqrt{\int_{\mathcal{D}} u^2 \, d\tau \int_{\mathcal{D}} \sum_{i=1}^{n} (p_i)^2 \, d\tau}$$

$$+ \sqrt{\int_{\mathcal{D}} u^2 \, d\tau \int_{\mathcal{D}} f^2 \, d\tau} + c \int_{\mathcal{D}} u^2 \, d\tau \geq 0.$$

(6.3.11)

With the notation introduced in Eqs. (6.3.1)–(6.3.4), and upon dividing both sides by $a > 0$, such a minorization reads as

$$U[\Gamma]P[\Gamma] - \frac{\alpha}{a} P^2[\mathcal{D}] + U[\mathcal{D}] \left(\frac{b}{a} P[\mathcal{D}] + \frac{1}{a} F[\mathcal{D}] + \frac{c}{a} U[\mathcal{D}] \right) \geq 0. \quad (6.3.12)$$

Now with the help of some $\alpha' > 0$ and $\beta > 0$ one can re-express (6.3.12) in the more convenient form

$$U[\Gamma]P[\Gamma] - \alpha'P^2[\mathcal{D}] + \beta U[\mathcal{D}](P[\mathcal{D}] + U[\mathcal{D}] + F[\mathcal{D}]) \geq 0. \qquad (6.3.13)$$

Here there is still room left for a final improvement, i.e.

$$U[\Gamma]P[\Gamma] - \lambda P^2[\mathcal{D}] + \mu U[\mathcal{D}](U[\mathcal{D}] + F[\mathcal{D}]) \geq 0, \qquad (6.3.14)$$

where λ and μ are suitable positive numbers.

Another non-trivial consequence of Eq. (6.1.1) was obtained by Caccioppoli upon diferentiation with respect to x^h, which yields an equation of the kind

$$\sum_{i,k=1}^{n} A^{ik} p_{hik} + \sum_{i,k=1}^{n} B^{hik} p_{ik} + \sum_{i=1}^{n} \frac{\partial B^i}{\partial x^h} p_i + \frac{\partial}{\partial x^h}(Cu) = \frac{\partial f}{\partial x^h}, \qquad (6.3.15)$$

having defined

$$p_{hik} \equiv \frac{\partial^3 u}{\partial x^h \partial x^i \partial x^k}. \qquad (6.3.16)$$

At this stage, multiplication of Eq. (6.3.15) by $p_h \, d\tau$ and subsequent integration and application of the Stokes' theorem yields the non-local equation

$$-\int_{\Gamma} p_h \sum_{i,k=1}^{n} A^{ik} p_{hi} \cos(\nu, x_k) d\sigma - \int_{\mathcal{D}} \sum_{i,k=1}^{n} A^{ik} p_{hi} p_{hk} \, d\tau - \int_{\mathcal{D}} p_h \sum_{i,k=1}^{n} \frac{\partial A^{ik}}{\partial x^k} p_{hi} \, d\tau$$

$$+\int_{\mathcal{D}} p_h \sum_{i,k=1}^{n} B^{hik} p_{ik} \, d\tau + \int_{\mathcal{D}} p_h \sum_{i=1}^{n} \frac{\partial B^i}{\partial x^h} p_i \, d\tau - \int_{\Gamma} p_h Cu \cos(\nu, x_h) d\sigma$$

$$-\int_{\mathcal{D}} p_{hh} Cu \, d\tau = -\int_{\Gamma} p_h f \cos(\nu, x_h) d\sigma - \int_{\mathcal{D}} p_{hh} f \, d\tau. \qquad (6.3.17)$$

If one now defines

$$R'[\mathcal{D}] \equiv \sqrt{R^2[\mathcal{D}] - \int_{\mathcal{D}} (p_{nn})^2 d\tau}, \quad R'[\Gamma] \equiv \sqrt{R^2[\Gamma] - \int_{\Gamma} (p_{nn})^2 d\sigma}, \qquad (6.3.18)$$

the Eq. (6.3.17) provides, with the help of majorizations and summation extended to the values $1, \ldots, n-1$ of h, an inequality of the form

$$P[\Gamma]R'[\Gamma] - \alpha_1 R'^2[\mathcal{D}] + \beta_1(P[\mathcal{D}] + U[\mathcal{D}] + F[\mathcal{D}])R'[\mathcal{D}]$$
$$+ \beta_2 P^2[\mathcal{D}] + \beta_0 P[\Gamma](U[\Gamma] + F[\Gamma]) \geq 0, \qquad (6.3.19)$$

where $\alpha_1, \beta_1, \beta_2, \beta_0$ are positive. In the intermediate steps, the algorithm strategy is as follows: a sum of squares is replaced with the square of a sum, or the other way around, with alteration by a bounded factor in the two cases; a product of factors is replaced by half-sum of two squares (majorization); a coefficient is suitably changed, whenever necessary to reinforce the inequality. If one further defines

$$T[\mathcal{D}] \equiv \sqrt{R'^2[\mathcal{D}] + (\beta_0)^2 \left(U^2[\mathcal{D}] + F^2[\mathcal{D}] \right)}, \tag{6.3.20}$$

$$T[\Gamma] \equiv \sqrt{R'^2[\Gamma] + (\beta_0)^2 \left(U^2[\Gamma] + F^2[\Gamma] \right)}, \tag{6.3.21}$$

one derives eventually from (6.3.19) the inequality

$$P[\Gamma]T[\Gamma] - \lambda_1 T^2[\mathcal{D}] + \mu_1(P^2[\mathcal{D}] + U^2[\mathcal{D}] + F^2[\mathcal{D}]) \geq 0, \tag{6.3.22}$$

with λ_1, μ_1 yet other positive numbers.

Let us now see how to apply the Leray lemma of Sect. 6.2 to the inequalities (6.3.14) and (6.3.22). For this purpose, let O be a point within D and Σ_ρ the hypersphere centred at O with radius ρ. Following Caccioppoli [14], we assume that Σ_r is contained in D, and we set for $\rho < r$

$$\varphi(\rho) \equiv P[\Gamma_\rho], \ \omega(\rho) \equiv U[\Gamma_\rho], \ \delta(\rho) \equiv \mu U[\Sigma_\rho](U[\Sigma_\rho] + F[\Sigma_\rho]), \tag{6.3.23}$$

Γ_ρ being the spherical hypersurface consisting of the frontier of Σ_ρ, while μ is the coefficient occurring in (6.3.14). This relation corresponds entirely to (6.2.1), and hence, upon defining

$$P^2[\Sigma_\rho] \equiv \int_0^\rho P^2[\Gamma_t]dt = \int_0^\rho \varphi^2(t)dt, \ U^2[\Sigma_\rho] \equiv \int_0^\rho U^2[\Gamma_t]dt, \tag{6.3.24}$$

we can re-express the result (6.2.2) with the notation appropriate to the elliptic equation (6.1.1), i.e.

$$
\begin{aligned}
P^2[\Sigma_\rho] &\leq \frac{\mu}{\lambda} U[\Sigma_r](U[\Sigma_r] + F[\Sigma_r]) + \frac{U^2[\Sigma_r]}{\lambda^2(r-\rho)^2} \\
&\leq \frac{U^2[\Sigma_r]}{(r-\rho)^2}\left(\frac{\mu}{\lambda} + \frac{1}{\lambda^2}\right) + \frac{\mu}{\lambda} U[\Sigma_r]F[\Sigma_r] \\
&\leq \gamma_1 \left(\frac{1}{(r-\rho)^2} U[\Sigma_r] + F[\Sigma_r]\right) U[\Sigma_r],
\end{aligned}
\tag{6.3.25}
$$

where we have defined

$$\frac{\mu}{\lambda} + \frac{1}{\lambda^2} \equiv \gamma_1, \tag{6.3.26}$$

from which $\frac{\mu}{\lambda} < \gamma_1$.

With analogous procedure one derives from (6.3.22) and Leray's lemma the inequality holding for all $r' \leq r$

$$T^2[\Sigma_\rho] \leq c_1 \left(P^2[\Sigma_{r'}] + U^2[\Sigma_{r'}] + F^2[\Sigma_{r'}] + \frac{1}{(r'-\rho)^2} \int_0^{r'} P^2[\Gamma_t] dt \right).$$

(6.3.27)

Now by virtue of (6.3.18) and (6.3.20) one finds

$$R^2[\Sigma_\rho] = R'^2[\Sigma_\rho] + \int_{\Sigma_\rho} (p_{nn})^2 d\tau \leq T^2[\Sigma_\rho] + \int_{\Sigma_\rho} (p_{nn})^2 d\tau, \qquad (6.3.28)$$

whereas p_{nn} is obtained straight away from Eq. (6.1.1). Hence (6.3.27) and (6.3.28) lead to

$$R^2[\Sigma_\rho] \leq c_2 \left(\frac{1}{(r'-\rho)^2} P^2[\Sigma_{r'}] + U^2[\Sigma_{r'}] + F^2[\Sigma_{r'}] \right).$$

(6.3.29)

In this inequality we can insert the majorization (6.3.25) evaluated at $\rho = r'$, finding therefore, for some positive number γ_2,

$$R^2[\Sigma_\rho] \leq \gamma_2 \left(\frac{1}{(r-\rho)^4} U^2[\Sigma_r] + \frac{1}{(r-\rho)^2} U[\Sigma_r] F[\Sigma_r] + F^2[\Sigma_r] \right). \quad (6.3.30)$$

The material in these first three sections exhausts what can be covered in a first lecture. The following section, which can be omitted in a first reading, continues the analysis of Caccioppoli's proof.

6.4 Caccioppoli's Proof of Integral Bounds: Part 2

Suppose now that Σ_r is not entirely contained in D. We can limit ourselves, which is enough for our purposes, to the case where Γ_r intersects S only in a portion S' belonging to the hyperplane $x_n = 0$ and containing the projection of Σ_r onto this. Let us denote by $\mathcal{D}_\rho$ the domain $D \cap \Sigma_\rho$, by Γ'_ρ the part of the frontier of $\mathcal{D}_\rho$ within D, and let us set

$$u(x^1, \ldots, x^{n-1}, 0) \equiv v(x^1, \ldots, x^{n-1}) \qquad (6.4.1)$$

when $(x^1, \ldots, x^{n-1})$ is varying within S'. We revert to the calculation that led us to Eq. (6.3.14) and we apply it to $\mathcal{D}_\rho$ rather than to $\mathcal{D}$. For this purpose, we multiply initially Eq. (6.1.1) by $(u - v) d\tau$ rather than by $u \, d\tau$. We therefore arrive at an inequality where V and Q for v are the counterpart of U and P for u, i.e.

$$(U[\Gamma'_\rho] + V[\Gamma'_\rho])P[\Gamma'_\rho] - \lambda' P^2[\mathcal{D}_\rho]$$
$$+ \mu' \left\{ U^2[\mathcal{D}_\rho] + V^2[\mathcal{D}_\rho] + Q^2[\mathcal{D}_\rho] + (U[\mathcal{D}_\rho] + V[\mathcal{D}_\rho])F[\mathcal{D}_\rho] \right\} \geq 0. \tag{6.4.2}$$

Of course, we have here

$$P^2[\mathcal{D}_\rho] \equiv \int_0^\rho P^2[\Gamma'_t]dt, \ U^2[\mathcal{D}_\rho] \equiv \int_0^\rho U^2[\Gamma'_t]dt, \tag{6.4.3}$$

while on the other hand it is clear that one can write, relying upon the definitions (6.3.1)–(6.3.6),

$$\int_0^\rho V^2[\Gamma'_t]dt = V^2[\mathcal{D}_\rho] \leq 2\rho \int_{S'_\rho} v^2 dx^1 ... dx^{n-1} \leq 2\rho \overline{U}^2[S'_\rho], \tag{6.4.4}$$

$$\int_0^\rho Q^2[\Gamma'_t]dt = Q^2[\mathcal{D}_\rho] \leq \eta_1 \overline{P}^2[S'_\rho], \tag{6.4.5}$$

where S'_ρ is the projection of Σ_ρ onto S', and η_1 is a constant depending on the local parametric representations adopted for S. In place of the inequality (6.3.25) one then gets

$$P^2[\mathcal{D}_\rho] \leq \gamma' \left[\frac{1}{(r-\rho)^2}(U^2[\mathcal{D}_r] + r\overline{U}^2[S'_r]) \right.$$
$$\left. + (U[\mathcal{D}_r] + \sqrt{r}\ \overline{U}[S'_r])F[\mathcal{D}_r] + r\overline{P}^2[S'_r] \right]. \tag{6.4.6}$$

In analogous way, on considering the bound that pertains to second derivatives, let us multiply the differentiated equation (6.3.15) by $\left(p_h - \frac{\partial v}{\partial x^h} \right) d\tau$, and let us integrate as we have done before, introducing the symbol Z, which is defined for v as R is defined for u. One then gets, with obvious meaning of the abbreviated notation, the inequality

$$(P+Q)R'[\Gamma'_\rho] - \alpha'_1 R'^2[\mathcal{D}_\rho] + \beta'_1(P+Q+Z+U+F)R'[\mathcal{D}_\rho]$$
$$+ \beta'_2 \left[(P+Q)P + (U+F)Z \right][\mathcal{D}_\rho] + \beta'_0(P+Q)(U+F)[\Gamma'_\rho] \geq 0. \tag{6.4.7}$$

From this inequality one obtains, upon setting

$$T \equiv \sqrt{R'^2 + (\beta'_0)^2(U^2 + F^2)}, \tag{6.4.8}$$

the inequality that corresponds to (6.3.22), i.e.

$$(P+Q)T[\Gamma'_\rho] - \lambda'_1 T^2[\mathcal{D}_\rho] + \mu'_1(P^2 + Q^2 + Z^2 + U^2 + F^2)[\mathcal{D}_\rho] \geq 0. \tag{6.4.9}$$

At this stage one proceeds as for the inequality (6.4.2), bounding $P^2[\mathcal{D}_\rho]$ with the help of (6.4.6) and pointing out that, by analogy with what we have done for Q, one can write

$$Z^2[\mathcal{D}_\rho] \leq \eta_2 \rho \overline{R}^2[S'_\rho].\tag{6.4.10}$$

Thus, one arrives eventually at the inequality which corresponds to (6.3.30), i.e.

$$
\begin{aligned}
R^2[\mathcal{D}_\rho] \leq \gamma'' \Bigg[&\frac{1}{(r-\rho)^4}(U^2[\mathcal{D}_r] + r\overline{U}^2[S'_r]) \\
+ &\frac{1}{(r-\rho)^2}(U[\mathcal{D}_r] + \sqrt{r}\,\overline{U}[S'_r])F[\mathcal{D}_r] + \frac{r}{(r-\rho)^2}\overline{P}^2[S'_r] \\
+ &r\overline{R}^2[S'_r] + F^2[\mathcal{D}_r] \Bigg].
\end{aligned}
\tag{6.4.11}
$$

In the applications of these formulae one is interested in the case when r and ρ are not too close, in such a way that ρ does not exceed a certain fraction of r. One can prove that, in such a case, one can neglect on the right-hand side of (6.4.11) the term containing $\overline{P}^2[S'_r]$ because, on the hypersphere S'_r, $r^2\overline{P}^2$ is majorized by a linear combination of $\overline{U}^2$ and $r^4\overline{R}^2$. In order to understand this one can remark that, for functions of a single variable, given $f : x \rightarrow f(x)$ in the open interval $]0, l[$, and defining

$$\int_0^l f^2(x)\mathrm{d}x \equiv A^2, \quad \int_0^l f'^2(x)\mathrm{d}x \equiv B^2, \quad \int_0^l f''^2(x)\mathrm{d}x \equiv C^2,\tag{6.4.12}$$

$l^2 B^2$ can be majorized by means of A^2 and $l^4 C^2$. Now upon denoting by μ the min of the absolute value f', one has

$$|f'| \leq \mu + \sqrt{l}C, \quad f^2 \leq \mu^2 + lC^2, \quad B^2 \leq \mu^2 l + l^2 C^2.\tag{6.4.13}$$

On the other hand, A does not increase if f is replaced by $\mu\left(\frac{l}{2} - x\right)$, so that

$$\frac{\mu^2 l^2}{12} \leq A^2, \quad l^2 B^2 \leq 12A^2 + l^4 C^2.\tag{6.4.14}$$

In our investigation, the condition (6.4.11) becomes therefore

$$
\begin{aligned}
R^2[\mathcal{D}_r] \leq \gamma'' \Bigg[&\frac{1}{(r-\rho)^4}(U^2[\mathcal{D}_r] + r\overline{U}^2[S'_r]) \\
+ &\frac{1}{(r-\rho)^2}(U[\mathcal{D}_r] + \sqrt{r}\,\overline{U}[S'_r])F[\mathcal{D}_r] + r\overline{R}^2[S'_r] + F^2[\mathcal{D}_r] \Bigg].
\end{aligned}
\tag{6.4.15}
$$

We are now in a position to prove the validity of (6.1.10), considering furthermore the general case of a domain $\mathcal{D}$ part of whose frontier intersects S. For this pupose, let us decompose the space in square domains Q with diagonal $\frac{\delta}{2}$, and for each of them let us consider the concentric hyperspheres Σ and Σ' with diameter $\frac{\delta}{2}$ and $\frac{3\delta}{2}$, respectively. The domains $Q_1, Q_2, \dots$ are taken to intersect $\mathcal{D}$ in the portions $\mathcal{D}_1, \mathcal{D}_2, \dots$, respectively.

If the distance of $\mathcal{D}$ from S is $\geq \delta$, one can majorize

$$J_{\mathcal{D}_i}^{(1)} \equiv P^2[\mathcal{D}_i], \quad J_{\mathcal{D}_i}^{(2)} \equiv R^2[\mathcal{D}_i] \tag{6.4.16}$$

by means of (6.3.25) and (6.3.30), and taking as hyperspheres majorizing Σ_ρ and Σ_r the Σ and Σ' of diameters $2\rho = \frac{\delta}{2}$ and $2r = \frac{3\delta}{2}$, respectively, that pertain to the domain Q_i containing $\mathcal{D}_i$. By summing the results for $i = 1, 2, \dots$ one obtains two bounds of the kind

$$J_{\mathcal{D}}^{(1)} = P^2[\mathcal{D}] \leq \frac{K_1'}{\delta^2}\left(U^2[D] + \delta^2 U[D]F[D]\right), \tag{6.4.17}$$

$$J_{\mathcal{D}}^{(2)} = R^2[\mathcal{D}] \leq \frac{K_2'}{\delta^4}\left(U^2[D] + \delta^2 U[D]F[D] + \delta^4 F^2[D]\right), \tag{6.4.18}$$

from which one gets

$$\|u\|_{D'} \leq K'\left[J_D + \int_D f^2 \, d\tau\right]. \tag{6.4.19}$$

At this stage Caccioppoli pointed out that, by virtue of the regularity assumed for S, one can decompose a suitable neighbourhood I of S on D in a certain number of domains $D_1, D_2, \dots$ contained in yet other domains $D_1', D_2', \dots$ respectively, in such a way that the following circumstances are verified:

(i) D_i' can undergo a continuous transformation of order 2 (i.e. invertible and defined by functions of class C^2) which turns the portion $S_i = S \cap D_i'$ of S into a field Π_i of the hyperplane ψ.
(ii) The image of D_i is contained in a hypersphere whose concentric hypersphere of radius twice as large is contained in the image of D_i', and has projection upon ψ contained in Π_i.

Let us now detach from $\mathcal{D}$ the domain $I \cap \mathcal{D}$, that we decompose into the portions $E_i \equiv D_i \cap \mathcal{D}$; to the remaining part E of $\mathcal{D}$ we apply the method described, and we do the same with the images of the E_i. We shall therefore arrive at separate inequalities for $P^2[E]$, $R^2[E]$ and for $P^2[E_i]$, $R^2[E_i]$, some of the latter being obtained by application of (6.4.6) or (6.4.15), all this without ruling out that E_i might have frontier points on S and hence its image might have frontier points on ψ. We shall thus find, as majorants of $J_{\mathcal{D}}^{(1)}$, $J_{\mathcal{D}}^{(2)}$, up to constant factors, the two expressions

$$\frac{\overline{U}^2[S']}{\delta} + \delta\overline{P}^2[S'] + \frac{U^2[D]}{\delta^2} + U[D]F[D] + \sqrt{\delta}\,\overline{U}[S']F[D],$$

$$\frac{\overline{U}^2[S']}{\delta^3} + \delta\overline{R}^2[S'] + \frac{U^2[D]}{\delta^4} + \frac{U[D]F[D]}{\delta^2} + F^2[D] + \frac{1}{\delta^{\frac{3}{2}}}\overline{U}[S']F[D],$$

where S' is a portion of S, and δ is sufficiently small but *limitable from below*. If we disregard the specifications reciprocal of S' and δ, we obtain a kind of bounds valid for every domain $\mathcal{D}$ and hence, which is the same, for D:

$$J_D^{(1)} = P^2[D] \le K_1\left(\overline{U}^2[S] + \overline{P}^2[S] + U^2[D] + F^2[D]\right), \tag{6.4.20}$$

$$J_D^{(2)} = R^2[D] \le K_2\left(\overline{U}^2[S] + \overline{R}^2[S] + U^2[D] + F^2[D]\right). \tag{6.4.21}$$

We note, with Caccioppoli, a simplification in (6.4.20), where F^2 occurs in place of $UF + \overline{U}F$.

Bearing instead in mind the variability of δ, one can obtain that S' is exterior, upon S, to a given portion S'' exterior to $\mathcal{D}$; this can be achieved by taking for δ a sufficiently small fraction, but always limitable from below, of the distance of S'' from $\mathcal{D}$. In other words, for every domain $\mathcal{D}$ whose distance from $S - S'$ is not less than δ, there exist the upper bounds

$$J_{\mathcal{D}}^{(1)} = P^2[\mathcal{D}] \le \frac{H_1}{\delta^2}\left(\delta\overline{U}^2[S'] + \delta^3\overline{P}^2[S'] + U^2[D] + \delta^4 F^2[D]\right), \tag{6.4.22}$$

$$J_{\mathcal{D}}^{(2)} = R^2[\mathcal{D}] \le \frac{H_2}{\delta^4}\left(\delta\overline{U}^2[S'] + \delta^5\overline{R}^2[S'] + U^2[D] + \delta^4 F^2[D]\right). \tag{6.4.23}$$

Note now that the γ coefficients of conditions (6.3.25), (6.4.6), (6.3.30), (6.4.11), (6.4.15) and K' of (6.4.17) and (6.4.18) depend on the maximum absolute values of coefficients of Eq. (6.1.1), and on the first derivatives of A^{ik}, B^i, the discriminant $\det A^{ik}$ having been set to 1. The K, H of (6.4.20)–(6.4.23) depend also on the curvatures of S or of a neighbourhood of S' on S, and on the diameter of D. Thus, the coefficients of the bounds studied so far can be assigned as functions of upper bounds for $|A^{ik}|$, $|B^i|$, $|C|$; the Lipschitz coefficients of A^{ik} and B^i; the Lipschitz coefficients on S of directional cosines of the normal to S; the diameter of D. Hence the bounds hold in general for functions u of class C^2, bounded coefficients A, B, C, and obeying Lipschitz conditions but for C, which can also fail to be continuous, and frontiers S with Lipschitz normal. This general case can be obtained as a limited case of another in which the functions A^{ik}, B^i are uniformly approximated by yet other functions a^{ik}, b^i everywhere smooth (possibly some polynomials), with equibounded first derivatives, while C is replaced by a bounded function c that approaches C on average; this scheme is considered within a domain D' contained in D and arbitrarily close to D, with frontier S' having continuous and equibounded curvatures. The bounds that pertain to the varied equation obtained in such a way for u approach in the limit those derived so far.

A remarkable partial extension of this result can be obtained, as far as the coefficients of (6.1.1) are concerned, by assuming that A^{ik}, B^i are just continuous, while C is taken to be bounded but not necessarily smooth. For this purpose, let us revert to the previous decomposition of the domain D, by means of square domains Q_i, into portions $\mathcal{D}_1, \mathcal{D}_2, \ldots$. For each $\mathcal{D}_h$, let us denote by $\mathcal{D}'_h$ the domain obtained by adjoining to it the surrounding domains. Let $A^{ik}_{(h)}$ and $B^i_{(h)}$ be the values taken by A^{ik}, B^i at a particular point of $\mathcal{D}_h$. The function u satisfies in $\mathcal{D}'_h$ an equation of the form

$$\sum_{i,k=1}^{n} A^{ik}_{(h)} p_{ik} + \sum_{i=1}^{n} B^i_{(h)} p_i + Cu = f_h = f + \varphi_h, \qquad (6.4.24)$$

where, on denoting by ε the maximal among the oscillations of the functions A^{ik}, B^i in the domains $\mathcal{D}'_h$, one has

$$|\varphi_h| \leq \varepsilon \left(\sum_{i,k=1}^{n} |p_{ik}| + \sum_{i=1}^{n} |p_i| \right). \qquad (6.4.25)$$

By applying the upper bounds (6.4.22) and (6.4.23) to this equation, replacing $\mathcal{D}$ by $\mathcal{D}_h$ and D by $\mathcal{D}'_h$, and summing with respect to h, the results take, by virtue of (6.4.25), the form

$$P^2[D] \leq \frac{H'_1}{\delta^2} \left(\delta \overline{U}^2[S] + \delta^3 \overline{P}^2[S] + U^2[D] + \delta^4 F^2[D] \right) + \lambda_1 \varepsilon^2 \left(P^2[D] + R^2[D] \right), \qquad (6.4.26)$$

$$R^2[D] \leq \frac{H'_2}{\delta^4} \left(\delta \overline{U}^2[S] + \delta^5 \overline{R}^2[S] + U^2[D] + \delta^4 F^2[D] \right) + \lambda_2 \varepsilon^2 \left(P^2[D] + R^2[D] \right), \qquad (6.4.27)$$

where H'_1, H'_2, λ_1, λ_2 are fixed. One then obtains immediately two inequalities of the form (6.4.20) and (6.4.21) as soon as δ is chosen in such a way that ε turns out to be sufficiently small. The bounds (6.4.20) and (6.4.21) can be therefore extended to equations with continuous coefficients.

Suppose now that in Eq. (6.1.1), written here in the form

$$E[u] = f, \qquad (6.4.28)$$

the coefficient of u is not positive, i.e. $C \leq 0$. Under such an assumption one arrives at the integral bounds

$$J_D \leq M \left[\overline{J}_S + \int_D f^2 d\tau \right], \qquad (6.4.29)$$

or

$$J_D \leq N \left[\max_S u^2 + \int_D f^2 d\tau \right],$$ (6.4.30)

by considering the differential expression which is the adjoint of $E[u]$, i.e.

$$G[u] \equiv \sum_{i,k=1}^{n} \frac{\partial^2}{\partial x^i \partial x^k} (A^{ik} u) - \sum_{i=1}^{n} \frac{\partial}{\partial x^i} (B^i u) + Cu.$$ (6.4.31)

We shall thus assume, in order to be able to apply to $E[u]$ and $G[u]$ the existence theorems for the Dirichlet problem, that the coefficients A^{ik} and B^i admit of third and second derivatives, respectively, and that S consists of regular manifolds of third order (i.e. representable by means of functions of class C^3).

Let v be the solution of the boundary-value problem

$$G[v] = 1, \quad v|_S = 0,$$ (6.4.32)

and let us denote by z a solution of

$$E[z] = g, \quad z|_S = 0.$$ (6.4.33)

From the so-called reciprocity theorem (the generalized Green lemma) one finds

$$\int_D z \, d\tau = \int_D vg \, d\tau,$$ (6.4.34)

from which it follows that for $g \geq 0$, since $z \leq 0$ by virtue of the hypothesis $C \leq 0$, one has

$$\int_D vg \, d\tau \leq 0.$$ (6.4.35)

Among the two inequalities

$$g_1 \geq g_2, \quad \int_D vg_1 \, d\tau \leq \int_D vg_2 \, d\tau,$$ (6.4.36)

the latter is therefore a consequence of the former, because $v \leq 0$.

On setting now $z \equiv u^2$, where u denotes the solution of Eq. (6.1.1) vanishing on S, one has

$$E[z] = 2fu + 2 \sum_{i,k=1}^{n} A^{ik} p_i p_k - Cu^2 \geq 2fu,$$ (6.4.37)

from which

$$\int_D u^2 \, d\tau = \int_D z \, d\tau \le 2 \int_D v f u \, d\tau \le 2\mathcal{V}\sqrt{\int_D f^2 d\tau \int_D u^2 d\tau}, \qquad (6.4.38)$$

$\mathcal{V}$ denoting the maximum of $|v|$. Therefore

$$\int_D u^2 d\tau \le 4\mathcal{V}^2 \int_D f^2 d\tau, \qquad (6.4.39)$$

and the inequality (6.4.29) is obtained, provided that u vanishes on S.

In order to establish it in general, one has to apply the reciprocity theorem by assuming that, on S, only v vanishes. The formula obtained in such a way, i.e.

$$\int_D z \, d\tau = \int_D v g \, d\tau - \int_S z \sum_{i,k=1}^n \frac{\partial}{\partial x^i}(A^{ik}v)\cos(\nu, x_k) d\sigma, \qquad (6.4.40)$$

provides in particular, for $g \ge 0$, the inequality

$$\int_D z \, d\tau \le - \int_S z \sum_{i,k=1}^n \frac{\partial}{\partial x^i}(A^{ik}v)\cos(\nu, x_k) d\sigma, \qquad (6.4.41)$$

which implies that, for $E[u] = 0$, $z = u^2$,

$$\int_D u^2 d\tau \le M \int_S u^2 d\sigma, \qquad (6.4.42)$$

where the coefficient M depends on the maximum values of $|v|$ and $\left|\frac{\partial v}{\partial x^i}\right|$.

If one now has to prove the inequality (6.4.30), the last remark is useless because, from the result for functions u vanishing on S, and from the known property that the solutions of $E[u] = 0$ are majorized by the values taken on the frontier, the inequality (6.4.30) follows for sure, the coefficient N being independent of derivatives of v.

Following Caccioppoli [14], we have therefore obtained, under the assumption $C \le 0$ in Eq. (6.1.1), the bounds

$$U^2[D] \le M\left(\overline{U}^2[S] + F^2[D]\right), \quad U^2[D] \le N\left(\max\overline{u}^2 + F^2[D]\right). \qquad (6.4.43)$$

In light of the first inequality in (6.4.43), the inequalities (6.4.20) and (6.4.21) become

$$J_D^{(1)} = P^2[D] \le K_1\left(\overline{U}^2[S] + \overline{P}^2[S] + F^2[D]\right), \qquad (6.4.44)$$

$$J_D^{(2)} = R^2[D] \le K_2\left(\overline{U}^2[S] + \overline{R}^2[S] + F^2[D]\right), \qquad (6.4.45)$$

whereas the second inequality in (6.4.43) yields, again by insertion into (6.4.20) and (6.4.21), the bounds

$$J_D^{(1)} = P^2[D] \le K_1\left(\max \bar{u}^2 + \overline{P}^2[S] + F^2[D]\right), \tag{6.4.46}$$

$$J_D^{(2)} = R^2[D] \le K_2\left(\max \bar{u}^2 + \overline{R}^2[S] + F^2[D]\right). \tag{6.4.47}$$

In analogous way, getting rid again of $U^2[D]$ by means of either the first or the second of (6.4.43), the inequalities (6.4.22) and (6.4.23) get simplified either in the form

$$J_{\mathcal{D}}^{(1)} = P^2[\mathcal{D}] \le \frac{H_1}{\delta^2}\left(\overline{U}^2[S] + F^2[D] + \delta^3\overline{P}^2[S']\right), \tag{6.4.48}$$

$$J_{\mathcal{D}}^{(2)} = R^2[\mathcal{D}] \le \frac{H_2}{\delta^4}\left(\overline{U}^2[S] + F^2[D] + \delta^5\overline{R}^2[S']\right), \tag{6.4.49}$$

or as follows:

$$J_{\mathcal{D}}^{(1)} = P^2[\mathcal{D}] \le \frac{H_1}{\delta^2}\left(\max \bar{u}^2 + F^2[D] + \delta^3\overline{P}^2[S']\right), \tag{6.4.50}$$

$$J_{\mathcal{D}}^{(2)} = R^2[\mathcal{D}] \le \frac{H_2}{\delta^4}\left(\max \bar{u}^2 + F^2[D] + \delta^5\overline{R}^2[S']\right). \tag{6.4.51}$$

In this way one obtains in particular the bounds (for D' within D)

$$\|u\|_D \le K\left[\|\bar{u}\|_S + \int_D f^2 d\tau\right], \quad \|u\|_{D'} \le K'\left[\overline{J}_S + \int_D f^2 d\tau\right], \tag{6.4.52}$$

$$\|u\|_D \le K\left[\|u\|_S^* + \int_D f^2 d\tau\right], \quad \|u\|_{D'} \le K'\left[\max_S u^2 + \int_D f^2 d\tau\right]. \tag{6.4.53}$$

The range of application of (6.4.46), (6.4.47) and (6.4.50), (6.4.51), in which there occurs the maximum of $|\bar{u}|$ rather than the root mean square value, is less wide than the range of the analogous (6.4.44), (6.4.45) and (6.4.48), (6.4.49), as far as the data of the solutions are concerned. It is therefore broader with respect to the equations, the coefficients of the bounds depending on fewer elements.

The coefficients of (6.4.46), (6.4.47) and (6.4.50), (6.4.51) can be assigned as functions of the maximum $\mathcal{V}$ of $|v|$, and of the elements discussed in the analysis leading to (6.4.24)–(6.4.27). Now also $\mathcal{V}$ depends on these elements, and furthermore on the maximum absolute values of second derivatives of the functions A^{ik}. Thus, as soon as one has checked that $\mathcal{V}$ does not depend on further elements, the bounds (6.4.46), (6.4.47) and (6.4.50), (6.4.51) hold for equations with bounded coefficients, with $\frac{\partial A^{ik}}{\partial x^k}$ and B^i being Lipschitz functions, and with frontiers S having a Lipschitz normal. This is certainly the case if also in $G[u]$ the coefficient Γ of u is not positive, because v turns out to be bounded with the coefficients of G (the discriminant det $\left\| A^{ik} \right\|$ being equal to 1, and the known term of the equation satisfied from v being equal to 1). If, in particular, the expression $E[u]$ is self-adjoint, then in order to assign the coefficients of (6.4.46), (6.4.47), (6.4.50), (6.4.51) it is enough to have bounded the functions A^{ik}, B^i, their Lipschitz coefficients, and C.

More generally, it is sufficient that, for some particular positive functions φ, Φ (bounded with their second derivatives by means of the coefficients of E, G) one has

$$E[\varphi] < 0, \quad G[\Phi] < 0, \tag{6.4.54}$$

for us to be able to revert to the case of negative C, Γ coefficients, on replacing u by $\frac{u}{\varphi}$ and v by $\frac{v}{\Phi}$.

If one makes assumptions on $E[u]$ only, it is important to remark that the integral $\int_D v \, d\tau$ can be bounded by the coefficients of u, because, from what we have seen before,

$$\int_D v \, d\tau = \int_D z \, d\tau \text{ when } E[z] = 1, z = 0 \text{ on } S.$$

At this stage, the general reader can proceed on his own to complete the study of the work in Caccioppoli [14]. We just emphasize here the geometric vision of Caccioppoli, with a clear description of many domains, sub-domains and their frontiers, whereas the modern treatments seem to have relegated these conceptual ingredients to the original papers. Hence we hope that our effort was not useless.

Chapter 7
Advanced Tools for the Caccioppoli-Leray Inequality

7.1 The Concept of Weak Solution

Classical analysis viewed the linear differential equations as local equations holding pointwise. For example, the inhomogeneous linear equation

$$Au = f \qquad (7.1.1)$$

was long studied pointwise, writing $u = u_0 + A^{-1} f$, where u_0 is a solution of the homogeneous equation $Au = 0$, and the inverse operator is an integral operator, with kernel given by the Green kernel of A, once that suitable boundary conditions have been assigned.

Let us now assume that Eq. (7.1.1) is considered on some open set Ω of $\mathbf{R}^n$. The desired function u is said to be a *weak solution* of (7.1.1) if, for all $\phi \in C_c^\infty(\Omega)$ (recall Sect. 5.1), one has the equality of scalar products

$$(u, A^\dagger \phi) = (f, \phi), \qquad (7.1.2)$$

where $A^\dagger$ is the (formal) adjoint of the linear operator A, for which $(u, A^\dagger \phi) = (Au, \phi)$. From now on, the partial differential equations of interest will be frequently regarded as weak equations, obtainable by taking the scalar product of the classical, local differential equation with a smooth function having compact support.

Note that, since $\phi \in C_c^\infty(\Omega)$ by hypothesis, the term (f, ϕ) on the right-hand side of (7.1.2) reduces to an integral over a finite-measure set, while the scalar product on the left-hand side suggests using integration by parts. This leads to the so-called divergence form of elliptic equations or elliptic systems.

© Springer International Publishing AG 2017
G. Esposito, *From Ordinary to Partial Differential Equations*, UNITEXT 106,
DOI 10.1007/978-3-319-57544-5_7

7.2 Caccioppoli-Leray for Elliptic Systems in Divergence Form

After having seen in Chap. 6 that ellipticity of linear equations makes it possible to obtain integral bounds on the solutions, we are now going to reformulate this property in modern language. For this purpose, denoting by Ω a bounded open set of $\mathbf{R}^n$, we consider the following system of equations in divergence form:

$$-\sum_{\alpha,\beta=1}^{n}\sum_{j=1}^{m}\partial_\alpha(A_{ij}^{\alpha\beta}\partial_\beta u^j) = f_i - \sum_{\alpha=1}^{n}\partial_\alpha F_i^\alpha, \ i = 1, ..., m, \tag{7.2.1}$$

where $u \in H^1_{loc}(\Omega; \mathbf{R}^m)$, $A_{ij}^{\alpha\beta} \in L^\infty(\Omega)$, $f_i \in L^2_{loc}(\Omega)$, $F_i^\alpha \in L^2_{loc}(\Omega)$. We are going to prove that

Theorem 7.1 *If the coefficients $A_{ij}^{\alpha\beta}$ satisfy the Legendre condition [4]*

$$\sum_{\alpha,\beta=1}^{n}\sum_{i,j=1}^{m}(A_{ij}^{\alpha\beta}\,\xi_\alpha^i\,\xi_\beta^j) \geq \lambda|\xi|^2, \ \forall\xi \in R^{m\times n}, \tag{7.2.2}$$

where $\lambda > 0$ and $R^{m\times n}$ is the space of $m \times n$ real matrices, and if

$$\sup_{x\in\Omega}\left|A_{ij}^{\alpha\beta}(x)\right| \leq \Lambda < \infty, \tag{7.2.3}$$

then there exists $c = c(\lambda, \Lambda) > 0$ such that, for every Euclidean ball $B_R(x_0)$ centred at x_0 of radius R, having closure $\overline{B_R(x_0)}$ contained in Ω, and $\forall k \in \mathbf{R}^m$, one has

$$c(\lambda, \Lambda)\int_{B_{\frac{R}{2}}(x_0)}|\nabla u|^2 dx \leq \frac{1}{R^2}\int_{B_R(x_0)}|u(x) - k|^2 dx + R^2\int_{B_R(x_0)}|f(x)|^2 dx$$

$$+ \int_{B_R(x_0)}|F(x)|^2 dx. \tag{7.2.4}$$

Proof Equation (7.2.1) can be written concisely in the form

$$-\operatorname{div}(A\,\operatorname{grad}u) = f - \operatorname{div}F. \tag{7.2.5}$$

Here we multiply both sides by a test function $\Phi \in C_c^\infty(B_R)$ as in Sect. 7.1, and we find

$$\int \Phi \operatorname{div} F = \int \Phi \sum_{\alpha=1}^{n} \frac{\partial}{\partial x^\alpha} F^\alpha = \int \sum_{\alpha=1}^{n} \left[\frac{\partial}{\partial x^\alpha} (\Phi F^\alpha) - F^\alpha \frac{\partial \Phi}{\partial x^\alpha} \right]$$

$$= - \int F \nabla \Phi, \tag{7.2.6}$$

where the divergence term in the integrand gives vanishing contribution to the integral because Φ is taken to belong to $C_c^\infty(B_R)$. Moreover, in analogous way one finds

$$- \int \Phi[\operatorname{div}(A \operatorname{grad} u)] = - \int \Phi \sum_{\alpha,\beta=1}^{n} \left(\frac{\partial}{\partial x^\alpha} \delta^{\alpha\beta} A \frac{\partial u}{\partial x^\beta} \right)$$

$$= - \int \sum_{\alpha,\beta=1}^{n} \left[\frac{\partial}{\partial x^\alpha} \left(\Phi A \delta^{\alpha\beta} \frac{\partial u}{\partial x^\beta} \right) - \frac{\partial \Phi}{\partial x^\alpha} A \delta^{\alpha\beta} \frac{\partial u}{\partial x^\beta} \right] \tag{7.2.7}$$

$$= \int A(\nabla\Phi)(\nabla u).$$

Now the simple but non-trivial idea is to exploit the arbitrariness in the choice of test function to take it as depending on the solution u itself, i.e.

$$\Phi = u\eta^2, \ \eta \in C_c^\infty(B_R), \ \eta = 1 \text{ in } B_{\frac{R}{2}}, \ |\nabla\eta| \le \frac{4}{R}, \tag{7.2.8}$$

η taking values in the closed interval $[0, 1]$. Without loss of generality, we further assume $x_0 = 0, k = 0$. The ansatz (7.2.8) can be inserted into the non-local equation resulting from (7.2.5) to (7.2.7), i.e.

$$\int_{B_R} A(\nabla u)(\nabla\Phi) - \int_{B_R} f\Phi - \int_{B_R} F\nabla\Phi = 0. \tag{7.2.9}$$

Since, in our case,

$$\nabla\Phi = \eta^2 \nabla u + 2\eta u \otimes \nabla\eta, \tag{7.2.10}$$

Equation (7.2.9) reads as

$$\int_{B_R} A(\nabla u)\eta^2(\nabla u) + 2 \int_{B_R} \eta A(\nabla u)(u \otimes \nabla\eta) - \int_{B_R} f u\eta^2$$

$$- \int_{B_R} F\eta^2 \nabla u - 2 \int_{B_R} \eta F u \otimes \nabla\eta = 0. \tag{7.2.11}$$

In this equation, the first integral is

$$\int_{B_R} A(\nabla u)\eta^2(\nabla u) = \int_{B_R} \eta^2 \sum_{\alpha,\beta=1}^{n} \sum_{i,j=1}^{m} A_{ij}^{\alpha\beta}(\partial_\alpha u^i)(\partial_\beta u^j) \geq \lambda \int_{B_R} \eta^2 |\nabla u|^2,$$

(7.2.12)

where we have exploited the Legendre condition (7.2.2) in order to obtain this lower bound, the role of ξ_α^i being played by $\partial_\alpha u^i$.

From now on, we shall find it useful to apply in the analysis the Young inequality, according to which, for p, q whose inverses add up to 1, one has

$$ab = (\varepsilon a)\frac{b}{\varepsilon} \leq \frac{1}{p}(\varepsilon a)^p + \frac{1}{q}\left(\frac{b}{\varepsilon}\right)^q.$$

(7.2.13)

For example, on taking $p = q = 2, a = \eta|\nabla u|, b = u$, the second integral in (7.2.11) can be majorized as follows [we are grateful to F. Pappalardo for checking the detailed application of (7.2.13)]:

$$2\int_{B_R} \eta A(\nabla u)(u \otimes \nabla \eta) \leq 2\int_{B_R} \eta|A||\nabla u||u||\nabla \eta|$$
$$\leq \frac{8}{R}\int_{B_R} \eta|\nabla u||u||A| \leq \frac{4}{R}\Lambda\left[\varepsilon \int_{B_R} \eta^2|\nabla u|^2 + \frac{1}{\varepsilon}\int_{B_R} |u|^2\right],$$

(7.2.14)

where we have exploited first the upper bound on the gradient of η in (7.2.8), then the upper bound on A in (7.2.3), and eventually the Young inequality (7.2.13) with ε replaced by $\sqrt{\varepsilon}$.

The third integral in (7.2.11) can be majorized on setting $\varepsilon = R, p = q = 2, a = |u|, b = |f|$ in the Young inequality (7.2.13), i.e.

$$\int_{B_R} \eta^2 \left|\sum_{i=1}^{m} f_i u^i\right| \leq \int_{B_R} |f||u| \leq \frac{1}{2R^2}\int_{B_R} |u|^2 + \frac{R^2}{2}\int_{B_R} |f|^2.$$

(7.2.15)

Along the same lines, we can exploit the Young inequality (7.2.13) with $\varepsilon = \frac{\sqrt{\lambda}}{2}, p = q = 2, a = \eta|\nabla u|, b = |F|$, hence finding

$$\int_{B_R} \eta^2 \left|\sum_{i=1}^{m} \sum_{\alpha=1}^{n} F_i^\alpha \partial_\alpha u^i\right| \leq \frac{\lambda}{4}\int_{B_R} \eta^2|\nabla u|^2 + \frac{1}{\lambda}\int_{B_R} |F|^2.$$

(7.2.16)

Last, but not least, one obtains for the fifth integral

$$
2 \int_{B_R} \eta F(u \otimes \nabla \eta) \leq 2 \int_{B_R} \eta \left| \sum_{\alpha=1}^n \sum_{i=1}^m |F_i^\alpha u^i| |\partial_\alpha \eta| \right| \leq \frac{8}{R} \int_{B_R} |F||u|
$$

$$
\leq \frac{4}{R} \left[R \int_{B_R} |F|^2 + \frac{1}{R} \int_{B_R} |u|^2 \right] \tag{7.2.17}
$$

$$
= 4 \int_{B_R} |F|^2 + \frac{4}{R^2} \int_{B_R} |u|^2,
$$

by virtue of the Young upper bound (7.2.13) when $\varepsilon = \sqrt{R}$, $p = q = 2$, $a = |F|$, $b = |u|$.

At this stage, the Legendre condition (7.2.2), jointly with the inequalities (7.2.12)–(7.2.17) lead to

$$
\lambda \int_{B_R} \eta^2 |\nabla u|^2 \leq \int_{B_R} \eta^2 A(\nabla u)(\nabla u)
$$

$$
= -2 \int_{B_R} \eta A(\nabla u)(u \otimes \nabla \eta) + \int_{B_R} f u \eta^2 + \int_{B_R} \eta^2 F \nabla u + 2 \int_{B_R} \eta F(u \otimes \nabla \eta)
$$

$$
\leq 2 \int_{B_R} \eta A(\nabla u)(u \otimes \nabla \eta) + \int_{B_R} f u \eta^2 + \int_{B_R} \eta^2 F \nabla u + 2 \int_{B_R} \eta F(u \otimes \nabla \eta)
$$

$$
\leq \left(\frac{4\Lambda}{R} \varepsilon \int_{B_R} \eta^2 |\nabla u|^2 + \frac{4\Lambda}{R\varepsilon} \int_{B_R} |u|^2 \right) + \left(\frac{1}{2R^2} \int_{B_R} |u|^2 + \frac{R^2}{2} \int_{B_R} |f|^2 \right)
$$

$$
+ \left(\frac{\lambda}{4} \int_{B_R} \eta^2 |\nabla u|^2 + \frac{1}{\lambda} \int_{B_R} |F|^2 \right) + \left(4 \int_{B_R} |F|^2 + \frac{4}{R^2} \int_{B_R} |u|^2 \right)
$$

$$
= \left(\frac{4\Lambda \varepsilon}{R} + \frac{\lambda}{4} \right) \int_{B_R} \eta^2 |\nabla u|^2 + \left(\frac{4\Lambda}{R\varepsilon} + \frac{1}{2R^2} + \frac{4}{R^2} \right) \int_{B_R} |u|^2
$$

$$
+ \frac{R^2}{2} \int_{B_R} |f|^2 + \left(4 + \frac{1}{\lambda} \right) \int_{B_R} |F|^2. \tag{7.2.18}
$$

Furthermore, one can exploit the arbitrariness of ε so as to choose it in such a way that

$$
\frac{4\Lambda \varepsilon}{R} = \frac{\lambda}{4}, \tag{7.2.19}
$$

which implies that

$$
\begin{aligned}
\left(\lambda - \frac{\lambda}{2}\right) \int_{B_R} \eta^2 |\nabla u|^2 &= \frac{\lambda}{2} \int_{B_R} \eta^2 |\nabla u|^2 \\
&\leq \left(\frac{\lambda}{4\varepsilon^2} + \frac{1}{2R^2} + \frac{4}{R^2}\right) \int_{B_R} |u|^2 + \frac{R^2}{2} \int_{B_R} |f|^2 \\
&+ \left(4 + \frac{1}{\lambda}\right) \int_{B_R} |F|^2.
\end{aligned}
\tag{7.2.20}
$$

Multiplication by 2 turns this into

$$
\lambda \int_{B_R} \eta^2 |\nabla u|^2 \leq \left(\frac{\lambda}{2\varepsilon^2} + \frac{9}{R^2}\right) \int_{B_R} |u|^2 + R^2 \int_{B_R} |f|^2 + \left(8 + \frac{2}{\lambda}\right) \int_{B_R} |F|^2.
\tag{7.2.21}
$$

Moreover, since $\eta = 1$ in the Euclidean ball $B_{\frac{R}{2}}$, one can write

$$
\lambda \int_{B_{\frac{R}{2}}} |\nabla u|^2 \leq \lambda \int_{B_R} \eta^2 |\nabla u|^2,
\tag{7.2.22}
$$

where the right-hand side can be majorized with the help of (7.2.21), hence leading to

$$
\lambda \int_{B_{\frac{R}{2}}} \eta^2 |\nabla u|^2 \leq \left(\frac{\lambda}{2\varepsilon^2} + \frac{9}{R^2}\right) \int_{B_R} |u|^2 + R^2 \int_{B_R} |f|^2 + \left(8 + \frac{2}{\lambda}\right) \int_{B_R} |F|^2.
\tag{7.2.23}
$$

By virtue of (7.2.19) we can here re-express

$$
\frac{1}{\varepsilon^2} = \frac{256\Lambda^2}{R^2 \lambda^2}.
\tag{7.2.24}
$$

Thus, upon defining

$$
c(\lambda, \Lambda) \equiv \frac{\lambda}{\left(9 + 128\frac{\Lambda^2}{\lambda}\right)},
\tag{7.2.25}
$$

one obtains eventually

$$c(\lambda, \Lambda) \int_{B_{\frac{R}{2}}} |\nabla u|^2 \leq \frac{1}{R^2} \int_{B_R} |u|^2 + \frac{R^2}{\left(9 + 128\frac{\Lambda^2}{\lambda}\right)} \int_{B_R} |f|^2$$

$$+ \frac{\left(8 + \frac{2}{\lambda}\right)}{\left(9 + 128\frac{\Lambda^2}{\lambda}\right)} \int_{B_R} |F|^2 \tag{7.2.26}$$

$$\leq \frac{1}{R^2} \int_{B_R} |u|^2 + R^2 \int_{B_R} |f|^2 + \int_{B_R} |F|^2.$$

Q.E.D.

7.3 Legendre versus Legendre-Hadamard Conditions

We have already encountered the Legendre condition in (7.2.2). This provides a concept of ellipticity for systems of equations. One therefore deals with functions

$$u : \Omega \subset \mathbf{R}^n \to \mathbf{R}^m,$$

with Greek letters $\alpha, \beta \in \{1, 2, ..., n\}$ used for the domain, and Latin letters $i, j \in \{1, 2, ..., m\}$ used for the codomain. The matrices $A_{ij}^{\alpha\beta}$ in (7.2.1) are rank-4 tensors. In the scalar case, ellipticity would be expressed by the condition

$$A \geq \lambda I, \quad \lambda > 0. \tag{7.3.1}$$

In the vector-valued case, one defines the Legendre condition (7.2.2), and in the system (7.2.1) $u \in H_0^{1,2}(\Omega; \mathbf{R}^m)$, while f_i and $F_i^\alpha \in L^2(\Omega)$.

Besides (7.3.1), it is useful to consider also the *coercivity condition*

$$\int_\Omega \langle A\nabla u, \nabla u \rangle dx \geq \lambda \int_\Omega |\nabla u|^2 dx, \quad u \in H_0^{1,2}(\Omega; \mathbf{R}^m). \tag{7.3.2}$$

This can be applied to test functions having the form $u_\tau(x) = \varphi(x)be^{i\tau x \cdot a}$ with $a \in \mathbf{R}^n$ and $b \in \mathbf{R}^m$, and implies the *Legendre-Hadamard condition*

$$\sum_{\alpha,\beta=1}^n \sum_{i,j=1}^m A_{ij}^{\alpha,\beta} \xi_\alpha^i \xi_\beta^j \geq \lambda |\xi|^2 \ \forall \xi = a \otimes b. \tag{7.3.3}$$

When the Legendre condition (7.2.2) is restricted to rank-1 matrices $\xi_\alpha^i = a_\alpha b^i$, it becomes the Legendre-Hadamard condition (7.3.3). If (7.3.3) is fulfilled, this does not imply that (7.2.2) holds. In Ambrosio [4], the author provides an example where $m = n = 2$ and the tensor $A_{ij}^{\alpha\beta}$ is defined by

$$A_{ij}^{\alpha\beta} \, \xi_\alpha^i \, \xi_\beta^j = \det(\xi) + \varepsilon|\xi|^2. \tag{7.3.4}$$

If such a quadratic form is restricted to diagonal matrices with eigenvalues t and $-t$, it equals $t^2(2\varepsilon - 1)$, and hence Legendre is not fulfilled if $\varepsilon < \frac{1}{2}$, whereas Legendre-Hadamard with $\lambda = \varepsilon$ is satisfied.

The Legendre-Hadamard condition provides however a sufficient condition for coercivity by virtue of a theorem of Garding, according to which, if (7.3.3) holds for some $\lambda > 0$ and $A_{ij}^{\alpha\beta} = A_{ji}^{\alpha\beta}$, then the lower bound (7.3.2) holds for all $u \in H^{1,2}(\mathbf{R}^n; \mathbf{R}^m)$.

7.4 Uniform, or Strong, or Uniform Strong, or Proper Ellipticity

A linear partial differential operator of order $2r$ with real or complex coefficients $a_\alpha(t)$, i.e.

$$\mathcal{M}u = \sum_{|\alpha|=0}^{2r} a_\alpha(t)D^\alpha u, \tag{7.4.1}$$

is said to be *elliptic* in a domain A, where the $a_\alpha(t)$ are defined, if its *characteristic polynomial* (or *leading symbol*) $M(t, \xi)$ satisfies the condition

$$M(t, \xi) = \sum_{|\alpha|=2r} a_\alpha(t)\xi^\alpha \neq 0, \tag{7.4.2}$$

for all non-vanishing vector ξ and $\forall t \in A$. Note that $M(t, \xi)$ contains only the highest values $2r$ of the multi-index $|\alpha|$, which ensures that the definition of ellipticity is coordinate-independent. The condition (7.4.2) is certainly verified if there exists a constant $a > 0$ such that

$$|M(t, \xi)| \geq a|\xi|^{2r}, \ \forall t \in A, \tag{7.4.3}$$

and in this case $\mathcal{M}$ is said to be *uniformly elliptic* in A [101].

Note that, if A is bounded and the coefficients a_α are continuous in A, the function $|M(t, \eta)|$ studied as t varies in A and η is varying on the spherical hypersurface $|\eta| = 1$ has, when (7.4.2) holds, a positive minimum a. Since M is a homogeneous polynomial of degree $2r$ in the variables ξ_i, one has

$$M\left(t, \frac{\xi}{|\xi|}\right) = \frac{M(t, \xi)}{|\xi|^{2r}}, \tag{7.4.4}$$

and hence from the lower bound $|M(t, \eta)| \geq a$ one derives the condition (7.4.3). Thus, if the domain A is bounded and the coefficients a_α, with $|\alpha| = 2r$, are continuous in A, the elliptic operator $\mathcal{M}$ is also uniformly elliptic.

Next, an elliptic operator $\mathcal{M}$ is said to be *strongly elliptic* in A if, for all $t \in A$, and for any real non-vanishing vector ξ, one has [101]

$$\mathrm{Re}\, M(t, \xi) \neq 0. \tag{7.4.5}$$

Furthermore, $\mathcal{M}$ is *uniformly strongly elliptic* in A if there exists an $a > 0$ such that, for all $t \in A$ and any real ξ, one has [101]

$$\mathrm{Re} M(t, \xi) \geq a|\xi|^{2r}. \tag{7.4.6}$$

Of course, every elliptic operator with real coefficients is also strongly elliptic. In analogous way to what we have done above, one proves that, if the domain A is bounded and the coefficients a_α with $|\alpha| = 2r$ are continuous in A, the operator $\mathcal{M}$, if strongly elliptic, is also uniformly strongly elliptic.

Another important concept is *proper ellipticity* of an operator. In order to define it, let ξ and η be two real, non-vanishing, mutually orthogonal vectors, and consider the algebraic equation in the variable z of degree $2r$ [101]:

$$M(t, \xi + z\eta) = 0. \tag{7.4.7}$$

If the operator $\mathcal{M}$ is elliptic, from (7.4.2) it follows that all roots of Eq. (7.4.7) have non-vanishing imaginary part. The operator $\mathcal{M}$ is then said to be *properly elliptic* if there are equally many roots of (7.4.7) with positive or negative imaginary part. If this condition is fulfilled, one denotes by $M^+(t, \xi, \eta, z)$ the polynomial having as roots only the roots of M with positive imaginary part.

If $n > 2$, every elliptic operator is also properly elliptic. Indeed, for fixed values of t and ξ let $P(\eta)$ and $N(\eta)$ be the numbers of roots of Eq. (7.4.7) with positive and negative imaginary part, respectively, for which one has $P(-\eta) = N(\eta)$. Moreover, since Eq. (7.4.7) does not admit by assumption real roots, while the complex roots depend in a continuous way on η, one has that P and N are constant when η is varying in a connected set. Since the set defined as

$$\{\eta : (\xi, \eta) = 0, |\eta| = 1\} \tag{7.4.8}$$

is connected if $n \geq 3$, one has in this case

$$P(\eta) = P(-\eta) \implies P(\eta) = N(\eta). \tag{7.4.9}$$

Note however that, if $n = 2$, the set (7.4.8) is not connected because it reduces to a pair of points, and hence the theorem stated does not hold. One can nevertheless prove that every strongly elliptic operator is properly elliptic also in the case $n = 2$. Indeed, if $n = 2$, there exist only two unit vectors η and $\eta' = -\eta$ orthogonal to ξ. In order to prove the theorem, it suffices to show that the two equations

$$M(t, \xi + z\eta) = 0, \; M(t, \xi - z\eta) = 0 \qquad (7.4.10)$$

have an equal number of roots with positive imaginary part. This is proved, in turn, if we show that, $\forall \tau \in [0, 1]$, the number of roots with positive imaginary part of the equation

$$\tau M(t, \xi + z\eta) + (1 - \tau)M(t, \xi - z\eta) = 0 \qquad (7.4.11)$$

remains constant. But this is clearly the case, since otherwise for at least a value of τ in the closed interval $[0, 1]$ the Eq. (7.4.11) would have a real root, which is impossible because, for real values of z, the left-hand side has non-vanishing real part. Q.E.D.

On denoting now by f a function defined in the domain A, let us consider the equation

$$\mathcal{M} u = f, \qquad (7.4.12)$$

which is said to be elliptic, or strongly elliptic, or properly elliptic if this is the nature of the operator $\mathcal{M}$, according to the previous definitions. The function u which solves the inhomogeneous equation (7.4.12) is required to belong to a suitable functional space. For example, if the coefficients of $\mathcal{M}$ and f are of class $C^{(0,\lambda)}(A)$, it is convenient to require that $u \in C^{(2r,\lambda)}(A)$ and that u should solve Eq. (7.4.12) everywhere within the domain A. If instead the coefficients of $\mathcal{M}$ are measurable and bounded, and the right-hand side $f \in L^p(A)$, one can require that u should be of class $H^{2r,p}(A)$ and should satisfy Eq. (7.4.12) almost everywhere within A.

Chapter 8
Aspects of Spectral Theory

8.1 Resolvent Set, Spectrum and Resolvent of a Linear Operator

Let A be a linear transformation defined on a sub-space $D(A)$ of a complex Hilbert space $\mathcal{H}$; the sub-space $D(A)$ is the domain of A. Let $R(A)$ be the range of A, i.e. the set

$$R(A) \equiv \{Af : f \in \mathcal{H}\}. \tag{8.1.1}$$

The *resolvent set* of A, denoted by $\rho(A)$, is the set of all complex numbers λ such that $\lambda I - A$ is one-to-one (I being the identity map), $R(\lambda I - A) = \mathcal{H}$, and the inverse $(\lambda I - A)^{-1}$ is a bounded linear transformation on $\mathcal{H}$. The *spectrum* of A, denoted by $\sigma(A)$, is the set of all complex numbers not lying in the resolvent set $\rho(A)$.

Theorem 8.1 *The resolvent set of A is open, and $(\lambda I - A)^{-1}$ is an analytic function on $\rho(A)$. If the linear map A is bounded, then $\rho(A)$ includes the set of all complex numbers λ having modulus bigger than the norm of A.*

Proof If λ_0 is an element of $\rho(A)$, then [1]

$$\begin{aligned}
\lambda I - A &= (\lambda_0 I - A) + (\lambda - \lambda_0)I \\
&= \left[I - (\lambda_0 - \lambda)I(\lambda_0 I - A)^{-1}\right](\lambda_0 I - A).
\end{aligned} \tag{8.1.2}$$

The operator in square brackets on the second line is invertible if the norm of $(\lambda_0 - \lambda)(\lambda_0 I - A)^{-1}$ is less than 1, i.e. if

$$|\lambda_0 - \lambda| < \left\|(\lambda_0 I - A)^{-1}\right\|^{-1}. \tag{8.1.3}$$

© Springer International Publishing AG 2017
G. Esposito, *From Ordinary to Partial Differential Equations*, UNITEXT 106,
DOI 10.1007/978-3-319-57544-5_8

Moreover, the Neumann expansion holds, according to which

$$\left[I - (\lambda_0 - \lambda)(\lambda_0 I - A)^{-1} \right]^{-1} = \sum_{k=0}^{\infty} (\lambda_0 - \lambda)^k (\lambda_0 I - A)^{-k}, \qquad (8.1.4)$$

where the negative power of $(\lambda_0 I - A)$ is defined as

$$(\lambda_0 I - A)^{-k} \equiv [(\lambda_0 I - A)^{-1}]^k. \qquad (8.1.5)$$

Thus, Eq. (8.1.2) implies that, when (8.1.3) holds, the complex number ρ belongs to the resolvent set $\rho(A)$, and the continuity of the inverse $(\lambda_0 I - A)^{-1}$ implies that

$$(\lambda I - A)^{-1} = \sum_{k=0}^{\infty} (\lambda_0 - \lambda)^k (\lambda_0 I - A)^{-k-1}. \qquad (8.1.6)$$

Such an inverse is therefore expressed in terms of a power series about any point $\lambda_0 \in \rho(A)$, i.e. $(\lambda I - A)$ is an analytic function of λ. Last, if A is bounded and the modulus of λ is bigger than the norm of A, one has

$$\lambda I - A = \lambda(I - \lambda^{-1} A), \ \left\| \lambda^{-1} A \right\| < 1. \qquad (8.1.7)$$

All this implies that λ belongs to the resolvent set. $\square$

The bounded linear transformation $(\lambda I - A)^{-1}$, defined on the resolvent set $\rho(A)$, is the *resolvent* of the linear operator A. By virtue of (8.1.6) one has the upper bound

$$\left\| (\lambda I - A)^{-1} \right\| \le \sum_{k=0}^{\infty} |\lambda_0 - \lambda|^k \left\| (\lambda_0 - A)^{-1} \right\|^{k+1}. \qquad (8.1.8)$$

Now one exploits the sum of the geometric series to find the upper bound

$$\left\| (\lambda I - A)^{-1} \right\| \le \frac{\left\| (\lambda_0 I - A)^{-1} \right\|}{1 - |\lambda_0 - \lambda| \left\| (\lambda_0 I - A)^{-1} \right\|}, \qquad (8.1.9)$$

where we bear in mind the inequality (8.1.3).

8.2 Modified Resolvent Set and Modified Resolvent

In the investigation of integral equations, one can encounter frequently equations of the form

$$f = g + \lambda A f, \qquad (8.2.1)$$

where one wants to solve for f in terms of g. If the operator $I - \lambda A$ is invertible, so that the inverse of λ belongs to the resolvent set $\rho(A)$, the solution of Eq. (8.2.1) is given by

$$f = (I - \lambda A)^{-1} g. \tag{8.2.2}$$

It is however useful not to consider the operator $(I - \lambda A)^{-1}$ itself, but, instead, the composition of A with such an inverse. A reason for doing this is that the range of the composition $A(I - \lambda A)^{-1}$ coincides with the range of A, and one has often a very precise information about the range of A. Note also the identity

$$I = (I - \lambda A)(I - \lambda A)^{-1} = (I - \lambda A)^{-1} - \lambda A(I - \lambda A)^{-1}, \tag{8.2.3}$$

from which follows the simple but non-trivial identity

$$(I - \lambda A)^{-1} = I + \lambda A(I - \lambda A)^{-1}. \tag{8.2.4}$$

This suggests defining, in turn, the modified resolvent set as follows.

The *modified resolvent set* $\rho_m(A)$ of the linear operator A is the set of all non-vanishing complex numbers λ such that $I - \lambda A$ is one-to-one, and $(I - \lambda A)^{-1}$ is a bounded linear transformation on the space X such that $R(I - \lambda A) = X$. If λ belongs to the modified resolvent set $\rho_m(A)$, the map

$$A_\lambda \equiv A(I - \lambda A)^{-1} \tag{8.2.5}$$

is the *modified resolvent* of A.

In light of the definition (8.2.5), the identity (8.2.4) may be written in the form

$$(I - \lambda A)^{-1} = I + \lambda A_\lambda. \tag{8.2.6}$$

One also notes that

$$\lambda \in \rho_m(A) \iff \lambda^{-1} \in \rho(A). \tag{8.2.7}$$

If the linear operator A is closed, the condition that $(\lambda I - A)^{-1}$ be bounded in the definition of resolvent set is redundant. Indeed, if $\lambda I - A$ is one-to-one and onto, then $(\lambda I - A)^{-1}$ is closed (A being closed) and defined everywhere on the Hilbert space $\mathcal{H}$. By virtue of the closed graph theorem the operator $(\lambda I - A)^{-1}$ is then bounded. In the same way, if A is closed, the boundedness condition for $(I - \lambda A)^{-1}$ in the definition of modified resolvent set is redundant.

One also remarks that if $D(A) = X$, and if λ belongs to the resolvent set $\rho(A)$, then

$$A(\lambda I - A)^{-1} = (\lambda I - A)^{-1} A. \tag{8.2.8}$$

Indeed, in any case, $(\lambda I - A)(\lambda I - A)^{-1} = I$, while in addition $(\lambda I - A)^{-1}(\lambda I - A)$ is the identity on $D(A)$. Thus, if $D(A) = \mathcal{H}$, then $(\lambda I - A)^{-1}(\lambda I - A) = I$. In analogous way, one finds that if $D(A) = \mathcal{H}$ and $\lambda \in \rho_m(A)$, then A and A_λ commute.

For modified resolvents, the following counterpart to Theorem 8.1 holds:

Theorem 8.2 *The modified resolvent set of A is open, and A_λ is an analytic function on $\rho_m(A)$. If $\lambda_0 \in \rho_m(A)$, then for $|\lambda - \lambda_0|$ less than the inverse of the norm of A_{λ_0}, $\lambda \neq 0$, $\lambda \in \rho_m(A)$ and*

$$A_\lambda = \sum_{k=0}^{\infty} (\lambda - \lambda_0)^k \, A_{\lambda_0}^{k+1}, \tag{8.2.9}$$

where the convergence is meant to be in the operator norm. Moreover,

$$\|A_\lambda\| \leq \frac{\|A_{\lambda_0}\|}{1 - |\lambda - \lambda_0| \, \|A_{\lambda_0}\|}, \quad |\lambda - \lambda_0| < \|A_{\lambda_0}\|^{-1}. \tag{8.2.10}$$

If A is bounded,

$$\left\{ \lambda : 0 < |\lambda| < \|A\|^{-1} \right\} \subset \rho_m(A). \tag{8.2.11}$$

Proof By adding and subtracting $\lambda_0 A$, one has first the identity

$$\begin{aligned} I - \lambda A &= I - \lambda_0 A + (\lambda_0 - \lambda) A \\ &= \left[I - (\lambda - \lambda_0) A_{\lambda_0} \right] (I - \lambda_0 A). \end{aligned} \tag{8.2.12}$$

Thus, if $|\lambda - \lambda_0|$ is majorized as in (8.2.10), one can write

$$(I - \lambda A)^{-1} = (I - \lambda_0 A)^{-1} \sum_{k=0}^{\infty} (\lambda - \lambda_0)^k A_{\lambda_0}^k. \tag{8.2.13}$$

By acting with A on both sides, one obtains the expansion (8.2.9), from which (8.2.10) follows.   □

In the applications, A is a differential operator, and there exists some number in the resolvent set of A. In the Dirichlet problem with zero Dirichlet data, the Hilbert space $\mathcal{H}$ is $L^2(\Omega)$.

8.3 Eigenvalues and Characteristic Values

A complex number λ is an *eigenvalue* of A if there exists a non-zero vector f such that

$$A f = \lambda f. \tag{8.3.1}$$

A complex number λ is said to be a *characteristic value* of A if

$$\lambda A g = g \tag{8.3.2}$$

for a non-zero vector g. By virtue of these definitions, every characteristic value of A is non-vanishing, and λ is a characteristic value of A if and only if its inverse λ^{-1} is an eigenvalue of A.

Given an eigenvalue λ of A, a non-zero vector f is a *generalized eigenvector* of A corresponding to λ if, for some positive integer k,

$$(\lambda I - A)^k f = 0. \tag{8.3.3}$$

The set of all generalized eigenvectors of A corresponding to λ, together with the origin in the Hilbert space $\mathcal{H}$, forms a sub-space of $\mathcal{H}$, whose dimension is the *multiplicity* of the eigenvalue λ. Similarly, if λ is a characteristic value of A, a non-zero vector f is a *generalized characteristic vector* of A corresponding to λ if, for some positive integer k,

$$(I - \lambda A)^k f = 0. \tag{8.3.4}$$

On denoting by $N(A)$ the null space of a transformation A, the vector f is a generalized eigenvector of T corresponding to λ if and only if $f \neq 0$ and f lies in the set

$$M(\lambda) = \cup_{k=1}^{\infty} N((\lambda I - T)^k). \tag{8.3.5}$$

The multiplicity of λ is therefore the dimension of this set. Such a set is a sub-space of the Hilbert space $\mathcal{H}$, because

$$N((\lambda I - T)^k) \subset N((\lambda I - T)^{k+1}), \ k \geq 1. \tag{8.3.6}$$

The set $M(\lambda)$ is said to be the *generalized eigenspace* of the operator T corresponding to λ.

8.4 Directions of Minimal Growth of the Resolvent

If θ is a real parameter and $a \geq 0$, one defines the set

$$\Xi(\theta, a) \equiv \left\{ re^{i\theta} : r > a \right\}. \tag{8.4.1}$$

Given a linear transformation A defined on a sub-space $D(A)$ of a Hilbert space $\mathcal{H}$, a vector $e^{i\theta}$ in the complex plane is a *direction of minimal growth* of the resolvent of A if, for some positive a,

$$\Xi(\theta, a) \subset \rho(A), \ \left\| (\lambda I - A)^{-1} \right\| = O(|\lambda|^{-1}) \text{ for } \lambda \in \Xi(\theta, a), \ |\lambda| \to \infty. \tag{8.4.2}$$

In analogous way, a vector $e^{i\theta}$ is a direction of minimal growth of the modified resolvent of A if, for some positive a,

$$\Xi(\theta, a) \subset \rho_m(A), \ A_\lambda = O(|\lambda|^{-1}) \text{ for } \lambda \in \Xi(\theta, a), \ |\lambda| \to \infty. \tag{8.4.3}$$

8.5 Decay Rate of the Resolvent Along Rays

In general, one cannot expect the resolvent, or modified resolvent, to decay faster along any ray than $O(|\lambda|^{-1})$, as is shown below.

Theorem 8.3 *Suppose that for some positive constant α and for some sequence of complex numbers λ_n in $\rho(A)$ such that $|\lambda_n| \to \infty$, the following estimate holds:*

$$\left\| (\lambda_n I - A)^{-1} \right\| \le \alpha |\lambda_n|^{-1}. \tag{8.5.1}$$

Then $\alpha \ge 1$. Correspondingly, if for some sequence of complex numbers in $\rho_m(A)$ such that $|\lambda_n| \to \infty$, the estimate

$$\left\| A_{\lambda_n} \right\| \le \alpha |\lambda_n|^{-1} \tag{8.5.2}$$

holds, then either $A = 0$ or $\alpha \ge 1$.

Proof By virtue of (8.1.9) one finds that $\lambda \in \rho(A)$ if $|\lambda - \lambda_n| < \alpha^{-1}|\lambda_n|$, and one obtains an estimate

$$\left\| (\lambda I - A)^{-1} \right\| \le \frac{\alpha |\lambda_n|^{-1}}{1 - |\lambda - \lambda_n| \alpha |\lambda_n|^{-1}} = \frac{1}{\alpha^{-1}|\lambda_n| - |\lambda - \lambda_n|}. \tag{8.5.3}$$

If α were less than 1, one could write it in the form

$$\alpha = \frac{1}{(1 + 2\varepsilon)} \tag{8.5.4}$$

for some $\varepsilon > 0$. Moreover, all circles

$$\{\lambda : |\lambda - \lambda_n| < (1 + \varepsilon)|\lambda_n|\} \tag{8.5.5}$$

are contained in $\rho(A)$, and for λ in such a circle, (8.5.3) implies

$$\left\| (\lambda I - A)^{-1} \right\| < \frac{1}{(1 + 2\varepsilon)|\lambda_n| - (1 + \varepsilon)|\lambda_n|} = \frac{1}{\varepsilon|\lambda_n|}. \tag{8.5.6}$$

Since such circles cover the complex plane, the inequality (8.5.6) implies that $(\lambda I - A)^{-1}$ exists and is uniformly bounded in the complex plane. On the other hand, we know from Theorem (8.2) that $(\lambda I - A)^{-1}$ is also analytic on $\rho(A)$, which coincides with the complex plane. The Liouville theorem implies therefore that $(\lambda I - A)^{-1}$ is constant, which in turn yields the constant nature of $(\lambda I - A)$, an absurd.

In the second part of the proof, by the same method above one shows that if $\alpha < 1$, then A_λ is constant. But since the norm of A_{λ_n} approaches 0, this implies that $A_\lambda = 0$ for all λ, i.e.

$$A(I - \lambda A)^{-1} = 0, \ \forall \lambda. \tag{8.5.7}$$

In this equation, application of the operator $I - \lambda A$ on the right yields eventually $A = 0$.
□

We refer again the reader to the work in Agmon [1] for the proof of other interesting properties, which are as follows.

Theorem 8.4 *The set of numbers θ such that $e^{i\theta}$ is a direction of minimal growth of the resolvent of A is open.*

Theorem 8.5 *If A is a self-adjoint operator on the Hilbert space $\mathcal{H}$, any non-real direction $e^{i\theta}$ is a direction of minimal growth of both the resolvent and the modified resolvent of A. Moreover, for $\arg(\lambda) = \theta$, one has the inequalities*

$$\left\| (\lambda I - A)^{-1} \right\| \leq \left| \frac{1}{\sin \theta} \right| |\lambda|^{-1}, \tag{8.5.8}$$

$$\| A_\lambda \| \leq \left| \frac{1}{\sin \theta} \right| |\lambda^{-1}|. \tag{8.5.9}$$

Theorem 8.6 *If A is a linear transformation with domain $D(A) \subset \mathcal{H}$, such that for all $f \in D(A)$*

$$\mathrm{Re}(Af, f) \geq \lambda_0 \| f \|^2 \tag{8.5.10}$$

for some real number λ_0, and if some complex number λ_1 lies in the resolvent set of A and has real part majorized by λ_0, then the half-plane

$$\{ \lambda : \mathrm{Re}(\lambda) < \lambda_0 \} \tag{8.5.11}$$

is contained in the resolvent set of A, and every direction $e^{i\theta}$ with θ in the open interval

$$\left] -\frac{\pi}{2}, \frac{3\pi}{2} \right[$$

is a direction of minimal growth of the resolvent of A.

8.6   Strongly Elliptic Boundary-Value Problems, and an Example

In Sect. 7.4 we have defined strong ellipticity for differential operators, and now we are going to extend this concept to boundary-value problems.

Let A be an elliptic differential operator with characteristic polynomial or leading symbol $A_m(x, \xi)$ and let $\mathcal{K}$ be a cone containing 0 such that, for $\xi \neq 0$, the spectrum of $A_m(x, \xi)$ lies in the complement of $\mathcal{K}$. Let $b^{(0)}(y, \zeta, D_r)$ be the leading symbol of the boundary operator $\mathcal{B}$ that expresses the boundary conditions under study. The boundary-value problem, i.e. the pair $(A, \mathcal{B})$, is then said to be *strongly elliptic* if, for

$$(0, 0) \neq (\zeta, \lambda) \in \partial \mathbf{R}_+^D \times \mathcal{K}, \tag{8.6.1}$$

the equations here written in terms of the geodesic distance r to the boundary:

$$A_m(y, 0; \zeta, D_r) f(r) = \lambda f(r), \tag{8.6.2}$$

$$\lim_{r \to \infty} f(r) = 0, \tag{8.6.3}$$

$$b^{(0)}(y; \zeta, D_r) f(r)\big|_{r=0} = g(\zeta), \tag{8.6.4}$$

have a unique solution.

For example, in the case of Dirichlet boundary conditions, with A an operator of Laplace type, one has the leading symbol

$$A_2(x, \xi) = |\xi|^2 = \sum_{\mu,\nu=0}^{n-1} (g^{-1})^{\mu\nu} \xi_\nu \xi_\mu, \tag{8.6.5}$$

and upon making the replacements demanded by (8.6.2), i.e.

$$\xi_j = \zeta_j \; \forall j = 1, ..., n-1; \; \xi_0 = -i\frac{d}{dr} \equiv D_r, \tag{8.6.6}$$

one obtains the ordinary differential equation

$$A_2(y, 0; \zeta, D_r) f(r) = \left(-\frac{d^2}{dr^2} + |\zeta|^2\right) f(r) = \lambda f(r). \tag{8.6.7}$$

The general solution of Eq. (8.6.7) reads as

$$f(r) = \alpha(\zeta)e^{-r\Lambda} + \beta(\zeta)e^{r\Lambda}, \tag{8.6.8}$$

having defined $\Lambda \equiv \sqrt{|\zeta|^2 - \lambda}$. The asymptotic condition (8.6.3) for $r \to \infty$ leads to $\beta = 0$, while Eq. (8.6.4) engenders $\alpha(\zeta) = g(\zeta)$, since the leading symbol of the boundary operator reduces to the identity in the Dirichlet case. One finds therefore that the Dirichlet boundary-value problem for an operator of Laplace type (i.e. Laplacian plus a potential term) is strongly elliptic with respect to the cone $\mathbf{C} - \mathbf{R}_+$.

The leading symbol nomenclature (rather than characteristic polynomial) is more appropriate, since ellipticity can be defined also for operators which are pseudo-differential rather than differential. For these operators, the symbol can be defined also on Riemannian manifolds, but it is not a polynomial [143]. The prescription

(8.6.6) reflects a simple but non-trivial property, i.e. the split of the cotangent bundle T^*M into the direct sum

$$T^*M = T^*(\partial M) \oplus T^*\mathbf{R}. \tag{8.6.9}$$

The functions f in Eq. (8.6.7) can be scalar functions or, more generally, smooth sections of a vector bundle over $\partial M \times \mathbf{R}_+$. The work in [5] has investigated in a systematic way the conditions for obtaining strongly elliptic boundary-value problems for the gauge theories of fundamental interactions studied in theoretical physics.

Chapter 9
Laplace Equation with Mixed Boundary Conditions

9.1 Uniqueness Theorems with Mixed Boundary Conditions

So far, we have said almost nothing about the basic Dirichlet and Neumann boundary-value problems for equations of elliptic type, apart from the strong ellipticity example of Sect. 8.6. We are now going to study in detail an example of mixed boundary conditions for partial differential equations of elliptic type, following the brilliant work in De Giorgi [33].

Let D be a regular domain of r-dimensional Euclidean space $\mathbf{R}^r$, with generic point $x = (x^1, \ldots, x^r)$, and suppose that two sets $\mathcal{F}_1 D$ and $\mathcal{F}_2 D$ are given, with empty intersection, whose sum coincides with the frontier $\mathcal{F}D$ of the domain D. Consider the differential equation [cf. Eq. (6.1.1)]

$$\left[\sum_{h,k=1}^{r} a^{hk}(x) \frac{\partial^2}{\partial x^h \partial x^k} + \sum_{h=1}^{r} b^h(x) \frac{\partial}{\partial x^h} + c(x) \right] u(x) = f(x) \qquad (9.1.1)$$

in the unknown function u, where a^{hk}, b^h, c, f are functions continuous in D. Moreover, the a^{hk} are taken to be such that Eq. (9.1.1) is of elliptic self-adjoint type, and over the whole of D one here assumes

$$c(x) \leq 0. \qquad (9.1.2)$$

Early work by Fichera [57] had shown that, given a function φ defined in the set $\mathcal{F}_2 D$, if the sets $D, \mathcal{F}_1 D, \mathcal{F}_2 D$, jointly with the functions $a^{hk}, b^h, c, f, \varphi$ satisfy some qualitative assumptions only mildly restrictive, there exists a unique function u defined in D and fulfilling therein the following conditions:

(i) The function u is of class C^2 in the set $(D - \mathcal{F}D)$ and satifies Eq. (9.1.1) at all points of such a set.

© Springer International Publishing AG 2017

G. Esposito, *From Ordinary to Partial Differential Equations*, UNITEXT 106,
DOI 10.1007/978-3-319-57544-5_9

(ii) The function u is continuous in $(D - \mathcal{F}_2 D)$ and verifies at all points of $\mathcal{F}_1 D$ the condition $u(x) = 0$.

(iii) On denoting by $\nu(\overline{x})$ the interior conormal to the frontier of D at the point $\overline{x}$, relative to the quadratic form

$$\sum_{h,k=1}^{r} a^{hk} \lambda_h \lambda_k,$$

one has, at all points $\overline{x}$ of $\mathcal{F}_2 D$,

$$\lim_{x \to \overline{x}} \left[\frac{du}{d\nu} \right]_{\nu(\overline{x})} = \varphi(\overline{x}). \qquad (9.1.3)$$

(iv) Existence and finiteness of the integral

$$\int_D \sum_{h=1}^{r} \left(\frac{\partial u}{\partial x^h} \right)^2 dx^1 ... dx^r.$$

(v) The function u satisfies two reciprocity relations with the Green function pertaining to the Dirichlet problem for a domain D' containing D.

In light of this theorem, Picone asked De Giorgi to study whether there exists a unique function u continuous in the whole domain D and fulfilling the first four conditions (i), (ii), (iii), (iv), but not the fifth.

9.2   The De Giorgi Family of Solutions

De Giorgi proved, by building an example, that there may exist infinitely many functions satisfying the conditions from (i) to (iv) of Sect. 9.1 and continuous in the codomain D. To begin, consider in the plane E_2 the domain characterized by the conditions

$$x^1 \in [-1, 1], \ x^2 \in \left[0, \sqrt{1 - (x^1)^2} \right], \qquad (9.2.1)$$

and denote by $\mathcal{F}_1 D$ the circumference with equations

$$x^2 = \sqrt{1 - (x^1)^2}, \ x^1 \in [-1, 1], \qquad (9.2.2)$$

with $\mathcal{F}_2 D$ the segment picked out by

$$x^1 \in [-1, 1], \ x^2 = 0. \qquad (9.2.3)$$

Interestingly, there exists a non-vanishing function in D satisfying the following conditions:

(i') u is continuous in $(D - \mathcal{F}D)$ together with all its partial derivatives, and is a harmonic function at all points of $(D - \mathcal{F}D)$:

$$\Delta_2 u = 0. \tag{9.2.4}$$

(ii') u is continuous in D and verifies

$$u(x) = 0, \ \forall x \in \mathcal{F}_1 D. \tag{9.2.5}$$

(iii') There exists a set N of vanishing linear measure contained in $\mathcal{F}_2 D$, such that u is continuous in $(D - N)$ together with its first derivatives and verifies on $(\mathcal{F}_2 D - N)$ the relation

$$\frac{du}{d\nu} = \frac{du}{dx^2} = 0, \tag{9.2.6}$$

where $\frac{du}{d\nu}$ denotes the derivative of u with respect to the normal to $\mathcal{F}D$ interior to D.
(iv') Finiteness of the integral

$$\int_D \left[\left(\frac{\partial u}{\partial x^1} \right)^2 + \left(\frac{\partial u}{\partial x^2} \right)^2 \right] dx^1 \, dx^2.$$

As a first step, one proves some lemmas that make it possible to prove, in turn, the existence of a function u satisfying the assumptions from (i') to (iv').

Lemma 9.1 *Given an open interval $T =]a, b[$ contained within the open interval $]-1, 1[$ and a positive number ε, it is always possible to build a pluri-interval $B \subset T$ such that (mis being the Lebesgue measure)*

$$2\mathrm{mis}(B) = \mathrm{mis}(T), \tag{9.2.7}$$

while for any point $x = (x^1, x^2) \in E_2$ one has the inequality

$$\left| 2 \int_B \log\left[(x^1 - \xi)^2 + (x^2)^2 \right] d\xi - \int_T \log\left[(x^1 - \xi)^2 + (x^2)^2 \right] d\xi \right| < \varepsilon. \tag{9.2.8}$$

Setting now $c = (b - a)$, for each positive integer n, we shall denote by B_n the set formed by the points ξ satisfying one of the relations

$$a + \frac{c(2h - 1)}{2n} \le \xi \le a + \frac{hc}{n} \ \text{for } h = 1, \ldots, n; \tag{9.2.9}$$

one has for sure

$$2\mathrm{mis}(B_n) = \mathrm{mis}(B). \tag{9.2.10}$$

9.3   De Giorgi's Clever use of the Characteristic Function of a Set

On denoting by φ the characteristic function of the set B, and by φ_n the characteristic function of the set B_n, i.e.

$$\varphi(\xi) = 1 \text{ if } \xi \in B, 0 \text{ if } \xi \notin B, \ \ \varphi_n(\xi) = 1 \text{ if } \xi \in B_n, 0 \text{ if } \xi \notin B_n, \tag{9.3.1}$$

we define the function ψ_n taking the value

$$\psi_n(\xi) \equiv 2\varphi_n(\xi) - \varphi(\xi), \tag{9.3.2}$$

from which one gets

$$\psi_n(\xi) = (-1)^k \text{ if } \xi \in \left]\frac{(k-1)c}{2n} + a, \frac{kc}{2n} + a\right[, \tag{9.3.3}$$

$$\psi_n(\xi) = 0 \text{ if } \xi < a, \text{ or } \xi > b. \tag{9.3.4}$$

If one now defines a new function Θ_n by integration of ψ_n over the real line, i.e.

$$\Theta_n(\xi) \equiv \int_{-\infty}^{\xi} \psi_n(t)\mathrm{d}t, \tag{9.3.5}$$

upon bearing in mind that the open interval $]a, b[$ is contained in the open interval $]-1, 1[$, as well as the values (9.3.3) and (9.3.4) taken by ψ_n, one finds the equalities and inequalities

$$\Theta_n(\xi) = 0, \text{ if } |\xi| \geq 1, \tag{9.3.6}$$

$$|\Theta_n(\xi)| \leq \frac{1}{n}. \tag{9.3.7}$$

Let us take now a point t in the interval $]-1, 1[$; on recalling when ψ_n vanishes according to (9.3.4), and bearing in mind that the interval $]-1, 1[$ contains the interval $]a, b[$, one obtains

$$\int_{-\infty}^{\infty} \psi_n(\xi) \log((t-\xi)^2) d\xi = \int_{-\infty}^{\infty} \psi_n(\xi+t) \log(\xi^2) d\xi$$
$$= \int_{-2}^{2} \psi_n(\xi+t) \log(\xi^2) d\xi. \tag{9.3.8}$$

A theorem of calculus of variations (cf. Part C) ensures that, given a function f that is square-integrable in the bounded interval $]\alpha, \beta[$, the functional

$$\int_{\alpha}^{\beta} f(\xi) y'(\xi) d\xi,$$

where $y'(\xi)$ is the standard notation for the first derivative of $y(\xi)$, is continuous in every set E of the space consisting of functions y for which the integrals

$$\int_{\alpha}^{\beta} (y'(\xi))^2 d\xi$$

are equi-bounded. [Recall that, if $\{f_n\}_{n \in N}$ is a sequence of functions defined in a subset of $\mathbf{R}$ containing the interval $[a, b]$, such functions are said to be equi-bounded in this interval if and only if there exists a positive M for which $|f_n(x)| \le M, \forall n \in N, \forall x \in [a, b]$.]

One now remarks that, on denoting by $\Theta'_n(\xi+t)$ the first derivative with respect to ξ of the function $\Theta_n(\xi+t)$, one has, by virtue of (9.3.5) and (9.3.7),

$$\Theta'_n(\xi+t) = \psi_n(\xi+t), \quad |\Theta_n(\xi+t)| < \frac{1}{n}. \tag{9.3.9}$$

On the other hand, $\log(\xi^2)$ is a square-integrable function in the interval $]-2, 2[$, whereas, by the very definition of $\psi_n(x)$, one has

$$\int_{-2}^{2} (\psi_n(t+\xi))^2 d\xi = \text{mis}(T) < 2, \tag{9.3.10}$$

no matter how one chooses the point t in the interval $]-1, 1[$.

We can then conclude, by virtue of (9.3.8), (9.3.9) and (9.3.10) and the above-mentioned theorem in calculus of variations, that, no matter how one takes a positive number ε, one can always find a positive integer $n(\varepsilon)$ such that, for $|t| \le 1$,

$$\left| \int_{-\infty}^{\infty} \psi_{n(\varepsilon)}(\xi) \log((\xi-t)^2) d\xi \right| < \varepsilon. \tag{9.3.11}$$

At this stage, let us consider the function

$$g(x) = g(x^1, x^2) = \int_{-\infty}^{\infty} \psi_{n(\varepsilon)}(\xi) \log\left[(x^1 - \xi)^2 + (x^2)^2 \right] d\xi. \tag{9.3.12}$$

Such a function is harmonic in $(\mathbf{R}^2 - \mathcal{F}_2 D)$ by construction, and it is continuous over the whole of $\mathbf{R}^2$. By virtue of (9.3.11) it satisfies, at all points of the set $\mathcal{F}_2 D$, the relation

$$|g(x)| < \varepsilon. \tag{9.3.13}$$

If we denote with $|x|$ the distance of the generic point $x = (x^1, x^2)$ from the origin of coordinates, and bear in mind (9.3.5)–(9.3.7) we find, integrating by parts,

$$\lim_{|x| \to \infty} g(x) = \lim_{|x| \to \infty} \int_{-\infty}^{\infty} \psi_{n(\varepsilon)} \log\left[(x^1 - \xi)^2 + (x^2)^2\right] d\xi$$

$$= \lim_{|x| \to \infty} 2 \int_{-\infty}^{\infty} \Theta_{n(\varepsilon)}(\xi) \frac{(x^1 - \xi)}{\left[(x^1 - \xi)^2 + (x^2)^2\right]} d\xi = 0. \tag{9.3.14}$$

Equation (9.3.14) ensures that, if the inequality (9.3.13) holds on $\mathcal{F}_2 D$, it must hold necessarily over the whole space $\mathbf{R}^2$ with Euclidean metric.

If one now sets $B = B_{n(\varepsilon)}$, one obtains by virtue of Eq. (9.3.12) and of the very definition of $\psi_{n(\varepsilon)}(\xi)$ the equality

$$\left| 2 \int_B \log\left[(x^1 - \xi)^2 + (x^2)^2\right] d\xi - \int_T \log\left[(x^1 - \xi)^2 + (x^2)^2\right] d\xi \right|$$

$$= \left| \int_{-\infty}^{\infty} \psi_{n(\varepsilon)}(\xi) \log\left[(x^1 - \xi)^2 + (x^2)^2\right] d\xi \right| = |g(x)|. \tag{9.3.15}$$

From (9.3.13) and (9.3.15) follows the inequality (9.2.8), and our theorem is proved. Q.E.D.

From Lemma 9.1 follows immediately a second Lemma, i.e.

Lemma 9.2 *Given at will a pluri-interval B within the interval $]-1, 1[$ and a positive number ε, it is always possible to build a pluri-interval $B^* \subset B$ such that, for any point $x = (x^1, x^2)$ of the Euclidean plane $\mathbf{R}^2$,*

$$\left| \int_B \log\left[(x^1 - \xi)^2 + (x^2)^2\right] d\xi - 2 \int_{B*} \log\left[(x^1 - \xi)^2 + (x^2)^2\right] d\xi \right| < \varepsilon, \tag{9.3.16}$$

and furthermore

$$2\mathrm{mis}(B^*) = \mathrm{mis}(B). \tag{9.3.17}$$

Let us now consider a pluri-interval B_0 within the interval $]-1, 1[$. By virtue of Lemma 9.2 it is certainly possible to build a sequence of pluri-intervals

$$B_0 \subset B_1 \subset \ldots \subset B_n \subset \ldots,$$

such that, for any choice of the index n, one gets

$$2\mathrm{mis}(B_{n+1}) = \mathrm{mis}(B_n),\tag{9.3.18}$$

and in addition one has, for every point $x = (x^1, x^2)$ contained in $\mathbf{R}^2$,

$$\left| 2\int_{B_{n+1}} \log\left[(x^1 - \xi)^2 + (x^2)^2\right] d\xi - \int_{B_n} \log\left[(x^1 - \xi)^2 + (x^2)^2\right] d\xi \right| < \frac{1}{4^n}.\tag{9.3.19}$$

If we set

$$w_n(x) = 2^n \int_{B_n} \log\left[(x^1 - \xi)^2 + (x^2)^2\right] d\xi,\tag{9.3.20}$$

the sequence of functions

$$w_0(x),\ w_1(x),\ \ldots,\ w_n(x),\ \ldots\tag{9.3.21}$$

converges uniformly over the whole plane $\mathbf{R}^2$ to a limit function

$$w(x) = \lim_{n\to\infty} w_n(x).\tag{9.3.22}$$

The function w turns out to be continuous over the whole plane $\mathbf{R}^2$ and harmonic in $(\mathbf{R}^2 - N)$, where N is here the set

$$N \equiv \left\{ x = (x^1, x^2) : x^1 \in \prod_{h=0}^{\infty} B_h,\ x^2 = 0 \right\}.\tag{9.3.23}$$

From the defining equations (9.3.23), bearing in mind that B_0 is taken to be a pluri-interval within $]-1, 1[$, one sees that N is a closed set contained in $\mathcal{F}_2 D$; moreover, (9.3.18) and (9.3.23) ensure that the set N has vanishing linear measure.

Let us consider now the circular domain of equation

$$(x^1)^2 + (x^2)^2 \le 1,\tag{9.3.24}$$

that we denote by D^*, while, for every real number α in the interval $]0, 1[$, we denote by D_α the domain defined by the inequalities

$$(x^1)^2 + (x^2)^2 \le 1,\ |x^2| \ge \alpha.\tag{9.3.25}$$

If we denote by $\mathcal{F}_1 D_\alpha$ the set identified by

$$(x^1)^2 + (x^2)^2 = 1,\ |x^2| \ge \alpha,\tag{9.3.26}$$

and by $\mathcal{F}_2 D_\alpha$ the set identified by

$$(x^1)^2 + (x^2)^2 < 1, \quad |x^2| = \alpha, \tag{9.3.27}$$

the frontier $\mathcal{F} D_\alpha$ of the domain D_α consists of the sum

$$\mathcal{F}_1 D_\alpha + \mathcal{F}_2 D_\alpha = \mathcal{F} D_\alpha. \tag{9.3.28}$$

Let us agree moreover that, given a point x belonging to $\mathcal{F} D^*$ (or to $\mathcal{F}_2 D_\alpha$), one denotes always with $\nu(x)$ the normal to $\mathcal{F} D^*$ (or to $\mathcal{F}_2 D_\alpha$) internal to D^* (or to D_α). We remark that, once a point y has been fixed in the Euclidean plane $\mathbf{R}^2$, and denoting by $|x - y|$ the distance from y of the generic point $x \in \mathbf{R}^2$, the gradient of $\log |x - y|$ (viewed as a function of the point x) has the direction and orientation of the half-line coming out from y and passing through x; upon bearing this in mind, one verifies immediately that, for $|\xi| < 1$, the following inequalities hold:

$$\frac{d}{d\nu} \log\left[(x^1 - \xi)^2 + (x^2)^2\right] < 0 \text{ if } x = (x^1, x^2) \in \mathcal{F} D^*, \tag{9.3.29}$$

$$\frac{d}{d\nu} \log\left[(x^1 - \xi)^2 + (x^2)^2\right] > 0 \text{ if } x = (x^1, x^2) \in \mathcal{F}_2 D_\alpha, \tag{9.3.30}$$

where $\frac{d}{d\nu}$ denotes derivation along the direction of the normal $\nu(x)$. From the definition (9.3.20) of w_n, and from the inequalities (9.3.29) and (9.3.30), one finds immediately the conditions

$$\frac{d}{d\nu} w_n(x) < 0 \text{ if } x \in \mathcal{F} D^*, \tag{9.3.31}$$

$$\frac{d}{d\nu} w_n(x) > 0 \text{ if } x \in \mathcal{F}_2 D_\alpha. \tag{9.3.32}$$

From Eq. (9.3.18) that relates Lebesgue measures of B_n and B_{n+1}, and from the definition (9.3.20) of w_n, there follows

$$\int_{\mathcal{F} D^*} \frac{dw_n}{d\nu} ds = -2\pi \text{mis}(B), \tag{9.3.33}$$

where ds is the standard notation for the line element on $\mathcal{F} D^*$. Since at the points of $\mathcal{F}_1 D_\alpha$ the normal to $\mathcal{F} D^*$ interior to D^* coincides with the normal to $\mathcal{F}_1 D_\alpha$ interior to D_α, from (9.3.31) to (9.3.33), recalling that $w_n(x)$ is harmonic in D_α, and the decomposition (9.3.28) of $\mathcal{F} D_\alpha$, one finds

$$\int_{\mathcal{F}_1 D_\alpha} \left| \frac{dw_n}{d\nu} \right| ds = - \int_{\mathcal{F}_1 D_\alpha} \frac{dw_n}{d\nu} ds < 2\pi \operatorname{mis}(B_0), \qquad (9.3.34)$$

$$\int_{\mathcal{F}_2 D_\alpha} \left| \frac{dw_n}{d\nu} \right| ds = \int_{\mathcal{F}_2 D_\alpha} \frac{dw_n}{d\nu} ds = - \int_{\mathcal{F}_1 D_\alpha} \frac{dw_n}{d\nu} ds. \qquad (9.3.35)$$

From (9.3.34) and (9.3.35) it follows that

$$\int_{\mathcal{F} D_\alpha} \left| \frac{dw_n}{d\nu} \right| < 4\pi \operatorname{mis}(B_0) < 8\pi. \qquad (9.3.36)$$

The uniform convergence of the sequence (9.3.21) implies that there exists a positive number M such that all functions of the sequence are majorized in absolute value by M, i.e.

$$|w_n(x)| < M \text{ if } |x| \le 1, \ n = 1, 2, \dots \ . \qquad (9.3.37)$$

By virtue of (9.3.36) and (9.3.37) one obtains

$$\int_{D_\alpha} \left[\left(\frac{\partial w_n}{\partial x^1} \right)^2 + \left(\frac{\partial w_n}{\partial x^2} \right)^2 \right] dx^1 \, dx^2 \le \int_{\mathcal{F} D_\alpha} \left| \frac{dw_n}{d\nu} \right| |w_n| ds < 8\pi M, \qquad (9.3.38)$$

and therefore, passing to the limit as $n \to \infty$, one finds the inequality

$$\int_{D_\alpha} \left[\left(\frac{\partial w_n}{\partial x^1} \right)^2 + \left(\frac{\partial w_n}{\partial x^2} \right)^2 \right] dx^1 \, dx^2 \le 8\pi M. \qquad (9.3.39)$$

From (9.3.39), passing to the limit as $\alpha \to 0$, one finds the inequality

$$\int_{D^*} \left[\left(\frac{\partial w_n}{\partial x^1} \right)^2 + \left(\frac{\partial w_n}{\partial x^2} \right)^2 \right] dx^1 \, dx^2 \le 8\pi M. \qquad (9.3.40)$$

If one now denotes by w^* the function harmonic in $(D^* - \mathcal{F} D^*)$ and continuous over the whole domain D^*, whose value $w^*(x)$ coincides with $w(x)$ on $\mathcal{F} D^*$, anf if we set therefore

$$u(x) \equiv w(x) - w^*(x), \qquad (9.3.41)$$

one verifies easily that the function u defined in such a way fulfills the conditions (i'), (ii'), (iii') and (iv') of Sect. 9.2.

In order to prove that conditions (i'), (ii') and (iii') are fulfilled, it is enough to remark that $u(x)$ is continuous in the domain D^*, vanishing on $\mathcal{F} D^*$, harmonic in

$(D - \mathcal{F}D^* - N)$, symmetric with respect to the line $x^2 = 0$, and bear in mind the definitions of the sets $D, \mathcal{F}_1 D, \mathcal{F}_2, D^*$.

In order to prove that $u(x)$ satisfies the condition (iv'), it is enough to bear in mind (9.3.40), (9.3.41) and note that, since the set N has empty intersection with $\mathcal{F}D^*$, $w(x)$ is continuous together with its first partial derivatives in $\mathcal{F}D^*$, and hence $w^*(x)$ has the same property in the whole circular domain D^*.

9.4   Perimeter of a Set and Reduced Frontier of Finite-Perimeter Sets

The sections from 9.1 to 9.3 exhaust the material covered in a lecture, but we cannot resist the temptation of presenting a modern application of characteristic functions, since we are aiming at encouraging an interdisciplinary view of mathematics. Hence we briefly enter here the realm of geometric measure theory.

If E is a subset of $\mathbf{R}^n$, with $n \geq 2$, and denoting by $\varphi(x, E)$ the characteristic function of Eq. (9.3.1), i.e.

$$\varphi(x, E) = 1 \text{ if } x \in E, \; 0 \text{ if } x \notin E,$$

and by $\star$ the convolution product

$$(f \star g)(x) \equiv \int f(x - \xi)g(\xi)\mathrm{d}\xi, \tag{9.4.1}$$

one considers $\forall \lambda > 0$ the 1-parameter family of functions

$$\varphi_\lambda(x) \equiv (\pi\lambda)^{-\frac{n}{3}}\exp\left(-\frac{|x|^2}{\lambda}\right) \star \varphi(x, E). \tag{9.4.2}$$

The *perimeter of the set* E is then meant to be the limit, either finite or infinite,

$$P(E) \equiv \lim_{\lambda \to 0} \int_{\mathbf{R}^n} |\mathrm{grad}\varphi_\lambda(x)| \; \mathrm{d}x. \tag{9.4.3}$$

As was proved by De Giorgi [34], the necessary and sufficient condition for finiteness of the perimeter $P(E)$ is the existence of a vector function $a(B)$, completely additive and bounded, defined for every set $B \subset \mathbf{R}^n$, verifying the generalized Gauss–Green formulae

$$\int_E (\mathrm{grad}g)\mathrm{d}x = -\int_{\mathbf{R}^n} g(x)\mathrm{d}a \tag{9.4.4}$$

for every function g of class C^1 on $\mathbf{R}^n$ and with compact support. In such a case one finds

$$P(E) = \int_{\mathbf{R}^n} |da|. \tag{9.4.5}$$

The meaning of Eq. (9.4.4) is that the function a is the gradient, in the sense of distribution theory, of the characteristic function $\varphi(x, E)$. One therefore writes

$$\int_B \operatorname{grad}\varphi(x, E) = a(B), \quad \int_B f(x)\operatorname{grad}\varphi(x, E) = \int_B f \, da. \tag{9.4.6}$$

If $\{E_h\}$ is a sequence of sets of $\mathbf{R}^n$, and E is a set of $\mathbf{R}^n$ verifying the condition

$$\lim_{h\to\infty} \int_{\mathbf{R}^n} |\varphi(x, E) - \varphi(x, E_h)| \, dx = \lim_{h\to\infty} \Big[\operatorname{mis}(E - E_h) + \operatorname{mis}(E_h - E)\Big] = 0, \tag{9.4.7}$$

one then has [34]

$$\min \lim_{h\to\infty} P(E_h) \geq P(E). \tag{9.4.8}$$

Among the sequences approximating a set E, those consisting of polygonal domains are of particular interest. A *polygonal domain* is indeed any set $E \subset \mathbf{R}^n$ which is the closure of an open set, and whose frontier $\mathcal{F}E$ is contained in the union of a finite number of hyperplanes of $\mathbf{R}^n$. A theorem ensures that, if the set E has finite perimeter, there exists a sequence $\{E_h\}$ of polygonal domains verifying the conditions

$$P(E) = \lim_{h\to\infty} P(E_h), \quad \lim_{h\to\infty} \Big[\operatorname{mis}(E - E_h) + \operatorname{mis}(E_h - E)\Big] = 0. \tag{9.4.9}$$

The sets approximated on average by means of polygonal domains having finite perimeter were introduced by Caccioppoli [16, 17]. By virtue of the theorems expressed by (9.4.7)–(9.4.9), such sets are all (and only those) *sets of finite perimeter*, and are therefore *Caccioppoli sets*.

Given a Caccioppoli set $E \subset \mathbf{R}^n$, the *reduced frontier* of E, denoted by $\mathcal{F}^*E$, is the set of points ξ verifying the following conditions:
(i) The integral of $|\operatorname{grad}\varphi(x, E)|$ over the open hypersphere $A_\rho(\xi)$ centred at ξ with radius ρ is positive, i.e.

$$\int_{A_\rho(\xi)} |\operatorname{grad}\varphi(x, E)| > 0, \quad \forall \rho > 0. \tag{9.4.10}$$

(ii) There exists the limit

$$\lim_{\rho \to 0} \frac{\int_{A_\rho(\xi)} \text{grad}\varphi(x, E)}{\int_{A_\rho(\xi)} |\text{grad}\varphi(x, E)|} = \nu(\xi). \tag{9.4.11}$$

(iii) The function ν has unit modulus, i.e.

$$|\nu(\xi)| = 1. \tag{9.4.12}$$

Interestingly, for any Caccioppoli set $E \subset \mathbf{R}^n$ and for any point $\xi \in \mathcal{F}^* E$, one finds [37]

$$\lim_{\rho \to 0} \rho^{1-n} \int_{A_\rho(\xi)} |\text{grad}\varphi(x, E)| = \omega_{n-1}, \tag{9.4.13}$$

$$\lim_{\rho \to 0} \rho^{-n} \text{mis}(A_\rho(\xi) \cap E) = \lim_{\rho \to 0} \rho^{-n} \text{mis}\left(A_\rho(\xi) - E\right) = \frac{\omega_n}{2}, \tag{9.4.14}$$

where ω_n is the measure of the hypersphere in $\mathbf{R}^n$ of unit radius.

The theory of finite-perimeter sets was initiated in the works by Caccioppoli [16, 17], but it was De Giorgi who put on firm ground the brilliant ideas of Caccioppoli in an impressive series of theorems. For an introduction to the appropriate branch of mathematics, i.e. geometric measure theory, we refer the reader to the beautiful books by Federer [56], De Giorgi et al. [37], Maggi [99].

Chapter 10
New Functional Spaces: Morrey and Campanato

10.1 Morrey and Campanato Spaces: Definitions and Properties

In the modern studies on elliptic equations, extensive use is made of Morrey and Campanato spaces, which are therefore presented here to the general reader.

If λ is a non-negative real number, $p \in [1, \infty[$, Ω is an open set of $\mathbf{R}^n$, $\Omega(x_0, r)$ is the part common to Ω and to the Euclidean ball centred at x_0 and of radius r, i.e.

$$\Omega(x_0, r) \equiv \Omega \cap B_r(x_0), \tag{10.1.1}$$

while the diameter of Ω is meant to be the sup of the Euclidean distance among points of Ω:

$$d_\Omega \equiv \sup_{x, y \in \Omega} d_E(x, y), \tag{10.1.2}$$

the function $f \in L^p(\Omega)$ is said to belong to the Morrey space $L^{p,\lambda}(\Omega)$ if (cf. [100, 102])

$$\sup_{r \in]0, d_\Omega[, \, x_0 \in \Omega} r^{-\lambda} \int_{\Omega(x_0, r)} |f|^p \, dx < \infty. \tag{10.1.3}$$

When this condition is fulfilled, one defines the Morrey norm of f according to

$$\|f\|_{L^{p,\lambda}} \equiv \left(\sup_{r \in]0, d_\Omega[, \, x_0 \in \Omega} r^{-\lambda} \int_{\Omega(x_0, r)} |f|^p \, dx \right)^{\frac{1}{p}}. \tag{10.1.4}$$

The Morrey functional spaces possess five basic properties, which are as follows [4].

© Springer International Publishing AG 2017
G. Esposito, *From Ordinary to Partial Differential Equations*, UNITEXT 106,
DOI 10.1007/978-3-319-57544-5_10

(i) $L^{p,\lambda}(\Omega; R)$ are Banach (i.e. vector spaces with a norm, where every Cauchy sequence converges to an element of the space), $\forall p \in [1, \infty[$ and $\forall \lambda \geq 0$.

(ii) $L^{p,0}(\Omega; R) = L^p(\Omega; R)$.

(iii) $L^{p,\lambda}(\Omega; R) = \{0\}$ if $\lambda > n$.

(iv) $L^{p,n}(\Omega; R) \sim L^\infty(\Omega; R)$.

(v) $L^{q,\mu}(\Omega; R) \subset L^{p,\lambda}(\Omega; R)$ if Ω is bounded, $q \geq p$ and $\frac{(n-\lambda)}{p} \geq \frac{(n-\mu)}{q}$.

The third and fourth property result from the Lebesgue differentiation theorem, while the fifth relies upon the Hölder inequality

$$\left(\int_{\Omega(x,r)} |f|^p dx \right) \leq \left(\int_{\Omega(x,r)} |f|^q dx \right)^{\frac{p}{q}} (\omega_n r^n)^{\left(1 - \frac{p}{q}\right)}$$

$$= C(n, p, q) \, \|f\|^p_{L^{q,m}} \, r^{\frac{\mu p}{q} + n\left(1 - \frac{p}{q}\right)}. \qquad (10.1.5)$$

This inequality suggests defining

$$\lambda_c \equiv \frac{\mu p}{q} + n\left(1 - \frac{p}{q}\right). \qquad (10.1.6)$$

In the fifth property expressed above, the condition upon λ reads therefore $\lambda \leq \lambda_c$.

Another class of functional spaces was introduced by Campanato, who gave important contributions to the study of elliptic systems in divergence form (and to parabolic equations and elasticity theory as well). If Ω is an open subset of $\mathbf{R}^n$, λ is a positive number and $1 \leq p < \infty$, a function $f \in L^p(\Omega)$ is said to belong to the *Campanato space* $\mathcal{L}^{p,\lambda}$ if (cf. [20, 21])

$$\|f\|^p_{\mathcal{L}^{p,\lambda}} \equiv \sup_{r \in]0, d_\Omega[, \, x_0 \in \Omega} r^{-\lambda} \int_{\Omega(x_0,r)} |f(x) - f_{x_0,r}|^p \, dx < \infty, \quad (10.1.7)$$

where d_Ω denotes again the diameter of Ω, while

$$f_{x_0,r} \equiv \frac{\int_{\Omega(x_0,r)} f(x) dx}{\mathrm{mis}\Omega(x_0, r)}. \qquad (10.1.8)$$

Three basic properties should be stressed, and they are as follows.

(i) The norm defined in (10.1.3) is, strictly, a seminorm, because the constants have vanishing $\mathcal{L}^{p,\lambda}$ norm. If the open set Ω is connected, however, the space $\mathcal{L}^{p,\lambda}$ is a Banach space modulo constants.

(ii) The space $\mathcal{L}^{q,\mu}$ is a proper subset of $\mathcal{L}^{p,\lambda}$ if the open set Ω is bounded, $p \leq q$ and

$$\frac{(n - \lambda)}{p} \geq \frac{(n - \mu)}{q}.$$

(iii) The space $C^{0,\alpha}$ of Chap. 5 is properly included into the space $\mathcal{L}^{p,n+\alpha p}$, because (denoting by $\mathcal{L}^n$ the Lebesgue measure in $\mathbf{R}^n$)

$$\int_{\Omega(x_0,r)} |f(x) - f_{x_0,r}|^p \mathrm{d}x \leq \left(\|f\|_{C^{0,\alpha}}\right)^p r^{\alpha p} \mathcal{L}^n(B(x_0,r)) = \left(\|f\|_{C^{0,\alpha}}\right)^p \omega_n r^{n+\alpha p}, \quad (10.1.9)$$

ω_n being, as usual, the volume of the unit ball in $\mathbf{R}^n$. Conversely, if a function f in the Campanato space $\mathcal{L}^{p,\lambda}$ has Hölder continuous representatives in its Lebesgue equivalence class, one can replace pointwise definition of Hölder spaces with an integral one.

Following again Ambrosio [4], we now state and prove a fundamental theorem that makes it precise in which sense the Morrey and Campanato spaces are equivalent.

Theorem 10.1 *If Ω is an open bounded set of $\mathbf{R}^n$, if the Lebesgue measure of $(\Omega \cap B_r(x_0))$ has the lower bound $c_* r^n$ for all x_0 in the closure of Ω and for all r in the open interval $]0, d_\Omega[$, and if $0 \leq \lambda < n$, then the following relation among Morrey and Campanato norm holds:*

$$\| \ \|_{L^{p,\lambda}} \cong \| \ \|_{\mathcal{L}^{p,\lambda}} + \| \ \|_{L^p}. \quad (10.1.10)$$

Proof By virtue of the Jensen inequality, one can write the majorization

$$\int_{\Omega(x_0,r)} |f_{x_0,r}|^p \mathrm{d}x \leq \int_{\Omega(x_0,r)} |f(x)|^p \mathrm{d}x, \quad (10.1.11)$$

and hence one can write the two majorizations

$$\int_{\Omega(x_0,r)} |f(x) - f_{x_0,r}|^p \mathrm{d}x \leq 2^{p-1} \left(\int_{\Omega(x_0,r)} |f(x)|^p \mathrm{d}x + \int_{\Omega(x_0,r)} |f_{x_0,r}|^p \mathrm{d}x \right)$$

$$\leq 2^p \int_{\Omega(x_0,r)} |f(x)|^p \mathrm{d}x, \quad (10.1.12)$$

which lead us from the Campanato to the Morrey norm.

Conversely, we would like to estimate

$$r^{-\lambda} \int_{\Omega(x_0,r)} |f(x)|^p \mathrm{d}x$$

with the sum of norms

$$\|f\|_{\mathcal{L}^{p,\lambda}} + \|f\|_{L^p}$$

for all $r \in]0, d_\Omega[$ and for all $x_0 \in \Omega$. Indeed, by virtue of the triangular inequality, and denoting by c a parameter depending on (c_*, n, p, λ), one obtains the majorizations

$$\int_{\Omega(x_0,r)} |f(x)|^p dx \leq 2^{p-1} \int_{\Omega(x_0,r)} |f(x) - f_{x_0,r}|^p dx + cr^n |f_{x_0,r}|^p$$

$$\leq c \left[r^\lambda \left(\|f\|_{\mathcal{L}^{p,\lambda}} \right)^p + r^n |f_{x_0,r}|^p \right]. \tag{10.1.13}$$

This intermediate step is simple but non-trivial, since it has made it possible to take out the summand $|f_{x_0,r}|^p$. The next task is now to obtain an estimate of such a term, by exploiting averages on concentric balls when the point x_0 of Ω is fixed, while

$$0 < r < \rho < d_\Omega.$$

Indeed, the following chain of majorizations is found to hold:

$$c_* \omega_n r^n |f_{x_0,r} - f_{x_0,\rho}|^p \leq \int_{\Omega(x_0,r)} |f_{x_0,r} - f(x) + f(x) - f_{x_0,\rho}|^p dx$$

$$\leq 2^{p-1} \left(\int_{\Omega(x_0,r)} |f_{x_0,r} - f(x)|^p dx + \int_{\Omega(x_0,r)} |f(x) - f_{x_0,\rho}|^p dx \right)$$

$$\leq 2^{p-1} \left(\|f\|_{\mathcal{L}^{p,\lambda}} \right)^p \left(r^\lambda + \rho^\lambda \right)$$

$$\leq 2^p \left(\|f\|_{\mathcal{L}^{p,\lambda}} \right)^p \rho^\lambda. \tag{10.1.14}$$

Hence one finds that

$$\left| f_{x_0,r} - f_{x_0,\rho} \right| \leq \left(\frac{2^p}{c_* \omega_n} \right)^{\frac{1}{p}} \|f\|_{\mathcal{L}^{p,\lambda}} \, r^{-\frac{n}{p}} \rho^{\frac{\lambda}{p}} \left(\rho^{\frac{n}{p}} \rho^{-\frac{n}{p}} \right)$$

$$= c \|f\|_{\mathcal{L}^{p,\lambda}} \left(\frac{\rho}{r} \right)^{\frac{n}{p}} \rho^{\frac{(\lambda-n)}{p}}. \tag{10.1.15}$$

Let us now fix $R > 0$, $r = 2^{-(k+1)} R$, $\rho = 2^{-k} R$, from which the ratio $\frac{\rho}{r}$ in (10.1.15) equals 2. Now we set $\tilde{c} \equiv c 2^{\frac{n}{p}}$, and we re-express (10.1.15) in the form

$$\left| f_{x_0, \frac{R}{2^{k+1}}} - f_{x_0, \frac{R}{2^k}} \right| \leq \tilde{c} \|f\|_{\mathcal{L}^{p,\lambda}} \left(\frac{R}{2^k} \right)^{\frac{(\lambda-n)}{p}}. \tag{10.1.16}$$

This inequality is then written N times, with $k = 0, 1, \ldots, N - 1$, i.e.

$$\left| f_{x_0, \frac{R}{2}} - f_{x_0,R} \right| \leq \tilde{c} \|f\|_{\mathcal{L}^{p,\lambda}} \, R^{\frac{(\lambda-n)}{p}}, \tag{10.1.17}$$

$$\left| f_{x_0, \frac{R}{4}} - f_{x_0, \frac{R}{2}} \right| \leq \tilde{c} \|f\|_{\mathcal{L}^{p,\lambda}} \left(\frac{R}{2} \right)^{\frac{(\lambda-n)}{p}}, \tag{10.1.18}$$

...

$$\left| f_{x_0, \frac{R}{2^N}} - f_{x_0, \frac{R}{2^{N-1}}} \right| \le \tilde{c} \, \|f\|_{\mathcal{L}^{p,\lambda}} \left(\frac{R}{2^{N-1}} \right)^{\frac{(\lambda-n)}{p}}. \tag{10.1.19}$$

Such inequalities can be added up, leading in turn to

$$\left| f_{x_0, \frac{R}{2^N}} - f_{x_0, R} \right| \le \tilde{c} \, \|f\|_{\mathcal{L}^{p,\lambda}} \, R^{\frac{(\lambda-n)}{p}} \left(1 + 2^{-\frac{(\lambda-n)}{p}} + \cdots + 2^{-\frac{(N-1)(\lambda-n)}{p}} \right)$$

$$= \tilde{c} \, \|f\|_{\mathcal{L}^{p,\lambda}} \, R^{\frac{(\lambda-n)}{p}} \frac{\left(1 - 2^{\frac{(n-\lambda)N}{p}} \right)}{\left(1 - 2^{\frac{(n-\lambda)}{p}} \right)}$$

$$\le \tilde{c} \, \|f\|_{\mathcal{L}^{p,\lambda}} \, R^{\frac{(\lambda-n)}{p}} 2^{\frac{N(n-\lambda)}{p}}$$

$$= \tilde{c} \, \|f\|_{\mathcal{L}^{p,\lambda}} \left(\frac{R}{2^N} \right)^{\frac{(\lambda-n)}{p}}. \tag{10.1.20}$$

At this stage, we choose $R \in]\frac{1}{2}d_\Omega, d_\Omega[$ and N such that $\frac{R}{2^N} = r$. By virtue of the triangular inequality, one finds

$$\left| f_{x_0, r} \right|^p \le 2^{p-1} \left(\left| f_{x_0, r} - f_{x_0, R} \right|^p + \left| f_{x_0, R} \right|^p \right), \tag{10.1.21}$$

where the terms on the right-hand side are majorized as follows (here c depends on d_Ω):

$$\left| f_{x_0, R} \right| \le c(d_\Omega) \, \|f\|_{L^p}, \tag{10.1.22}$$

$$\left| f_{x_0, r} - f_{x_0, R} \right|^p \le c \left(\|f\|_{\mathcal{L}^{p,\lambda}} \right)^p r^{\lambda-n}. \tag{10.1.23}$$

Hence one obtains, from (10.1.13) and (10.1.21)–(10.1.23), the upper bound

$$r^{-\lambda} \int_{\Omega(x_0, r)} |f(x)|^p dx \le c \left[\left(\|f\|_{\mathcal{L}^{p,\lambda}} \right)^p + r^{n-\lambda} \left| f_{x_0, r} \right|^p \right]$$

$$= c \left(\|f\|_{\mathcal{L}^{p,\lambda}} \right)^p + c r^{n-\lambda} 2^{p-1} \left[c \left(\|f\|_{\mathcal{L}^{p,\lambda}} \right)^p r^{\lambda-n} + c^p \left(\|f\|_{L^p} \right)^p \right]$$

$$\le c \left[\left(\|f\|_{\mathcal{L}^{p,\lambda}} \right)^p + (d_\Omega)^{n-\lambda} \left(\|f\|_{L^p} \right)^p \right]. \tag{10.1.24}$$

This agrees with (10.1.7), and therefore the desired proof is completed. $\square$

10.2   Functions of Bounded Mean Oscillation, and an Example

The Campanato space $\mathcal{L}^{1,n}$ plays an important role in modern analysis, and consists of the functions of *bounded mean oscillation*, abbreviated sometimes as BMO. By definitions, this is the space of all real-valued functions f defined on Ω such that there exists a constant C satisfying the inequality

$$\int_{\Omega(x_0,r)} |f(x) - f_{x_0,r}| \, dx \leq Cr^n, \ \forall r \in]0, d_\Omega[, \ \forall x_0 \in \Omega. \qquad (10.2.1)$$

A simple but non-trivial example is obtained when Ω is the open interval $]0, 1[$ and f is the logarithm function: $f(x) = \log(x)$. For any value $a, r > 0$ one finds indeed

$$\int_a^{a+r} |\log(t) - \log(a+r)| \, dt = \int_a^{a+r} \Big[\log(a+r) - \log(t)\Big] dt$$

$$= (a+r-a)\log(a+r) - \Big[t\log(t) - t\Big]_{t=a}^{t=a+r}$$

$$= r\log(a+r) - \Big[(a+r)\log(a+r) - a\log(a)\Big] + (a+r-a)$$

$$= r + a\log\left(\frac{a}{(a+r)}\right) \leq r. \qquad (10.2.2)$$

Thus, the logarithm is a function of bounded mean oscillation on the open interval $]0, 1[$.

10.3   Glancing Backwards: The Spaces $W_{\text{Loc}}^{1,p}$ and $W^{1,p}$

In this chapter devoted to abstract spaces we find it appropriate to make contact with the modern nomenclature on Sobolev spaces, first defined in Chap. 5. Thus, at the risk of slight repetitions, we focus on what follows, inspired by Lieb and Loss [95].

Locally summable functions are an important class of distributions, but one can refine that class by studying functions whose distributional derivatives are also locally summable functions. This class is denoted by $W_{\text{loc}}^{1,1}(\Omega)$. Moreover, just as $L_{\text{loc}}^p(\Omega)$ is related to $L_{\text{loc}}^1(\Omega)$, one can also define the class of functions $W_{\text{loc}}^{1,p}(\Omega)$ for each $1 \leq p \leq \infty$, i.e.

$$W_{\text{loc}}^{1,p} \equiv \Big\{ f : \Omega \to \mathbf{C}, \ f \in L_{\text{loc}}^p(\Omega) \text{ and } \partial_i f, \text{ as a distribution in } \mathcal{D}'(\Omega),$$

$$\text{is a } L_{\text{loc}}^p(\Omega) - \text{function for } i = 1, \ldots, n \Big\}. \qquad (10.3.1)$$

This set of functions, $W_{\text{loc}}^{1,p}(\Omega)$, forms a vector space but not a normed one. One has the inclusion $W_{\text{loc}}^{1,r}(\Omega) \subset W_{\text{loc}}^{1,p}(\Omega)$ if $r > p$.

One can also define $W^{1,p}(\Omega) \subset W_{\text{loc}}^{1,p}(\Omega)$ in analogous way, i.e.

$$W^{1,p}(\Omega) \equiv \left\{ f : \Omega \to \mathbf{C} : f \text{ and } \partial_i f \text{ are in } L^p(\Omega) \text{ for } i = 1, \ldots, n \right\}. \quad (10.3.2)$$

The space $W^{1,p}(\Omega)$ can be made into a normed space, by defining the norm (cf. Eq. (10.3.1))

$$\| f \|_{W^{1,p}(\Omega)} \equiv \left\{ \left(\| f \|_{L^p(\Omega)} \right)^p + \sum_{j=1}^{n} \left(\| \partial_j f \|_{L^p(\Omega)} \right)^p \right\}^{\frac{1}{p}}, \quad (10.3.3)$$

and this space is complete, i.e. every Cauchy sequence in this norm has a limit in $W^{1,p}(\Omega)$. This follows from the completeness of $L^p(\Omega)$ jointly with the definition of distributional derivative, i.e. if $f^j \to f$ and $\partial_i f^j \to g_i$, then it follows that $g_i = \partial_i f$ in $\mathcal{D}'(\Omega)$. The spaces $W^{1,p}(\Omega)$ are the Sobolev spaces of the modern literature. The superscript 1 in $W^{1,p}(\Omega)$ denotes the fact that the first derivatives of f are pth-power summable functions.

Strong convergence of a sequence of functions in $W^{1,p}(\Omega)$ means that the sequence converges strongly to f in $L^p(\Omega)$, and the n sequences

$$\left\{ \partial_1 f^j \right\}, \ldots, \left\{ \partial_n f^j \right\},$$

formed from the derivatives of f^j, converge in $L^p(\Omega)$ to the n functions $\partial_1 f, \ldots, \partial_n f$ in $L^p(\Omega)$. In the case of $W_{\text{loc}}^{1,p}(\Omega)$ one requires convergence only on every compact subset of Ω.

Weak convergence in $W^{1,p}(\Omega)$ means that, for each $L^q(\Omega)$-function g, one requires that

$$\int g \left(f^j - f \right) \to 0 \text{ and, for each } i, \int g \left(\partial_i f^j - \partial_i f \right) \to 0, \quad (10.3.4)$$

as $j \to \infty$, with the understanding that $\frac{1}{p} + \frac{1}{q} = 1$. For the space $W_{\text{loc}}^{1,p}(\Omega)$ one requires this only for each g with compact support.

Similar definitions apply to the spaces $W^{m,p}(\Omega)$ and $W_{\text{loc}}^{m,p}(\Omega)$ with $m > 1$. The first m derivatives of these functions are $L^p(\Omega)$-functions and, similarly to (10.3.3),

$$\left(\| f \|_{W^{m,p}(\Omega)} \right)^p \equiv \left(\| f \|_{L^p(\Omega)} \right)^p + \sum_{j=1}^{n} \left(\| \partial_j f \|_{L^p(\Omega)} \right)^p$$

$$+ \cdots + \sum_{j_1=1}^{\infty} \cdots \sum_{j_m=1}^{\infty} \left(\| \partial_{j_1} \cdots \partial_{j_m} f \|_{L^p(\Omega)} \right)^p. \quad (10.3.5)$$

10.4 How Sobolev Discovered His Functional Spaces

The material in this section would require at least a separate lecture, and we hope that the general reader will find it useful and enlightening. Sobolev's work uses indeed with great creativity all tools of real and functional analysis. In his seminal paper, translated in English in Sobolev [130], Sobolev begins by recalling a previous result by himself [128], according to which the space $L_2^{(\nu)}$ of functions of n variables whose derivatives up to the order ν are square-integrable, is a sub-space of the space $C^{(\nu-[\frac{n}{2}]-1)}$ of functions which have continuous derivatives up to the order $\nu-[\frac{n}{2}]-1$ inclusive. Convergence of a certain sequence in $L_2^{(\nu)}$ automatically induces convergence in $C^{(\nu-[\frac{n}{2}]-1)}$. In Sobolev [130], the author succeeded in proving an imbedding theorem of the space $L_p^{(\nu)}$ into a space $L_q^{(\nu-l)}$, where q depends on p, l, n.

After generalizing some inequalities of F. Riesz, Sobolev introduced what was then a new definition of derivative. For this purpose, he considered a function ϕ of the variables $x^1, \ldots, x^n$ which is summable in a domain D, and a domain D' lying, along with its boundary, inside D. He also considered any function $\psi(x^1, \ldots, x^n)$ having continuous derivatives up to the order

$$\alpha \equiv \alpha_1 + \cdots \alpha_n$$

inclusive, and vanishing outside of D'. If the function ϕ itself has continuous derivatives up to the order α, the equation

$$\int_D \cdots \int \left(\phi \frac{\partial^\alpha \psi}{\partial(x^1)^{\alpha_1} \ldots \partial(x^n)^{\alpha_n}} + (-1)^{\alpha-1} \psi \frac{\partial^\alpha \phi}{\partial(x^1)^{\alpha_1} \ldots \partial(x^n)^{\alpha_n}} \right) dx^1 \ldots dx^n = 0 \quad (10.4.1)$$

is then valid. Sobolev viewed this equation as the *definition of derivative* of ϕ, that he denoted by

$$\frac{\partial^\alpha \phi}{\partial(x^1)^{\alpha_1} \ldots \partial(x^n)^{\alpha_n}}.$$

This is a function of the n-tuple of variables $x^1, \ldots, x^n$ which is summable over any bounded subset of the domain D and satisfies Eq. (10.4.1) for all ψ.

Such a derivative is unique, because upon assuming for a moment the existence of two such derivatives

$$\left[\frac{\partial^\alpha \phi}{\partial(x^1)^{\alpha_1} \ldots \partial(x^n)^{\alpha_n}} \right]_1 \quad \text{and} \quad \left[\frac{\partial^\alpha \phi}{\partial(x^1)^{\alpha_1} \ldots \partial(x^n)^{\alpha_n}} \right]_2$$

one finds for any ψ (from now on, the symbol $\int_D \cdots \int$ denotes n integrals over the domain D of n-dimensional Euclidean space, no matter whether we write the subscript D for the first or the last integral)

$$\int_D \cdots \int \psi \left\{ \left[\frac{\partial^\alpha \phi}{\partial(x^1)^{\alpha_1} \ldots \partial(x^n)^{\alpha_n}} \right]_1 - \left[\frac{\partial^\alpha \phi}{\partial(x^1)^{\alpha_1} \ldots \partial(x^n)^{\alpha_n}} \right]_2 \right\} dx^1 \ldots dx^n = 0, \quad (10.4.2)$$

which implies that these two derivatives are equal to each other everywhere, up to a zero-measure set.

One can provide simple examples showing that the derivative in the above weak sense may exist when the function does not have a derivative almost everywhere. Conversely, the existence of the derivative almost everywhere does not necessarily imply that the weak derivative exists. Sobolev considered a function ϕ that can be split into the sum [130]

$$\phi(x^1, x^2) = f_1(x^1) + f_2(x^2), \tag{10.4.3}$$

where neither f_1 nor f_2 is differentiable. Then the classical derivative of ϕ, once with respect to x^1 and once with respect to x^2, does not exist, whereas, in the weak sense,

$$\frac{\partial^2 \phi}{\partial x^1 \partial x^2} = 0 \tag{10.4.4}$$

at all points. Indeed, one has

$$\int_D \cdots \int \phi \frac{\partial^2 \psi}{\partial x^1 \partial x^2} dx^1 \ldots dx^n = \int_D \cdots \int f_1(x^1) \frac{\partial}{\partial x^2} \left(\frac{\partial \psi}{\partial x^1} \right) dx^1 \ldots dx^n$$
$$+ \int_D \cdots \int f_2(x^2) \frac{\partial}{\partial x^1} \left(\frac{\partial \psi}{\partial x^2} \right) dx^1 \ldots dx^n = 0, \tag{10.4.5}$$

because both terms can be integrated by parts. On the other hand, let us consider the choice

$$\phi = F(x^1), \tag{10.4.6}$$

where F is assumed to have a summable derivative almost everywhere, but is not absolutely continuous. In such a case, the Sobolev weak derivative does not exist.

In order to prove existence of the weak derivative, it is useful to use the following result:

Theorem 10.2 *If one can approximate a given summable function ϕ by a sequence of continuously differentiable functions*

$$\phi_k(x^1, \ldots, x^n) \qquad (k = 0, 1, \ldots) \tag{10.4.7}$$

such that, for any function ψ, continuous and vanishing outside the domain D' which is interior to the domain D, the equation

$$\lim_{k \to \infty} \int_D \cdots \int \phi_k \psi \, dx^1 \ldots dx^n = \int \cdots \int \phi \psi \, dx^1 \ldots dx^n \tag{10.4.8}$$

holds, and if, in addition, for any bounded subset D' of the domain D

$$\int_{D'} \cdots \int \left| \frac{\partial^\alpha \phi}{\partial(x^1)^{\alpha_1} \ldots \partial(x^n)^{\alpha_n}} \right|^p dx^1 \ldots dx^n \leq A, \qquad (10.4.9)$$

where $p > 1$, then the weak derivative $\frac{\partial^\alpha \phi}{\partial(x^1)^{\alpha_1} \ldots \partial(x^n)^{\alpha_n}}$ exists and satisfies the majorization

$$\int_{D'} \cdots \int \left| \frac{\partial^\alpha \phi}{\partial(x^1)^{\alpha_1} \ldots \partial(x^n)^{\alpha_n}} \right|^p dx^1 \ldots dx^n \leq A. \qquad (10.4.10)$$

Proof The appropriate tool is the theorem on weak compactness of the unit sphere of the space of functions which are p-th power summable [130, 131]. By virtue of this weak compactness, there exists a function $\omega_{\alpha_1,\ldots,\alpha_n}$ and a sub-sequence

$$\frac{\partial^\alpha \phi_{k_i}}{\partial(x^1)^{\alpha_1} \ldots \partial(x^n)^{\alpha_n}}, \quad i = 1, 2, \ldots, \qquad (10.4.11)$$

such that, for any function ψ which is p'-th power summable on D', having set $p' \equiv \frac{p}{(p-1)}$, the equation

$$\lim_{i \to \infty} \int_{D'} \cdots \int \psi \frac{\partial^\alpha \phi_{k_i}}{\partial(x^1)^{\alpha_1} \ldots \partial(x^n)^{\alpha_n}} dx^1 \ldots dx^n = \int_{D'} \cdots \int \psi \omega_{\alpha_1,\ldots,\alpha_n} dx^1 \ldots dx^n \quad (10.4.12)$$

holds. Furthermore, one has

$$\int_{D'} \cdots \int \left| \omega_{\alpha_1,\ldots,\alpha_n} \right|^p dx^1 \ldots dx^n \leq A. \qquad (10.4.13)$$

Let us write the identity

$$\int \cdots \int_{D'} \psi \frac{\partial^\alpha \phi_{k_i}}{\partial(x^1)^{\alpha_1} \ldots \partial(x^n)^{\alpha_n}} dx^1 \ldots dx^n$$
$$= (-1)^\alpha \int \cdots \int_{D'} \phi_{k_i} \frac{\partial^\alpha \psi}{\partial(x^1)^{\alpha_1} \ldots \partial(x^n)^{\alpha_n}} dx^1 \ldots dx^n, \qquad (10.4.14)$$

where ψ is a continuously differentiable function that vanishes outside D'. On passing to the limit in this equation one obtains

$$\int \cdots \int_{D'} \psi \omega_{\alpha_1,\ldots,\alpha_n} dx^1 \ldots dx^n = (-1)^\alpha \int \cdots \int_{D'} \phi \frac{\partial^\alpha \psi}{\partial(x^1)^{\alpha_1} \ldots \partial(x^n)^{\alpha_n}} dx^1 \ldots dx^n. \quad (10.4.15)$$

Thus, the function $\omega_{\alpha_1,\ldots,\alpha_n}$ is the desired derivative of ϕ. $\square$

Sobolev went on by considering some domain D in $\mathbf{R}^n$, such that any of its interior or boundary points can be reached by the point of a moving spherical sector of constant magnitude and shape. In D, he considered a function $\phi(x^1, \ldots, x^n)$ satisfying the following assumptions:

(i) Given $q^* \geq 1$, the function $|\phi(x^1, \ldots, x^n)|^{q^*}$ is summable on D;

(ii) The function $\phi(x^1, \ldots, x^n)$ admits all derivatives of order l in the domain D;

(iii) All derivatives of order l are p-th power summable, with $p \geq 1$, i.e.

$$\int \cdots \int_D \left| \frac{\partial^l \phi}{\partial (x^1)^{\alpha_1} \ldots \partial (x^n)^{\alpha_n}} \right|^p dx^1 \ldots dx^n \leq A^p, \; l \leq n. \quad (10.4.16)$$

Furthermore, various relations among p, n and l are conceivable, i.e. [130, 131]

$$1 < p < \frac{n}{l}, \; 1 = p < \frac{n}{l}, \; 1 < p = \frac{n}{l}, \; 1 = p = \frac{n}{l}, \; p > \frac{n}{l}. \quad (10.4.17)$$

In Sobolev [130], the author restricted himself to those cases when

$$\frac{1}{q^*} > \frac{1}{p} - \frac{1}{n}, \quad (10.4.18)$$

and he proved a basic result.

Theorem 10.3 *If $1 < p < \frac{n}{l}$, then the function ϕ is q-th power summable on the domain D, where*

$$\frac{1}{q} = \frac{1}{p} - \frac{l}{n}, \quad (10.4.19)$$

and in addition one has

$$\left[\int \cdots \int_D |\phi|^q dx^1 \ldots dx^n \right]^{\frac{1}{q}} \leq N A^p + L \left[\int \cdots \int_D |\phi|^{q^*} dx^1 \ldots dx^n \right]^{\frac{1}{q^*}}, \quad (10.4.20)$$

where the constants N and L depend on the shape of the domain D and on the exponents p, q^*, k, n and l, but do not depend on the form of the function ϕ.

Proof One begins by considering the case when ϕ is continuous with its derivatives up to the order l. At each point $y^1, \ldots, y^n$ of the domain D, one considers the system of polar coordinates defined by the formulae

$$x^i - y^i = r(\sin \theta_1)(\sin \theta_2) \ldots (\sin \theta_{i-1})(\cos \theta_i), \; i = 1, 2, \ldots, n-2, \quad (10.4.21)$$

$$x^{n-1} - y^{n-1} = r(\sin \theta_1)(\sin \theta_2) \ldots (\sin \theta_{n-2})(\cos \phi), \quad (10.4.22)$$

$$x^n - y^n = r(\sin \theta_1)(\sin \theta_2) \ldots (\sin \theta_{n-2})(\sin \phi). \quad (10.4.23)$$

The spherical sector that reaches the point $y^1, \ldots, y^n$ is taken to have the equation

$$r = h, \quad \theta_1 = \alpha. \tag{10.4.24}$$

Consider now the function

$$\psi(\xi) = 1 \text{ if } \xi \leq \frac{1}{3}, \tag{10.4.25a}$$

$$\psi(\xi) = -\frac{1}{\pi}\text{arc tg sh}\frac{\left(\xi - \frac{1}{2}\right)}{\left(\frac{2}{3} - \xi\right)\left(\xi - \frac{1}{3}\right)} + \frac{1}{2}, \quad \xi \in \left[\frac{1}{3}, \frac{2}{3}\right], \tag{10.4.25b}$$

$$\psi(\xi) = 0 \text{ if } \xi \geq \frac{2}{3}. \tag{10.4.25c}$$

This function is chosen in such a way that it has continuous derivatives of any order at all points. One can use it to construct the function Ω by the formula

$$\Omega \equiv \left\{ \phi\frac{\partial^{l-1}}{\partial r^{l-1}} - \frac{\partial\phi}{\partial r}\frac{\partial^{l-2}}{\partial r^{l-2}} + \cdots \right.$$
$$\left. \cdots + \frac{\partial^{l-1}\phi}{\partial r^{l-1}}(-1)^{l-1} \right\}\left[\frac{r^{l-1}}{(l-1)!}\psi\left(\frac{r}{h}\right)\right]. \tag{10.4.26}$$

From these definitions one finds the identities

$$\Omega(0, \theta_1, \ldots, \theta_{n-2}, \phi) = \phi(y^1, \ldots, y^n), \quad \Omega(h, \theta_1, \ldots, \theta_{n-2}, \phi) = 0, \tag{10.4.27}$$

$$\frac{\partial\Omega}{\partial r} = \left\{ \phi\frac{\partial^l}{\partial r^l} + (-1)^{l-1}\frac{\partial^l\phi}{\partial r^l} \right\}\left[\frac{r^{l-1}}{(l-1)!}\psi\left(\frac{r}{h}\right)\right], \tag{10.4.28}$$

which yield the simple and useful formula

$$\phi(y^1, \ldots, y^n) = \int_h^0 \frac{\partial\Omega}{\partial r}dr. \tag{10.4.29}$$

In this formula one can first multiply both sides by the function

$$\psi\left(\frac{\theta_1}{\alpha}\right)(\sin\theta_1)^{n-2}(\sin\theta_2)^{n-3}\ldots(\sin\theta_{n-2})d\theta_1\ldots d\theta_{n-2}d\phi$$

$$=\frac{1}{r^{n-1}}\psi\left(\frac{\theta_1}{\alpha}\right)\frac{D(x^1,\ldots,x^n)}{D(r,\theta_1,\ldots,\theta_{n-2},\phi)}d\theta_1\ldots d\theta_{n-2}d\phi,\qquad(10.4.30)$$

and then integrate between the limits

$$\theta_1\in[0,\alpha],\ \theta_i\in[0,\pi]\ (i=2,3,\ldots,n-2),\ \phi\in[-\pi,\pi].\quad(10.4.31)$$

This procedure yields the value

$$\phi(y^1,\ldots,y^n)=\int\ldots\int_{V_0}\psi\left(\frac{\theta_1}{\alpha}\right)\frac{\partial\Omega}{\partial r}r^{1-n}dx^1\ldots dx^n,\qquad(10.4.32)$$

where V_0 denotes the sector reaching the point $y^1,\ldots,y^n$. In this formula the integrand contains $\frac{\partial\Omega}{\partial r}$, which is provided by (10.4.28). If we bear in mind that the l-th derivative of ϕ with respect to r is a linear combination of various derivatives of order l of ϕ with respect to the variables $x^1,\ldots,x^n$ with bounded coefficients, we obtain [130, 131]

$$\psi\left(\frac{\theta_1}{\alpha}\right)r^{1-n}\frac{\partial\Omega}{\partial r}=\mu\phi+\sum\mu_{\alpha_1,\ldots,\alpha_n}\frac{\partial^l\phi}{\partial(x^1)^{\alpha_1}\ldots\partial(x^n)^{\alpha_n}},\qquad(10.4.33)$$

where μ is a bounded continuous function, and the functions $\mu_{\alpha_1,\ldots,\alpha_n}$ satisfy the majorization

$$\left|\mu_{\alpha_1,\ldots,\alpha_n}\right|\le\frac{A}{r^{n-l}}\psi\left(\frac{r}{h}\right).\qquad(10.4.34)$$

The continuity of the function μ follows from the property

$$\frac{\partial^l}{\partial r^l}\left[\frac{r^{l-1}}{(l-1)!}\phi\left(\frac{r}{h}\right)\right]=0,\ \text{if}\ r\le\frac{h}{3}.\qquad(10.4.35)$$

Moreover, one can exploit the Minkowski inequality to find

$$\left[\int\ldots\int_D|\phi(Q)|^q dQ\right]^{\frac{1}{q}}\le\left[\int\ldots\int_D\left[\int\ldots\int_{r\le h}|\phi|\,|\mu|dP\right]^q dQ\right]^{\frac{1}{q}}$$

$$+\sum\left[\int\ldots\int_D\left[\int\ldots\int_{r\le h}\left|\frac{\partial^l\phi}{\partial(x^1)^{\alpha_1}\ldots\partial(x^n)^{\alpha_n}}\right|\,|\mu_{\alpha_1,\ldots,\alpha_n}|dP\right]^q dQ\right]^{\frac{1}{q}}.$$

$$\qquad(10.4.36)$$

At this stage, one exploits a fundamental theorem about integrals of potential type proved in Sobolev [130, 131]. If a function $f(x^1, \ldots, x^n) = f(P)$ is given which is p-th power summable, with $p > 1$, one writes (defining $dP \equiv dx^1 \ldots dx^n$)

$$\int \ldots \int |f(P)|^p dP \equiv B^p, \tag{10.4.37}$$

and one defines the λ-potential of f according to

$$U(Q) \equiv \int \ldots \int r^{-\lambda} f(P) dP, \quad \lambda < \frac{n}{p'}, \tag{10.4.38}$$

with the understanding that Q is some point with coordinates $(y^1, \ldots, y^n)$ and r is the Euclidean distance

$$r \equiv \sqrt{\sum_{i=1}^{n} (x^i - y^i)^2}. \tag{10.4.39}$$

The theorem needed is as follows.

Theorem 10.4 *The function $U(Q)$ is q-th power summable on the whole space $y^1, \ldots, y^n$, where*

$$\frac{1}{q} \equiv \frac{\lambda}{n} - \frac{1}{p'} > 1. \tag{10.4.40}$$

Moreover, the following inequality holds:

$$\int \ldots \int [U(Q)]^q dQ \leq K B^q, \tag{10.4.41}$$

where K is some constant.

By virtue of Theorem 10.4, the terms under the summation sign in (10.4.36) can be estimated by means of the inequality

$$\left[\int \ldots \int_D \left[\int \ldots \int_{r \leq h} \left| \frac{\partial^l \phi}{\partial (x^1)^{\alpha_1} \ldots \partial (x^n)^{\alpha_n}} \right| |\mu_{\alpha_1, \ldots, \alpha_n}| dP \right]^q dQ \right]^{\frac{1}{q}} \leq M A. \tag{10.4.42}$$

As far as the first line of (10.4.36) is concerned, one notes that, adding and subtracting q^* within the exponent q,

$$\left[\int \ldots \int_{r \leq h} |\phi| |\mu| dP \right]^q = \left[\int \ldots \int_{r \leq h} |\phi| |\mu| dP \right]^{q^*} \left[\int \ldots \int_{r \leq h} |\phi| |\mu| dP \right]^{q - q^*}. \tag{10.4.43}$$

The integral of $|\phi|\,|\mu|$ can be in addition majorized according to

$$\int \cdots \int_{r\leq h} |\phi|\,|\mu| \mathrm{d}P \leq \left[\int \cdots \int_{r\leq h} |\mu|^{q^{*'}} \mathrm{d}P\right]^{\frac{1}{q^{*'}}} \left[\int \cdots \int_{r\leq h} |\phi|^{q^*} \mathrm{d}P\right]^{\frac{1}{q^*}}$$

$$= M_2 \left[\int \cdots \int_{r\leq h} |\phi|^{q^*} \mathrm{d}P\right]^{\frac{1}{q^*}}, \qquad (10.4.44)$$

from which Eq. (10.4.43) leads to the majorization

$$\left[\int \cdots \int_{r\leq h} |\phi|\,|\mu| \mathrm{d}P\right]^{q} \leq M_3 \int \cdots \int_{r\leq h} |\phi|^{q^*} \mathrm{d}P \left[\int \cdots \int_{r\leq h} |\phi|^{q^*} \mathrm{d}P\right]^{\frac{(q-q^*)}{q^*}}$$

$$\leq M_3 \left[\int \cdots \int_{r\leq h} |\phi|^{q^*} \mathrm{d}P\right]^{\frac{(q-q^*)}{q^*}} \int \cdots \int_{r\leq h} |\phi|^{q^*} \mathrm{d}P, \qquad (10.4.45)$$

and hence also (by integrating n times over D and then taking the $\frac{1}{q}$ power)

$$\left[\int \cdots \int_{D} \left[\int \cdots \int_{r\leq h} |\phi|\,|\mu| \mathrm{d}P\right]^{q} \mathrm{d}Q\right]^{\frac{1}{q}}$$

$$\leq M_4 \left[\int \cdots \int_{D} |\phi|^{q^*} \mathrm{d}P\right]^{\frac{(q-q^*)}{qq^*}} \left[\int \cdots \int_{D} \left[\int \cdots \int_{r\leq h} |\phi|^{q^*} \mathrm{d}P\right] \mathrm{d}Q\right]^{\frac{1}{q}}$$

$$= M_4 \left[\int \cdots \int_{D} |\phi|^{q^*} \mathrm{d}P\right]^{\frac{1}{q^*}-\frac{1}{q}} \left[\int \cdots \int_{D} |\phi|^{q^*} \left[\int \cdots \int_{r\leq h} \mathrm{d}Q\right] \mathrm{d}P\right]^{\frac{1}{q}}$$

$$\leq M_5 \left[\int \cdots \int_{D} |\phi|^{q^*} \mathrm{d}P\right]^{\frac{1}{q^*}}. \qquad (10.4.46)$$

By virtue of (10.4.42), (10.4.46) and (10.4.36) one therefore obtains the desired inequality (10.4.20), which gets proved, so far, for continuous functions. It is now useful to recall theorems that make it possible to approximate a given function by means of continuously differentiable functions.

If ϕ is a function summable on every finite subset of n-dimensional Euclidean space, one can construct a new function defined by [130, 131]

$$\phi_k(x^1, \ldots, x^n) = \chi k^n \int \cdots \int \phi(y^1, \ldots, y^n) \psi(kr) \mathrm{d}y^1 \ldots \mathrm{d}y^n, \qquad (10.4.47)$$

where ψ is the function defined in Eq. (10.4.25), and the constant χ is defined as the inverse of the integral of ϕ over n-dimensional Euclidean space, i.e.

$$\chi \equiv \frac{1}{\int \cdots \int \phi(r) \mathrm{d}x^1 \ldots \mathrm{d}x^n}. \qquad (10.4.48)$$

The ϕ_k are said to be a *sequence of averaged functions* for ϕ. Their properties are as follows.

(i) At all points where

$$\lim \frac{\int \dots \int_{r \leq \tau} \phi(y^1, \dots, y^n) dy^1 \dots dy^n}{\int \dots \int_{r \leq \tau} dy^1 \dots dy^n} = \phi(x^1, \dots, x^n), \quad (10.4.49)$$

the ϕ_k satisfy the limit relation

$$\lim_{k \to \infty} \phi_k(x^1, \dots, x^n) = \phi(x^1, \dots, x^n). \quad (10.4.50)$$

Indeed, a theorem of Lebesgue ensures that (10.4.49) holds almost everywhere. One first proves the formula [130, 131]

$$\phi_k = \chi k^n \int_0^1 \left\{ \int_{C(\psi > \lambda)} \dots \int \phi(y^1, \dots, y^n) dy^1 \dots dy^n \right\} d\lambda, \quad (10.4.51)$$

where $C(\psi > \lambda)$ is the sphere in which the value $\psi(kr) > \lambda$. In order to prove this formula, one notes that the integral on the right-hand side of (10.4.47) can be evaluated with as much accuracy as desired in the form of the sum

$$\int \dots \int \phi(y^1, \dots, y^n) \psi(kr) dy^1 \dots dy^n = \sum_{i=0}^N \lambda_i \left[\int_{\psi(kr) \geq \lambda_i} \dots \int \phi \, dy^1 \dots dy^n \right.$$

$$\left. - \int_{\psi(kr) \geq \lambda_{i+1}} \dots \int \phi \, dy^1 \dots dy^n \right] + \int_{\psi(kr)=1} \dots \int \phi \, dy^1 \dots dy^n + \epsilon, \quad (10.4.52)$$

where the λ's satisfy

$$0 = \lambda_0 < \lambda_1 < \lambda_2 < \dots < \lambda_{N-1} < \lambda_N = 1, \quad (10.4.53)$$

and the difference $(\lambda_{i+1} - \lambda_i)$ is uniformly as small as desired for sufficiently large N. The equality (10.4.52) can be re-expressed in the convenient way

$$\int \dots \int \phi(y^1, \dots, y^n) \psi(kr) dy^1 \dots dy^n = \epsilon(N) + \lambda_0 \int_{r \leq k} \dots \int \phi \, dy^1 \dots dy^n$$

$$+ \sum_{i=1}^N (\lambda_i - \lambda_{i-1}) \int_{\psi(kr) \geq \lambda_i} \dots \int \phi \, dy^1 \dots dy^n, \quad (10.4.54)$$

from which Eq. (10.4.51) is proved by a passage to the limit. In analogous way, one finds

$$\frac{1}{k^n\chi} = \int_0^1 \left[\int_{C(\psi(kr)\geq\lambda)} \cdots \int dy^1 \dots dy^n \right].$$ (10.4.55)

From the limit (10.4.49) one obtains

$$\int_{r\leq\tau} \cdots \int \phi(y^1,\dots,y^n) dy^1 \dots dy^n = \left[\phi(x^1,\dots,x^n) + \epsilon(\tau)\right] \int_{r\leq\tau} \cdots \int dy^1 \dots dy^n.$$ (10.4.56)

with the understanding that both $\epsilon(\tau)$ and τ approach 0. Upon inserting (10.4.56) and (10.4.55) into (10.4.51), one obtains

$$\phi_k = \frac{\int_0^1 \left\{ \left[\phi(x^1,\dots,x^n) + \epsilon_k(\lambda)\right] \int_{\psi(kr)\geq\lambda} \cdots \int dy^1 \dots dy^n \right\} d\lambda}{\int_0^1 \left[\int_{\psi(kr)\geq\lambda} \cdots \int dy^1 \dots dy^n \right] d\lambda},$$ (10.4.57)

where $\epsilon_k(\lambda) = \epsilon(\tau)$, $\psi(k\tau) \geq \lambda$, and hence

$$|\phi_k - \phi| \leq \frac{\int_0^1 \left[|\epsilon_k(\lambda)| \int_{\psi(kr)\geq\lambda} \cdots \int dy^1 \dots dy^n \right] d\lambda}{\int_0^1 \left[\int_{\psi(kr)\geq\lambda} \cdots \int dy^1 \dots dy^n \right] d\lambda},$$ (10.4.58)

i.e. the limit (10.4.50) is indeed proved.

(ii) If the function ϕ is p-th power summable, in that

$$\int_\infty \cdots \int |\phi|^p dx^1 \dots dx^n < \infty, \quad p \geq 1,$$ (10.4.59)

the same holds for the function ϕ_k, i.e.

$$\int_\infty \cdots \int |\phi_k|^p dx^1 \dots dx^n < \infty,$$ (10.4.60)

and in addition the following limit holds:

$$\lim_{k\to\infty} \int_\infty \cdots \int |\phi_k - \phi|^p dx^1 \dots dx^n = 0.$$ (10.4.61)

Proof The property just stated clearly holds if the function ϕ is bounded and differs from 0 only on a finite domain. Moreover, if the absolute value of ϕ is majorized by a constant M, one can write the same majorization for ϕ_k as well, i.e.

$$|\phi_k| \leq \chi k^n \int \cdots \int M\psi(kr) dy^1 \dots dy^n = M.$$ (10.4.62)

The summability property (10.4.60) is therefore obvious, and the limit (10.4.61) follows from the Lebesgue theorem on the admissibility to pass the limit under the integral sign for bounded functions.

In the general case, one begins by pointing out that the function ϕ may be always expressed as the sum of two functions $\overline{\phi}^{(N)}$ and $\widetilde{\phi}^{(N)}$

$$\phi = \overline{\phi}^{(N)} + \widetilde{\phi}^{(N)}, \tag{10.4.63}$$

where

$$\overline{\phi}^{(N)} = \phi \text{ if } |\phi| \leq N, \ \overline{\phi}^{(N)} = N \text{ if } |\phi| > N, \tag{10.4.64}$$

and of course

$$\int \ldots \int \left| \widetilde{\phi}^{(N)} \right|^p \, dx^1 \ldots dx^n < \epsilon. \tag{10.4.65}$$

The averages for $\overline{\phi}^{(N)}$ and $\widetilde{\phi}^{(N)}$ are $\overline{\phi}_k^{(N)}$ and $\widetilde{\phi}_k^{(N)}$, respectively, and can be studied separately. Of course, the average for ϕ is just the sum of such averages, i.e.

$$\phi_k = \overline{\phi}_k^{(N)} + \widetilde{\phi}_k^{(N)}. \tag{10.4.66}$$

Following again Sobolev [130, 131], consider the integral

$$\left[\int_\infty \ldots \int |\phi - \phi_k|^p \, dx^1 \ldots dx^n \right]^{\frac{1}{p}} \leq \left[\int_\infty \ldots \int \left| \overline{\phi}^{(N)} - \overline{\phi}_k^{(N)} \right|^p \, dx^1 \ldots dx^n \right]^{\frac{1}{p}}$$
$$+ \left[\int_\infty \ldots \int \left| \widetilde{\phi}^{(N)} \right|^p \, dx^1 \ldots dx^n \right]^{\frac{1}{p}} + \left[\int_D \ldots \int \left| \widetilde{\phi}_k^{(N)} \right|^p \, dx^1 \ldots dx^n \right]^{\frac{1}{p}}.$$
$$\tag{10.4.67}$$

The number N is chosen in such a way that

$$\left[\int_\infty \ldots \int \left| \widetilde{\phi}^{(N)} \right|^p \, dx^1 \ldots dx^n \right]^{\frac{1}{p}} < \frac{\epsilon}{3K}, \tag{10.4.68}$$

where the constant K is obtained from the product

$$K \equiv \chi \int_{r \leq 1} \ldots \int dx^1 \ldots dx^n > 1. \tag{10.4.69}$$

As far as the last integral occurring in (10.4.67) is concerned, i.e.

$$\int_D \cdots \int \left| \widetilde{\phi}_k^{(N)} \right|^p \, dx^1 \ldots dx^n,$$

one finds, for the modulus considered in the integrand,

$$\left| \widetilde{\phi}_k^{(N)} \right| = \chi k^n \left| \int_{r \leq \frac{1}{k}} \cdots \int \widetilde{\phi}^{(N)}(y^1, \ldots, y^n) \psi(kr) dy^1 \ldots dy^n \right|$$

$$\leq \chi k^n \left[\int_{r \leq \frac{1}{k}} \cdots \int \left| \widetilde{\phi}^{(N)}(y^1, \ldots, y^n) \right|^p \, dy^1 \ldots dy^n \right]^{\frac{1}{p}} \left[\int \cdots \int |\psi(kr)|^{p'} dy^1 \ldots dy^n \right]^{\frac{1}{p'}}$$

$$\leq \left(\chi k^n \right)^{1 - \frac{1}{p'}} \left[\int_{r \leq \frac{1}{k}} \cdots \int \left| \widetilde{\phi}^{(N)}(y^1, \ldots, y^n) \right|^p \, dy^1 \ldots dy^n \right]^{\frac{1}{p}}, \tag{10.4.70}$$

where

$$\frac{1}{p'} + \frac{1}{p} = 1. \tag{10.4.71}$$

The inequality (10.4.70) can be therefore re-expressed in the form

$$\left| \widetilde{\phi}_k^{(N)} \right|^p \leq \chi k^n \int_{r \leq \frac{1}{k}} \cdots \int \left| \widetilde{\phi}^{(N)}(y^1, \ldots, y^n) \right|^p \, dy^1 \ldots dy^n. \tag{10.4.72}$$

Now we can integrate the left-hand side of (10.4.72) and majorize it as follows:

$$\int_\infty \cdots \int \left| \widetilde{\phi}_k^{(N)} \right|^p \, dx^1 \ldots dx^n$$

$$\leq \int_\infty \cdots \int \chi k^n \left[\int_{r \leq \frac{1}{k}} \cdots \int \left| \widetilde{\phi}^{(N)}(y^1, \ldots, y^n) \right|^p \, dy^1 \ldots dy^n \right] dx^1 \ldots dx^n$$

$$= \int_\infty \cdots \int \left| \widetilde{\phi}^{(N)}(y^1, \ldots, y^n) \right|^p \left(\chi k^n \int_{r \leq \frac{1}{k}} \cdots \int dx^1 \ldots dx^n \right) dy^1 \ldots dy^n$$

$$\leq K \int_\infty \cdots \int \left| \widetilde{\phi}^{(N)} \right|^p \, dy^1 \ldots dy^n$$

$$\leq \frac{\epsilon}{3}, \tag{10.4.73}$$

where, on the last line, the inequality (10.4.68) has been exploited.

Last, on the first line of (10.4.67), for k sufficiently large the integral on the right-hand side can be majorized according to

$$\left[\int \cdots \int \left| \overline{\phi}^N - \overline{\phi}_k^{(N)} \right|^p \, dx^1 \ldots dx^n \right]^{\frac{1}{p}} \leq \frac{\epsilon}{3}, \tag{10.4.74}$$

and hence (10.4.67) reads eventually as

$$\left[\int_\infty \cdots \int |\phi - \phi_k|^p \ dx^1 \ldots dx^n \right]^{\frac{1}{p}} < \frac{\epsilon}{3} + \frac{\epsilon}{3} + \frac{\epsilon}{3} = \epsilon. \qquad (10.4.75)$$

By virtue of the inequality (10.4.75), the relations (10.4.60) and (10.4.61) follow in the general case. □

As a corollary, if for the function ϕ the integral

$$\int_D \cdots \int |\phi|^p \ dx^1 \ldots dx^n$$

taken over any bounded domain D is bounded, one can then substitute for the infinite domain everywhere, in the formulation of the theorem, any finite domain.

(iii) The derivative of the averaged function is equal to the averaged function of the derivative, if the latter is understood in the sense specified above.

Proof Since $\psi(kr)$ depends only on the differences of the coordinates $(x^i - y^i)$, one finds

$$\frac{\partial^\alpha \phi_k}{\partial(x^1)^{\alpha_1} \ldots \partial(x^n)^{\alpha_n}} = \chi k^n \int \cdots \int \phi(y^1, \ldots, y^n) \frac{\partial^\alpha \psi(kr)}{\partial(x^1)^{\alpha_1} \ldots \partial(x^n)^{\alpha_n}} dy^1 \ldots dy^n$$

$$= \chi k^n \int \cdots \int \phi(y^1, \ldots, y^n)(-1)^\alpha \frac{\partial^\alpha \psi(kr)}{\partial(y^1)^{\alpha_1} \ldots \partial(y^n)^{\alpha_n}} dy^1 \ldots dy^n$$

$$= \chi k^n \int \cdots \int \frac{\partial^\alpha \phi}{\partial(y^1)^{\alpha_1} \ldots \partial(y^n)^{\alpha_n}} \psi(kr) dy^1 \ldots dy^n, \qquad (10.4.76)$$

where, on the last line, we have exploited the definition of generalized derivative. The desired property is therefore proved. □

(iv) If there exist derivatives of the averaged functions

$$\frac{\partial^\alpha \phi_k}{\partial(x^1)^{\alpha_1} \ldots \partial(x^n)^{\alpha_n}}$$

which have, over some domain D, an integral uniformly bounded with respect to k, i.e.

$$\int \cdots \int \left| \frac{\partial^\alpha \phi_k}{\partial(x^1)^{\alpha_1} \ldots \partial(x^n)^{\alpha_n}} \right|^p dx^1 \ldots dx^n \leq A, \ p > 1, \qquad (10.4.77)$$

then the derivative

$$\frac{\partial^\alpha \phi}{\partial(x^1)^{\alpha_1} \ldots \partial(x^n)^{\alpha_n}}$$

of the function ϕ exists and is p-th power summable on D, and satisfies in turn the inequality

$$\int_D \cdots \int \left| \frac{\partial^\alpha \phi}{\partial (x^1)^{\alpha_1} \ldots \partial (x^n)^{\alpha_n}} \right|^p dx^1 \ldots dx^n \leq A. \qquad (10.4.78)$$

Furthermore, from properties (ii) and (iii) follows, as a corollary, the following limit:

$$\lim_{k \to \infty} \int_D \cdots \int \left| \frac{\partial^\alpha \phi_k}{\partial (x^1)^{\alpha_1} \ldots \partial (x^n)^{\alpha_n}} - \frac{\partial^\alpha \phi}{\partial (x^1)^{\alpha_1} \ldots \partial (x^n)^{\alpha_n}} \right|^p dx^1 \ldots dx^n = 0.$$
$$(10.4.79)$$

From (10.4.78) follow (10.4.77) and (10.4.79).

The fundamental theorem in the general case. Let ϕ be any function q^*-th power summable on the domain D, and let the derivatives of order l of ϕ have bounded p-th power integrals. One can now construct a sequence of averaged functions ϕ_k. By virtue of the results proved so far, for this sequence all the hypotheses we are interested in remain valid, since the averaged function is q^*-th power summable, and its derivatives of order l are p-th power summable on the whole interior of the domain. One can establish the inequality (10.4.20) for the averaged functions. After such a step, a passage to the limit yields this inequality in the general case. $\square$

For the analysis of the remaining sub-cases in Eq. (10.4.17), we refer the reader to the last page in Sobolev [130].

Chapter 11
Pseudo-Holomorphic and Polyharmonic Frameworks

11.1 Local Theory of Pseudo-Holomorphic Functions

As is well known, the theory of Laplace's equation in the two-dimensional Euclidean plane $\mathbf{R}^2$ is essentially equivalent to the theory of complex-analytic functions. Moreover, one can also associate a theory of generalized analytic functions, also called pseudo-analytic or pseudo-holomorphic, with every linear elliptic equation of second order for an unknown function of two independent variables. This is the local theory we are going to outline in this section, while referring the reader to Bers [10] and Vekua [135] for a much more extensive treatment.

Let us assume that Φ is a harmonic function. The resulting Laplace equation is here written in Cartesian coordinates (x, y) for the Euclidean plane $\mathbf{R}^2$, i.e.

$$\triangle \Phi \equiv \Phi_{xx} + \Phi_{yy} = 0. \tag{11.1.1}$$

On defining

$$u \equiv \frac{\partial \Phi}{\partial x}, \quad v \equiv -\frac{\partial \Phi}{\partial y}, \tag{11.1.2}$$

the functions u and v solve the Cauchy-Riemann equations

$$\frac{\partial u}{\partial x} - \frac{\partial v}{\partial y} = 0, \quad \frac{\partial u}{\partial y} + \frac{\partial v}{\partial x} = 0, \tag{11.1.3}$$

so that the complex gradient $w \equiv u + iv$ is a holomorphic function of the complex variable $z \equiv x + iy$, because from Eq. (11.1.3)

$$\frac{\partial w}{\partial x} = -i\frac{\partial w}{\partial y}, \tag{11.1.4}$$

while the real-valued functions u and v are harmonic, because, from Eq. (11.1.3), one finds

© Springer International Publishing AG 2017
G. Esposito, *From Ordinary to Partial Differential Equations*, UNITEXT 106,
DOI 10.1007/978-3-319-57544-5_11

$$\triangle u = \left(\frac{\partial^2}{\partial x \partial y} - \frac{\partial^2}{\partial y \partial x} \right) v = 0, \tag{11.1.5}$$

$$\triangle v = \left(-\frac{\partial^2}{\partial x \partial y} + \frac{\partial^2}{\partial y \partial x} \right) u = 0. \tag{11.1.6}$$

On the other hand, one can find a conjugate harmonic function Ψ with the help of the Cauchy-Riemann equations

$$\frac{\partial \Phi}{\partial x} - \frac{\partial \Psi}{\partial y} = 0, \quad \frac{\partial \Phi}{\partial y} + \frac{\partial \Psi}{\partial x} = 0. \tag{11.1.7}$$

In this system, the function Φ is the real part of the holomorphic function $W \equiv \Phi + i\Psi$. The functions W and w are related by

$$w(z) = \frac{dW}{dz}. \tag{11.1.8}$$

Let us now consider the general form of a linear elliptic equation in two independent variables, i.e.

$$\left(a_{11}\frac{\partial^2}{\partial \xi^2} + 2a_{12}\frac{\partial^2}{\partial \eta \partial \xi} + a_{22}\frac{\partial^2}{\partial \eta^2} + a_1 \frac{\partial}{\partial \xi} + a_2 \frac{\partial}{\partial \eta} + a_0 \right) \phi = 0, \tag{11.1.9}$$

where the coefficients $a_{ij}(\xi, \eta)$ are taken to have Hölder continuous partial derivatives, and the existence of a positive solution $\phi_0(\xi, \eta)$ is also assumed. Moreover, we pass to new variables

$$x = x(\xi, \eta), \quad y = y(\xi, \eta), \tag{11.1.10}$$

by means of a map which is taken to be a homeomorphism satisfying the Beltrami equations associated with the metric

$$a_{22}d\xi \otimes d\xi - a_{12}(d\xi \otimes d\eta + d\eta \otimes d\xi) + a_{11}d\eta \otimes d\eta.$$

In order to help the general reader, we here recall that, if

$$g_{11}(x, y)dx \otimes dx + g_{12}(dx \otimes dy + dy \otimes dx) + g_{22}(x, y)dy \otimes dy$$

is a Riemannian metric defined in a domain D, a function $w = u + iv$ in D is said to be *conformal with respect to this metric* if u and v satisfy the Beltrami equations (here $g \equiv \sqrt{g_{11}g_{22} - (g_{12})^2}$)

$$g \begin{pmatrix} \frac{\partial u}{\partial x} \\ \frac{\partial u}{\partial y} \end{pmatrix} = \begin{pmatrix} -g_{12} & g_{11} \\ -g_{22} & g_{12} \end{pmatrix} \begin{pmatrix} \frac{\partial v}{\partial x} \\ \frac{\partial v}{\partial y} \end{pmatrix}, \tag{11.1.11}$$

which can be also written in the form

$$\frac{\partial w}{\partial x} + i\frac{\partial w}{\partial y} = \mu\left(\frac{\partial w}{\partial x} - i\frac{\partial w}{\partial y}\right), \tag{11.1.12}$$

the complex-valued function μ being given by

$$\mu \equiv \frac{(g_{11} - g_{22} + 2ig_{12})}{(g_{11} + g_{22} + 2g)}. \tag{11.1.13}$$

Now we write the function ϕ solving Eq. (11.1.9) in the factorized form

$$\phi = \Phi\,\phi_0, \tag{11.1.14}$$

where ϕ_0 is the positive solution of Eq. (11.1.9) whose existence has been assumed. Under the action of the homeomorphism (11.1.10), the linear operator acting on ϕ transforms in such a way that Eq. (11.1.9) is turned into

$$\left(\Delta + \alpha(x, y)\frac{\partial}{\partial x} + \beta(x, y)\frac{\partial}{\partial y}\right)\Phi(x, y) = 0. \tag{11.1.15}$$

By virtue of Eq. (11.1.15), the definitions (11.1.2) lead now to the first-order linear system

$$\frac{\partial u}{\partial x} - \frac{\partial v}{\partial y} = \Delta\Phi = -\alpha\frac{\partial\Phi}{\partial x} - \beta\frac{\partial\Phi}{\partial y} = -\alpha u + \beta v, \tag{11.1.16}$$

$$\frac{\partial u}{\partial y} + \frac{\partial v}{\partial x} = \left(\frac{\partial^2}{\partial y\partial x} - \frac{\partial^2}{\partial x\partial y}\right)\Phi = 0, \tag{11.1.17}$$

which is a special case of the Hilbert-Carleman system

$$\frac{\partial u}{\partial x} - \frac{\partial v}{\partial y} = \alpha_{11}u + \alpha_{12}v, \tag{11.1.18}$$

$$\frac{\partial u}{\partial y} + \frac{\partial v}{\partial x} = \alpha_{21}u + \alpha_{22}v. \tag{11.1.19}$$

With the help of complex notation, Eqs. (11.1.18) and (11.1.19) can be re-expressed in the form (see chapter by Bers in Courant and Hilbert [28])

$$\frac{\partial w}{\partial \overline{z}} = a(z)w + b(z)\overline{w}, \tag{11.1.20}$$

where the a and b functions are defined by

$$4a \equiv \alpha_{11} + \alpha_{22} + i(\alpha_{21} - \alpha_{12}), \tag{11.1.21}$$

$$4b \equiv \alpha_{11} - \alpha_{22} + i(\alpha_{21} + \alpha_{12}). \tag{11.1.22}$$

A continuously differentiable solution of Eq. (11.1.20) is said to be a *pseudo-analytic function of the first kind* defined by the system (11.1.18) and (11.1.19), or an $[a, b]$-pseudo-analytic function.

The complex gradient $\frac{\partial \Phi}{\partial x} + i \frac{\partial \Phi}{\partial y}$ of every solution of Eq. (11.1.15) is an $[a, \overline{a}]$-pseudo-analytic function, with $4a = -\alpha - i\beta$. Conversely, given Eq. (11.1.15), one can find under suitable assumptions two real-valued functions $\tau(x, y)$ and $\sigma(x, y) > 0$ obeying the equations of Hilbert-Carleman type

$$\frac{\partial \sigma}{\partial x} - \frac{\partial \tau}{\partial y} = \alpha \sigma, \tag{11.1.23}$$

$$\frac{\partial \sigma}{\partial y} + \frac{\partial \tau}{\partial x} = \beta \sigma. \tag{11.1.24}$$

Indeed, if one considers the system

$$\sigma \frac{\partial \Phi}{\partial x} + \tau \frac{\partial \Phi}{\partial y} = \frac{\partial \Psi}{\partial y}, \tag{11.1.25}$$

$$\sigma \frac{\partial \Phi}{\partial y} - \tau \frac{\partial \Phi}{\partial x} = -\frac{\partial \Psi}{\partial x}, \tag{11.1.26}$$

differentiation of Eq. (11.1.25) with respect to x and of Eq. (11.1.26) with respect to y, with subsequent addition of the resulting equations, leads to the equation

$$\sigma \triangle \Phi + \left(\frac{\partial \sigma}{\partial x} - \frac{\partial \tau}{\partial y} \right) \frac{\partial \Phi}{\partial x} + \left(\frac{\partial \sigma}{\partial y} + \frac{\partial \tau}{\partial x} \right) \frac{\partial \Phi}{\partial y} = 0, \tag{11.1.27}$$

because the terms involving mixed derivatives of Φ and Ψ add up to 0. By virtue of Eqs. (11.1.23) and (11.1.24), Eq. (11.1.27) is turned into

$$\sigma(x, y) \left(\triangle + \alpha(x, y) \frac{\partial}{\partial x} + \beta(x, y) \frac{\partial}{\partial y} \right) \Phi(x, y) = 0, \tag{11.1.28}$$

which is then fully equivalent to Eq. (11.1.15), since $\sigma(x, y)$ is taken to be positive. Thus, Eq. (11.1.15) is equivalent to the elliptic system (11.1.25) and (11.1.26). A function $\Phi + i\Psi$ corresponding to a solution of the elliptic system (11.1.25) and (11.1.26) with Hölder continuous coefficients is said to be a *pseudo-analytic function of the second kind* associated with this system. In other words, *every solution of Eq. (11.1.15) is the real part of a pseudo-analytic function of the second kind.*

Note that the theory here outlined is *local*, because it relies on first-order systems of partial differential equations which generalize the Cauchy-Riemann equations by including terms proportional to the undifferentiated unknown functions of the independent variables (x, y). Recall now the basic formulae of complex analysis

$$\frac{\partial w}{\partial z} = \frac{1}{2}\left(\frac{\partial w}{\partial x} - i\frac{\partial w}{\partial y}\right), \quad \frac{\partial w}{\partial \bar{z}} = \frac{1}{2}\left(\frac{\partial w}{\partial x} + i\frac{\partial w}{\partial y}\right). \tag{11.1.29}$$

Two solutions $F(z)$ and $G(z)$ of Eq. (11.1.20) are said to form a pair of *generators* if

$$F(z) \neq 0\ \forall z, \quad G(z) \neq 0\ \forall z, \quad \mathrm{Im}\left(\frac{G}{F}\right) > 0. \tag{11.1.30}$$

For example, for the Cauchy-Riemann equations, the functions 1 and i form a pair $(1, i)$ of generators, and the same is true for the pair (e^z, ie^z).

If (F, G) is a generating pair for Eq. (11.1.20), every complex-valued function $w(z)$ may be written in a unique way in the form

$$w(z) = \phi(z)F(z) + \psi(z)G(z), \tag{11.1.31}$$

where the functions ϕ and ψ are real-valued. It is convenient to associate with every function w written in the form (11.1.31) the function

$$\omega(z) \equiv \phi(z) + i\psi(z). \tag{11.1.32}$$

If the function w is pseudo-analytic of the first kind, the function ω is pseudo-analytic of the second kind.

Since two arbitrary smooth functions F, G satisfying the conditions (11.1.30) form a generating pair for some equation of the form (11.1.20), they are said to be (F, G)-pseudo-analytic functions.

Theorem 11.1 *The function w in (11.1.31) is (F, G)-pseudo-analytic if and only if*

$$\frac{\partial \phi}{\partial \bar{z}}F + \frac{\partial \psi}{\partial \bar{z}}G = 0. \tag{11.1.33}$$

Proof Since both F and G are taken to solve Eq. (11.1.20), one finds from (11.1.31)

$$\begin{aligned}
\frac{\partial w}{\partial \bar{z}} &= \frac{\partial \phi}{\partial \bar{z}}F + \phi\frac{\partial F}{\partial \bar{z}} + \frac{\partial \psi}{\partial \bar{z}}G + \psi\frac{\partial G}{\partial \bar{z}} \\
&= \phi(aF + b\overline{F}) + \psi(aG + b\overline{G}) + \frac{\partial \phi}{\partial \bar{z}}F + \frac{\partial \psi}{\partial \bar{z}}G \\
&= aw + b\overline{w} + \frac{\partial \phi}{\partial \bar{z}}F + \frac{\partial \psi}{\partial \bar{z}}G \\
&= \frac{\partial w}{\partial \bar{z}} + \frac{\partial \phi}{\partial \bar{z}}F + \frac{\partial \psi}{\partial \bar{z}}G,
\end{aligned} \tag{11.1.34}$$

which implies the validity of Eq. (11.1.33). Now define the real-valued functions $\tau(z)$ and $\sigma(z) > 0$ by the relation

$$\sigma(z) - \mathrm{i}\tau(z) = \mathrm{i}\frac{F(z)}{G(z)}. \tag{11.1.35}$$

Then Eq. (11.1.33) becomes identical with Eqs. (11.1.25) and (11.1.26), hence justifying the nomenclature used so far.

Remark A holomorphic function is a function of complex variable which is continuous and admits first derivative. As was proved by [71], this is sufficient to prove continuity of the first derivative as well. One then thinks of a holomorphic function as a differentiable function whose real and imaginary parts obey the Cauchy-Riemann equations. Since the functions called pseudo-analytic in the literature obey, in the local theory, a generalization of the Cauchy-Riemann equations, as we have shown in detail, we prefer to call them, from now on, *pseudo-holomorphic functions*.

11.2   Global Theory of Pseudo-Holomorphic Functions

In his papers [18, 19] on the foundations for a general theory of pseudo-holomorphic functions, Caccioppoli began by stressing how impressive is the richness and variety of results one obtains by the only hypothesis of complex differentiability of a holomorphic function, i.e., from the geometric point of view, the conformal nature of the associated transformation in the plane. He therefore asked himself whether some of these results can be deduced from more general (i.e. qualitative) assumptions, according to which the Cauchy-Riemann equations expressing complex differentiability are replaced by differential inequalities (see below). The equalities among angles of the conformal representation would then be replaced by simple bounds, and the uniqueness of the analytic continuation would be lost. In return, a novel perspective on qualitative features would arise, including the global behaviour, the theorems of Picard, Schottky, Landau, Bloch, normal families and uniformization theory [18, 19]. The basic concepts one relies upon are as follows, while we refer the reader to the beautiful retirement address by Hedrick [77] for an even broader overview on generic functions (neither holomorphic nor pseudo-holomorphic) of complex variable.

We here denote again by u and v the real and imaginary part of a function

$$w(z) \equiv u(x, y) + \mathrm{i}v(x, y) \tag{11.2.1}$$

defined in a field A, a bounded open set of the z plane such that all its points are internal points. Let the functions $u(x, y)$, $v(x, y)$ satisfy the following assumptions:

(i) $u(x, y)$ and $v(x, y)$ are absolutely continuous in x and y for almost all values of y and x respectively, while their first derivatives with respect to x and y are square-integrable in every internal portion of A.

(ii) If

$$J = \frac{\partial(u, v)}{\partial(x, y)} = \frac{\partial u}{\partial x}\frac{\partial v}{\partial y} - \frac{\partial u}{\partial y}\frac{\partial v}{\partial x} \tag{11.2.2}$$

is the Jacobian of the map $(x, y) \to (u, v)$, $\Phi(x, y)$ is the upper limit (no confusion with the notation for the Laplacian should arise)

$$\Phi(x, y) \equiv \overline{\lim_{\Delta z \to 0}} \left| \frac{\Delta w}{\Delta z} \right|, \tag{11.2.3}$$

$\varphi(x, y)$ is the lower limit

$$\varphi(x, y) \equiv \underline{\lim_{\Delta z \to 0}} \left| \frac{\Delta w}{\Delta z} \right|, \tag{11.2.4}$$

then $J \geq 0$ almost everywhere in A, and there exists a positive real number $\mu \in]0, 1]$ such that

$$\varphi(x, y) \geq \mu \Phi(x, y) \tag{11.2.5}$$

almost everywhere in A. The function w is here said to be *globally pseudo-holomorphic*. Every value of $\mu \leq 1$ corresponds to a class C_μ of pseudo-holomorphic functions; in particular, if $\mu = 1$, C_1 is the class of holomorphic functions.

11.3 Upper and Lower Bound for the Increment Ratio

Outside of the holomorphic framework, the increment ratio $\frac{\Delta w}{\Delta z}$ has indeed a rich structure because, upon defining

$$m \equiv \frac{\Delta y}{\Delta x} \tag{11.3.1}$$

one has

$$\frac{\Delta w}{\Delta z} = \frac{\frac{\partial u}{\partial x} + i\frac{\partial v}{\partial x} + m\left(\frac{\partial u}{\partial y} + i\frac{\partial v}{\partial y}\right)}{(1 + im)} + \frac{(\varepsilon_1 + i\varepsilon_2) + m(\varepsilon_3 + i\varepsilon_4)}{(1 + im)}, \tag{11.3.2}$$

where ε_i tends to zero as $\Delta z \to 0$, for all $i = 1, 2, 3, 4$. Multiplication and division by $(1 - im)$ on the right-hand side yields therefore

$$\frac{\Delta w}{\Delta z} = A + iB + O(\varepsilon), \tag{11.3.3}$$

where

$$A \equiv \frac{\left(\frac{\partial u}{\partial x} + m\frac{\partial u}{\partial y}\right) + m\left(\frac{\partial v}{\partial x} + m\frac{\partial v}{\partial y}\right)}{(1 + m^2)}, \tag{11.3.4}$$

$$B \equiv \frac{\left(\frac{\partial v}{\partial x} + m\frac{\partial v}{\partial y}\right) - m\left(\frac{\partial u}{\partial x} + m\frac{\partial u}{\partial y}\right)}{(1 + m^2)}, \tag{11.3.5}$$

and hence

$$\lim_{\Delta z \to 0} \left|\frac{\Delta w}{\Delta z}\right|^2 = A^2 + B^2. \tag{11.3.6}$$

Now a patient calculation shows that, in the expression $(1 + m^2)^2(A^2 + B^2)$, a cancellation occurs and some reassembling can be made, so that eventually [77]

$$A^2 + B^2 = (1 + m^2)^{-1}(E + 2Fm + Gm^2) \equiv r(m), \tag{11.3.7}$$

where, according to a standard notation, we have set

$$E \equiv \left(\frac{\partial u}{\partial x}\right)^2 + \left(\frac{\partial v}{\partial x}\right)^2, \ G \equiv \left(\frac{\partial u}{\partial y}\right)^2 + \left(\frac{\partial v}{\partial y}\right)^2, \ F \equiv \frac{\partial u}{\partial x}\frac{\partial u}{\partial y} + \frac{\partial v}{\partial x}\frac{\partial v}{\partial y}. \tag{11.3.8}$$

To study the maximum or minimum of $A^2 + B^2$ as a function of the real variable m, we have to evaluate its first derivative, which vanishes if m solves the algebraic equation of second degree

$$Fm^2 + (E - G)m - F = 0, \tag{11.3.9}$$

solved by

$$m = -\frac{(E - G)}{2F} \pm \frac{1}{2F}\sqrt{(E - G)^2 + 4F^2} = m_{\pm}. \tag{11.3.10}$$

It is clear from Eq. (11.3.10) that for any generic E, F and G, m_+ is positive whereas m_- is negative. Once we take second derivative of the function $A^2 + B^2$ with respect to m, we obtain after some simplifications the expression

$$\frac{\partial^2}{\partial m^2}(A^2 + B^2) = -\frac{2F}{m(1 + m^2)} \quad \text{at } m = m_{\pm} \tag{11.3.11}$$

and hence it is clear that $\frac{\partial^2}{\partial m^2}(A^2 + B^2)|_{m_+} < 0$ whereas $\frac{\partial^2}{\partial m^2}(A^2 + B^2)|_{m_-} > 0$.
Interestingly, setting for convenience

$$\omega \equiv \sqrt{(E - G)^2 + 4F^2} = \sqrt{(E + G)^2 - 4J^2}, \tag{11.3.12}$$

one finds from (11.3.7) that [77]

$$r_+ = r(m_+) = \frac{E + 2Fm_+ + Gm_+^2}{(1 + m_+^2)} = \frac{(E + G)}{2} + \frac{\omega}{2}, \tag{11.3.13}$$

$$r_- = r(m_-) = \frac{E + 2Fm_- + Gm_-^2}{(1 + m_-^2)} = \frac{(E + G)}{2} - \frac{\omega}{2}. \tag{11.3.14}$$

In other words, the maximum and minimum values of r are themselves solutions of the quadratic equation [77]

$$\rho^2 - (E + G)\rho + J^2 = 0, \tag{11.3.15}$$

and the upper and lower limit (11.2.3) and (11.2.4) turn out to obey the identities [18, 19]

$$2\Phi^2 = E + G + \omega, \quad 2\varphi^2 = E + G - \omega, \tag{11.3.16}$$

$$\Phi^2 + \varphi^2 = E + G, \quad \Phi^2 \varphi^2 = \frac{1}{4}[(E + G)^2 - \omega^2] = J^2 = EG - F^2, \tag{11.3.17}$$

jointly with the inequalities

$$\mu J \le \varphi^2 \le \Phi^2 \le \frac{J}{\mu}, \quad J \ge \frac{\mu}{(1 + \mu^2)}(\Phi^2 + \varphi^2). \tag{11.3.18}$$

A *quasiconformal map* is a homeomorphism like the one in Eq. (11.2.1) with positive Jacobian (11.2.2). Such a map takes infinitesimal circles into infinitesimal ellipses, and is called quasiconformal if the eccentricity of these ellipses is uniformly bounded. This condition can be expressed by either of the three equivalent differential inequalities in (11.3.18). This property is conformally invariant: if $w = w(z)$ has it, so does the function $U(\zeta) = F\{w[f(\zeta)]\}$, where F and f are conformal mappings. Bers calls a function $w(z)$ quasi-conformal if it is of the form

$$w = f[\chi(z)], \tag{11.3.19}$$

where $\zeta = \chi(z)$ is a homeomorphism of $|z| \le 1$ onto $|\zeta| \le 1$ with $\chi(0) = 0, \chi(1) = 1$, which satisfies a uniform Hölder condition together with its inverse, and where $f(\zeta)$ is an analytic function of the complex variable $\zeta, |\zeta| < 1$.

11.4 Decomposition Theorem for Biharmonic Functions

The kernel of powers of the Laplacian is a topic with a well-established tradition both in pure mathematics and in mathematical physics [3, 6, 100], including the search for conformally invariant supplementary conditions in the gauge theories of fundamental interactions [51]. To begin with, we consider the biharmonic equation in $\mathbf{R}^3$, i.e.

$$\triangle \triangle u = 0. \tag{11.4.1}$$

Of course, every harmonic function is also biharmonic and, more generally, belongs to the kernel of arbitrary powers n of the Laplacian, whereas the converse does not hold. For a given choice of n, we shall look for functions which are in the kernel of Δ^n but not in the kernel of Δ^m, for all positive $m < n$.

Following Bergman and Schiffer [9], let φ and ψ be two arbitrary harmonic functions, then

$$f(P) = r^2 \varphi(P) + \psi(P), \quad r^2 \equiv \sum_{i=1}^{3}(x^i)^2, \tag{11.4.2}$$

is a solution of the biharmonic equation (11.4.1). Indeed, the action of Δ on f evaluated at the point P yields

$$\Delta f(P) = 6\varphi(P) + 4\sum_{i=1}^{3} x^i \frac{\partial \varphi(P)}{\partial x^i}, \tag{11.4.3}$$

where the expression on the right-hand side

$$\sum_{i=1}^{3} x^i \frac{\partial \varphi}{\partial x^i} = r \frac{\partial \varphi}{\partial r} \tag{11.4.4}$$

is a harmonic function if φ is harmonic, as one can prove by differentiation. Thus, Δf is a harmonic function, being the sum of two harmonic functions, and one can write

$$\Delta(\Delta f(P)) = 0. \tag{11.4.5}$$

Theorem 11.2 *Every biharmonic function can be written in the form* (11.4.2).

Proof For this purpose, given a biharmonic function $f(P)$, consider

$$D(P) \equiv \Delta f(P). \tag{11.4.6}$$

The function D is, by definition, a harmonic function, because

$$\Delta D(P) = \Delta^2 f(P) = 0. \tag{11.4.7}$$

By virtue of (11.4.3), one obtains from (11.4.6) the differential equation

$$6\varphi + 4r \frac{\partial \varphi}{\partial r} = D, \tag{11.4.8}$$

which has the particular integral

$$\varphi(P) = \frac{1}{4} r^{-\frac{3}{2}} \int_{0}^{r} \tau^{\frac{1}{2}} D(P(\tau)) \, d\tau, \tag{11.4.9}$$

with the understanding that the integration path is taken along a fixed radius vector. The harmonic nature of such a $\varphi(P)$ is proved by remarking that, for any harmonic function $D(P)$ (including our $D(P)$), the function

$$\Phi(P) \equiv r^{-k} \int_0^r \tau^{k-1} D(P(\tau)) \, d\tau, \ k > 0 \qquad (11.4.10)$$

is also harmonic. In order to prove this, it is sufficient to focus on homogeneous harmonic polynomials, which can be written in the form

$$H_n(P) = r^n h(\alpha, \beta), \qquad (11.4.11)$$

where α, β are the spherical coordinates on the unit 2-sphere. Every harmonic function can be indeed expanded into spherical harmonics $h(\alpha, \beta)$. Upon writing D in (11.4.10) as on the right-hand side of (11.4.11), one finds

$$\Phi_n = \frac{H_n}{(n+k)}, \qquad (11.4.12)$$

hence Φ_n is harmonic because H_n is harmonic, and our $\varphi(P)$ in (11.4.9) is harmonic with $k = \frac{3}{2}$ in (11.4.10). This implies in turn that the function ψ defined by

$$\psi(P) \equiv f(P) - r^2 \varphi(P) \qquad (11.4.13)$$

is harmonic because

$$\triangle \psi = D - \triangle(r^2 \varphi) = 0, \qquad (11.4.14)$$

by virtue of (11.4.9). The desired split (11.4.2) for biharmonic functions is therefore proved. $\qquad \square$

11.5 Boundary-Value Problems for the Biharmonic Equation

Hereafter, following again Bergman and Schiffer [9], we study functions u, v of class C^4 in a domain A of the Euclidean space $\mathbf{R}^2$, bounded by a curve $\gamma = \partial A$ where the functions of interest are of class C^3. On denoting by σ a real parameter in the closed interval $[-1, 1]$, one defines the bilinear integral

$$p(u, v) \equiv \int \int_A \left[(\triangle u)(\triangle v) - (1 - \sigma) \left(\frac{\partial^2 u}{\partial (x^1)^2} \frac{\partial^2 v}{\partial (x^2)^2} + \frac{\partial^2 u}{\partial (x^2)^2} \frac{\partial^2 v}{\partial (x^1)^2} \right. \right.$$
$$\left. \left. - 2 \frac{\partial^2 u}{\partial x^1 \partial x^2} \frac{\partial^2 v}{\partial x^1 \partial x^2} \right) \right] dx^1 \, dx^2. \qquad (11.5.1)$$

This integral defines therefore a bilinear form such that

$$p(u, v) = p(v, u), \quad p(u, u) \geq 0. \tag{11.5.2}$$

On denoting by $\mathbf{n}$ the inner normal along the boundary curve $\partial \mathcal{A}$, one can define the functionals

$$M(u) \equiv \sigma \, \triangle u + (1 - \sigma) \sum_{j,k=1}^{2} \cos(\mathbf{n}, \mathbf{x}^j) \cos(\mathbf{n}, \mathbf{x}^k) \frac{\partial^2 u}{\partial x^j \partial x^k} \tag{11.5.3}$$

and

$$V(u) \equiv \frac{\partial}{\partial n}(\triangle u) + (1 - \sigma)\frac{\partial}{\partial s}\left(\sum_{j,k=1}^{2} (-1)^k \cos(\mathbf{n}, \mathbf{x}^j) \cos(\mathbf{n}, \mathbf{x}^{3-k}) \frac{\partial^2 u}{\partial x^j \partial x^k} \right). \tag{11.5.4}$$

The functional $M(u)$ admits a physical interpretation as the bending moment around an axis which is tangent to the boundary curve $\partial \mathcal{A}$. The functional $V(u)$ is found to be equal to

$$V = X_3 - \frac{\partial M}{\partial s}, \tag{11.5.5}$$

where X_3 is the magnitude of the force on the edge of the plate acting in the x_3-direction, while M is the moment around an axis which is orthogonal to the boundary curve.

The conceivable boundary-value problems for the biharmonic equation consist in looking for solutions of the fourth-order equation (11.4.1) in the domain $\mathcal{A}$ which satisfy, on the boundary curve $\gamma = \partial \mathcal{A}$, one of the following four sets of conditions [9]:

$$u(P) = f_1(P), \quad \frac{\partial u}{\partial n}(P) = f_2(P), \tag{11.5.6}$$

$$u(P) = f_1(P), \quad M(u(P)) = f_2(P), \tag{11.5.7}$$

$$M(u(P)) = f_1(P), \quad V(u(P)) = f_2(P), \tag{11.5.8}$$

$$\frac{\partial u}{\partial n}(P) = f_1(P), \quad V(u(P)) = f_2(P). \tag{11.5.9}$$

If a solution of Eq. (11.4.1) with either of the boundary conditions (11.5.6)–(11.5.9) exists, the positive-definite character of the energy integral $p(u, u)$ may be applied to prove uniqueness theorems. One shows in this way that the solution is determined uniquely with (11.5.6) or (11.5.7), up to an arbitrary linear function with (11.5.8), and up to an arbitrary constant with (11.5.9).

Theorem 11.3 *If a solution u of the biharmonic equation (11.4.1) satisfies the homogeneous boundary conditions $u = 0$, $\frac{\partial u}{\partial n} = 0$ on the boundary curve γ, then u vanishes identically throughout the domain $\mathcal{A}$.*

Proof First, one applies the identity

$$p(u, v) = \int\int_{\mathcal{A}} (\Delta^2 u)v \; dx^1 \; dx^2 - \int_{\partial A} \left(M(u)\frac{\partial v}{\partial n} - V(u)v \right) ds \qquad (11.5.10)$$

with $v = u$. This leads to $p(u, u) = 0$, hence the resulting u is a harmonic function. Since it is taken to vanish on the boundary curve, u must indeed vanish identically in $\mathcal{A}$.

Theorem 11.4 *If u is a solution of the biharmonic equation (11.4.1) which on the boundary curve satisfies $u = 0$, $M(u) = 0$, then u vanishes identically in $\mathcal{A}$.*

Proof One follows the same method of Theorem 11.3, i.e. application of (11.5.10) with $v = u$ yields $p(u, u) = 0$, and u, being a harmonic function that vanishes on the boundary curve, must vanish identically in the domain $\mathcal{A}$.

Theorem 11.5 *If u solves the biharmonic equation (11.4.1) and satisfies on the boundary curve $M(u) = 0$, $V(u) = 0$, then, if $\sigma \in [-1, 1[$, u must be a linear function, whereas, if $\sigma = 1$, u is an arbitrary harmonic function.*

Proof In either case for σ, one has $p(u, u) = 0$. If $\sigma \in [-1, 1[$, u must be linear, whereas if one takes for u any linear function, then $M(u) = 0$, $V(u) = 0$. Moreover, if $\sigma = 1$, u may be harmonic, and if u is any harmonic function one has $M(u) = 0$ and $V(u) = 0$.

Theorem 11.6 *If u is a solution of the biharmonic equation which on the boundary curve satisfies the homogeneous conditions $\frac{\partial u}{\partial n} = 0$, $V(u) = 0$, then the function u reduces to a constant.*

Proof Since u must be harmonic, the vanishing of its normal derivative on the boundary curve γ implies that u is a constant, and for an arbitrary constant the condition $V(u) = 0$ is indeed satisfied on γ.

11.6 Fundamental Solution of the Biharmonic Equation

Since the integral $p(u, v)$ is symmetric, i.e. $p(u, v) = p(v, u)$, by virtue of the identity (11.5.10) one obtains the Rayleigh-Green identity

$$\int\int_{\mathcal{A}} \left((\Delta^2 u)v - u \, \Delta^2 v \right) dx^1 \; dx^2$$
$$= \int_{\partial A} \left(M(u)\frac{\partial v}{\partial n} - V(u)v - M(v)\frac{\partial u}{\partial n} + V(v)u \right) ds, \qquad (11.6.1)$$

which holds for all functions u and v which are of class C^4 in $\mathcal{A}$ and of class C^3 on $\partial\mathcal{A}$. Now one sets

$$S(P, Q) \equiv r^2 \log(r), \quad r^2 \equiv (x^1 - y^1)^2 + (x^2 - y^2)^2, \quad P \equiv (x^1, x^2), \quad Q \equiv (y^1, y^2).$$
$$(11.6.2)$$

When viewed as a function of each of its points, S is regular except for $P = Q$. Furthermore, differentiation shows that

$$\triangle S = 4(\log(r) + 1), \quad \triangle^2 S = 0, \quad \forall P \neq Q. \tag{11.6.3}$$

In light of the theorem of Sect. 11.4 on the general form of biharmonic functions, the form of $S(P, Q)$ is easily understood. Following the nomenclature used by Bergman and Schiffer [9], we say that S is a *fundamental singularity*. The identity (11.6.1) and the fundamental singularity S make it possible to express a biharmonic function in terms of the boundary data u, $\frac{\partial u}{\partial n}$, $V(u)$ and $M(u)$.

Theorem 11.7 *For all biharmonic functions for which the identity (11.6.1) holds, one has*

$$u(Q) = \frac{1}{8\pi} \int \mathcal{A} \left(-M_P(u(P)) \frac{\partial S(P, Q)}{\partial n_P} + V_P(u(P)) S(P, Q) \right.$$
$$\left. + M_P(S(P, Q)) \frac{\partial u(P)}{\partial n_P} - V_P(S(P, Q)) u(P) \right) ds_P. \tag{11.6.4}$$

Proof Let us remove a small circle of radius r centred at Q from the domain $\mathcal{A}$, denoting the circumference by C_r. From the biharmonic equation (11.4.1), and from the identity (11.6.1) where v is set equal to the fundamental singularity S,

$$\int_{\mathcal{A}+C_r} \left(-M(u) \frac{\partial S}{\partial n} + V(u)S + M(S) \frac{\partial u}{\partial n} - V(S)u \right) ds = 0. \tag{11.6.5}$$

One can now check from (11.6.2) that

$$\frac{\partial^2 S}{\partial x^i \partial x^j} = 2 \frac{(x_i - y_i)(x_j - y_j)}{r^2} \quad \text{if } i \neq j, \tag{11.6.6}$$

and hence the formulae (11.5.3) for $M(u)$ and (11.5.4) for $V(u)$ imply that

$$\lim_{r \to 0} \int_{C_r} V(S)u \, ds = 8\pi u(Q), \tag{11.6.7}$$

while the limits of the integrals around C_r of the other three terms of (11.6.5) are each zero.

In general, a solution $G(P, Q)$ of the biharmonic equation (11.4.1) is said to be a *fundamental solution* if $G(P, Q) + S(P, Q)$ is a regular solution of (11.4.1) in

$\mathcal{A}$ when viewed either as a function of the first argument P or as a function of the second argument Q. Thus, the integral representation (11.6.4) remains valid when $S(P, Q)$ is replaced by any fundamental solution $G(P, Q)$.

It can be proved that, if the domain $\mathcal{A}$ has a sufficiently smooth boundary, there exists a fundamental solution $G(P, Q)$ which satisfies the two boundary conditions

$$G(P, Q) = 0, \quad \frac{\partial G(P, Q)}{\partial n_p} = 0, \quad Q \in \mathcal{A}, \ P \in \partial \mathcal{A}. \tag{11.6.8}$$

The fundamental solution is symmetric, i.e. $G(P, Q) = G(Q, P)$, and, upon inserting the boundary conditions (11.6.8) into the integral representation (11.6.4), two terms drop out and one finds

$$u(Q) = \frac{1}{8\pi} \int_{\partial \mathcal{A}} \left(M_P(G(P, Q)) \frac{\partial u(P)}{\partial n} - V_P(G(P, Q)) u(P) \right) ds_P. \tag{11.6.9}$$

In particular, for the unit circle, the Green's function is given by

$$G(z_1, z_2) = |z_1 - z_2|^2 \log \left| \frac{\bar{z}_1 z_2 - 1}{z_1 - z_2} \right| - \frac{1}{2} \left(|z_1|^2 - 1 \right) \left(|z_2|^2 - 1 \right). \tag{11.6.10}$$

11.7 Mean-Value Property for Polyharmonic Functions

In Sect. 3.4 we proved the standard mean-value theorem for harmonic functions, and now, following the work of [97], we are going to state how these properties are extended to polyharmonic functions.

Let Ω be an open set of $\mathbf{R}^n$, x a point of Ω and $R \in]0, \text{dist}(x, \partial \Omega)[$. The *solid* and *spherical means* of a continuous function $u \in C^0(\Omega)$ are defined by

$$M(u; x, R) \equiv \frac{1}{\sigma(n) R^n} \int_{B(x, R)} u(y) dy, \tag{11.7.1}$$

and

$$N(u; x, R) \equiv \frac{1}{n \sigma(n) R^{n-1}} \int_{S(x, R)} u(y) dS(y), \tag{11.7.2}$$

respectively, where $\sigma(n)$ is the volume of the unit ball in $\mathbf{R}^n$, equal, as we know, to $\frac{\pi^{\frac{n}{2}}}{\Gamma(\frac{n}{2}+1)}$, while dS denotes the surface measure on $S(x, R)$. There exist relations between solid and spherical means, made precise by the following result

Lemma 11.1 *Let u be a continuous function on Ω. Then for any $x \in \Omega$ and R in the open interval $]0, \text{dist}(x, \partial \Omega)[$, one has*

$$\left(\frac{R}{n}\frac{\partial}{\partial R} + 1\right) M(u; x, R) = N(u; x, R). \tag{11.7.3}$$

Moreover, if u is of class C^2 on Ω, one has

$$\frac{n}{R}\frac{\partial}{\partial R} N(u; x, R) = M(\triangle u; x, R). \tag{11.7.4}$$

The proof of (11.7.3) is obtained by the evaluation of $M(u; x, R)$ in spherical coordinates, whereas the proof of (11.7.4) results from the Green formula [96].

Let now m be a positive integer. A function u of class C^{2m} on Ω is said to be *m-harmonic* on Ω if it is a solution of the equation

$$\triangle^m u = \triangle(\triangle^{m-1} u) = 0. \tag{11.7.5}$$

These polyharmonic functions belong therefore, by definition, to the kernel of positive powers of the Laplacian on open sets of $\mathbf{R}^n$. Two theorems are of interest for us, and are as follows.

Theorem 11.8 *Mean-value property: Let m be a positive or zero integer. If the function u of class $(2m + 2)$ on Ω is $(m + 1)$-harmonic on Ω, then for any $x \in \Omega$ and R in the open interval $]0, \text{dist}(x, \partial\Omega)[$ one has [96, 106]*

$$M(u; x, R) = \sum_{k=0}^{m} \frac{\triangle^k u(x)}{4^k \left(\frac{n}{2} + 1\right)_k k!} R^{2k}, \tag{11.7.6}$$

and

$$N(u; x, R) = \sum_{k=0}^{m} \frac{\triangle^k u(x)}{4^k \left(\frac{n}{2}\right)_k k!} R^{2k}, \tag{11.7.7}$$

where

$$(a)_k \equiv a(a + 1) \cdots (a + k - 1) \tag{11.7.8}$$

is the standard notation for the Pochhammer symbol.

Theorem 11.9 *Converse to the mean-value property: Let m be a positive or zero integer. If the function u is of class C^{2m} on Ω and, for all $x \in \Omega$ and R small enough, the formula (11.7.6) or (11.7.7) holds, then [96, 106]*

$$\triangle^{m+1} u = 0 \text{ in } \Omega. \tag{11.7.9}$$

Interestingly, the mean-value characterization of harmonic functions can be used to define harmonic functions on metric measure spaces. In other words, if (X, ρ, μ) is a metric measure space with a metric ρ and a Borel regular measure μ which is positive on open sets and finite on bounded sets, then a continuous function $u : \Omega \to \mathbf{R}$ on an open set $\Omega \subset X$ is said to be *harmonic* on Ω if, for every $x \in \Omega$ and any closed ball $B(x, R) \subset \Omega$ one has [66]

$$u(x) = \frac{1}{\mu(B(x, R))} \int_{B(x,R)} u(y) d\mu(y). \qquad (11.7.10)$$

If the measure is continuous with respect to the metric, then harmonic functions on metric measure spaces satisfy the maximum principle (cf. Sect. 3.5) and other important properties [66]. Harmonic functions can be defined also on groups, homogeneous spaces and graphs [26, 144].

On a metric measure space (X, ρ, μ) as above, if Ω is an open subset of X, the *solid mean* of a continuous function on Ω is defined, for any $x \in \Omega$ and $R \in]0, \text{dist}(x, \partial\Omega)[$, by [97]

$$M_X(u; x, R) \equiv \frac{1}{\mu(B_\rho(x, R))} \int_{B_\rho(x,R)} u(y) d\mu(y). \qquad (11.7.11)$$

The function $u \in C^0(\Omega)$ is said to be (X, ρ, μ)-polyharmonic on Ω if there exist functions $u_k \in C^0(\Omega)$ for $k = 0, 1, \dots, m$ with some positive integer m and $\varepsilon \in C^0(\Omega, \mathbf{R}_+)$ such that [97]

$$M_X(u; x, R) = \sum_{k=0}^{m} u_k(x) R^k, \qquad (11.7.12)$$

for any $x \in \Omega$ and $R \in]0, \varepsilon(x)[$.

Part III
Calculus of Variations

Chapter 12
The Euler Equations

12.1 Statement of the Problem

Following [36], we shall develop the theory on the real line, although many of the results can be extended to $\mathbf{R}^n$. Let T be the open interval $]a, b[$ contained in $\mathbf{R}$, and assume we are given a function $f(x, y, y')$ defined on the Cartesian product $T \times \mathbf{R} \times \mathbf{R}$, taken to be of class C^∞ for the time being. One then considers the functional

$$I(y) \equiv \int_a^b f(x, y(x), y'(x))dx \qquad (12.1.1)$$

defined for the functions $y \in C^1(T)$, where $C^1(T)$ is the space of functions continuous on the closure $\overline{T}$ of T, having first derivative continuous on T, i.e.

$$C^1(T) \equiv \{ f \in C^0(\overline{T}) : f' \in C^0(T) \}. \qquad (12.1.2)$$

Let $\mathcal{F}$ be the class of functions y such that

$$\mathcal{F} \equiv \{ y \in C^1(T) : y(a) = c_1 \in \mathbf{R}, \ y(b) = c_2 \in \mathbf{R} \}. \qquad (12.1.3)$$

One con then consider the following problems:

Problem (a) To prove the existence of a function $\overline{y}$ in the class $\mathcal{F}$ that minimizes the functional I of Eq. (12.1.1).

Problem (b) Uniqueness of such a function.

Problem (c) Regularization of such a function.

© Springer International Publishing AG 2017
G. Esposito, *From Ordinary to Partial Differential Equations*, UNITEXT 106,
DOI 10.1007/978-3-319-57544-5_12

12.2   The Euler Integral Condition

For the time being, let us look only for necessary conditions for a function $\overline{y}$ of the class $\mathcal{F}$ to be minimizing, in other words such that

$$I(\overline{y}) \leq I(y), \ \forall y \in \mathcal{F}. \tag{12.2.1}$$

This inequality is equivalent to imposing that, having defined

$$C_0^1(T) \equiv \left\{ y \in C^1(T) : y(a) = y(b) = 0 \right\}, \tag{12.2.2}$$

one can write

$$I(\overline{y}) \leq I(\overline{y} + w), \ \forall w \in C_0^1(T). \tag{12.2.3}$$

Upon fixing now a certain $w \in C_0^1(T)$, one finds that

$$I(\overline{y} + \lambda w) = \int_a^b f(x, \overline{y} + \lambda w, \overline{y}' + \lambda w') dx \tag{12.2.4}$$

is a function continuous and differentiable of the real variable λ. Thus, if $\overline{y}$ is minimizing, such a function has a minimum at $\lambda = 0$; it is therefore a necessary condition that

$$\frac{d}{d\lambda} I(\overline{y} + \lambda w) = 0 \text{ when } \lambda = 0. \tag{12.2.5}$$

By virtue of the identity

$$\frac{d}{d\lambda} I(\overline{y} + \lambda w) = \int_a^b \left[\frac{\partial f}{\partial y}(x, \overline{y} + \lambda w, \overline{y}' + \lambda w')w + \frac{\partial f}{\partial y'}(x, \overline{y} + \lambda w, \overline{y}' + \lambda w')w' \right] dx, \tag{12.2.6}$$

it turns out that the previous condition can be written in the form

$$\int_a^b \left[\frac{\partial f}{\partial y}(x, \overline{y}, \overline{y}')w + \frac{\partial f}{\partial y'}(x, \overline{y}, \overline{y}')w' \right] dx = 0, \tag{12.2.7}$$

for all $w \in C_0^1(T)$.

Equation (12.2.7) is known as the *Euler integral condition*. It is a necessary condition for the problem (a) when the class $\mathcal{F}$ consists of functions belonging to the space $C^1(T)$ defined in (12.1.2).

12.3 The Euler Differential Condition

In particular, if the class $\mathcal{F}$ is included in $C^2(T)$, one can derive from Eq. (12.2.7) a pointwise condition equivalent to it. In such a case, indeed, the function of the x variable

$$f(x, \overline{y}(x), \overline{y}'(x))$$

is of class C^1 on T, so that, integrating by parts and remembering that w vanishes at the ends of the interval, one obtains

$$\int_a^b w(x) \left\{ \frac{\partial f}{\partial y}(x, \overline{y}(x), \overline{y}'(x)) - \frac{d}{dx} \left[\frac{\partial f}{\partial y'}(x, y(x), y'(x)) \right] \right\} dx = 0, \quad (12.3.1)$$

for all $w \in C_0^1(T)$, and hence, given the arbitrariness of w, one finds eventually

$$\frac{\partial f}{\partial y}(x, \overline{y}(x), \overline{y}'(x)) = \frac{d}{dx} \left[\frac{\partial f}{\partial y'}(x, \overline{y}(x), \overline{y}'(x)) \right], \quad (12.3.2)$$

for all $x \in T$. Equation (12.3.2) is known as the *Euler differential condition*; the functions $\overline{y} \in \mathcal{F}$ which satisfy this equation are said to be *minimal* (recall, however, that Eq. (12.3.2) is only a necessary condition).

Let us consider now the second derivative of the function (12.2.4). Bearing in mind the expression of the first derivative, and the fact that, when $\lambda = 0$, one has a minimum, one obtains

$$\frac{d^2}{d\lambda^2} \left[I(\overline{y} + \lambda w) \right]_{\lambda=0} = \int_a^b \left\{ \frac{\partial^2 f}{\partial y'^2}(x, \overline{y}(x), \overline{y}'(x))(w'(x))^2 \right.$$
$$+ \frac{\partial^2 f}{\partial y \partial y'}(x, \overline{y}(x), \overline{y}'(x))w(x)w'(x)$$
$$\left. + \frac{\partial^2 f}{\partial y^2}(x, \overline{y}(x), \overline{y}'(x))w^2(x) \right\} dx \geq 0, \quad (12.3.3)$$

for every function $w \in C_0^1(T)$. It is therefore useful to study functionals of the kind

$$I_w \equiv \int_a^b \left[\alpha(x)(w'(x))^2 + \beta(x)w(x)w'(x) + \gamma(x)w^2(x) \right] dx, \quad (12.3.4)$$

for $w \in C_0^1(T)$. Before doing that, it is appropriate to consider another kind of problem, i.e. a variational problem with constraints.

12.4   Variational Problem with Constraints

Given the functional $I(y)$ of Eq. (12.1.1), we here look for functions that minimize $I(y)$ within the framework of functions of class $C_0^1(T)$ such that also the following condition (constraint) is verified:

$$J(y) \equiv \int_a^b g(x, y(x), y'(x)) \mathrm{d}x = 0, \qquad (12.4.1)$$

where g is a sufficiently regular function. We now provide necessary conditions for a function y to solve the problem with constraints; such conditions correspond to the Euler integral and differential condition, respectively.

Suppose that $\overline{y}$ is a minimizing function for the problem: let v_1 and v_2 be two functions of class $C_0^1(T)$, and $(\lambda_1, \lambda_2) \equiv \lambda$ a pair of real numbers such that, having defined

$$y_\lambda \equiv \overline{y} + \lambda_1 v_1 + \lambda_2 v_2, \qquad (12.4.2)$$

one finds

$$J(y_\lambda) = 0. \qquad (12.4.3)$$

By definition of minimizing function, and writing $y_\lambda' \equiv \frac{\mathrm{d}}{\mathrm{d}x} y_\lambda$, one then finds that

$$\int_a^b f(x, y_\lambda, y_\lambda') \mathrm{d}x \geq \int_a^b f(x, \overline{y}, \overline{y}') \mathrm{d}x. \qquad (12.4.4)$$

Let us now set

$$K_i \equiv \left[\frac{\partial J(y_\lambda)}{\partial \lambda_i} \right]_{\lambda=0}, \quad i = 1, 2. \qquad (12.4.5)$$

Two circumstances may then occur:

$$(K_1)^2 + (K_2)^2 = 0, \qquad (12.4.6)$$

or

$$(K_1)^2 + (K_2)^2 > 0. \qquad (12.4.7)$$

Let us consider for the time being the second case, i.e. (12.4.7); let us suppose, to fix the ideas, that K_1 does not vanish. One can then solve the implicit function $J(y_\lambda) = 0$, expressing λ_1 as a function of λ_2 in a neighbourhood

$$\{|\lambda| < \delta\}.$$

If $\lambda_1 = \varphi(\lambda_2)$, one then obtains

$$\left[\varphi'(\lambda_2)\right]_{\lambda_2=0} = -\frac{K_2}{K_1}. \tag{12.4.8}$$

By replacing, in Eq. (12.4.2), λ_1 with its value $\varphi(\lambda_2)$, and noting also that $\lambda_1 = \varphi(\lambda_2) = 0$ if $\lambda_2 = 0$, because y is minimizing, one has to find

$$\left[\frac{\mathrm{d}}{\mathrm{d}\lambda_2} I\left(y_{\varphi(\lambda_2),\lambda_2}\right)\right]_{\lambda_2=0} = 0. \tag{12.4.9}$$

In other words, one obtains

$$\left[-\frac{\partial}{\partial\lambda_1} I(y_\lambda)\frac{K_2}{K_1} + \frac{\partial}{\partial\lambda_2} I(y_\lambda)\right]_{\lambda=0} = 0. \tag{12.4.10}$$

In light of the definition (12.4.5), one can re-express this equation in the form

$$\left[\frac{\partial}{\partial\lambda_1} I(y_\lambda)\frac{\partial}{\partial\lambda_2} J(y_\lambda) - \frac{\partial}{\partial\lambda_2} I(y_\lambda)\frac{\partial}{\partial\lambda_1} J(y_\lambda)\right]_{\lambda=0} = 0. \tag{12.4.11}$$

We note that the condition (12.4.11) is also satisfied if (12.4.6) holds. The condition (12.4.11) is equivalent to the vanishing of the determinant

$$\det\begin{pmatrix} \frac{\partial I(y_\lambda)}{\partial\lambda_1} & \frac{\partial I(y_\lambda)}{\partial\lambda_2} \\ \frac{\partial J(y_\lambda)}{\partial\lambda_1} & \frac{\partial J(y_\lambda)}{\partial\lambda_2} \end{pmatrix} = 0 \text{ at } \lambda = 0. \tag{12.4.12}$$

This implies that there exists a non-zero vector $\mu \equiv (\mu_1, \mu_2)$ such that one can write

$$\mu_1\left[\frac{\partial I}{\partial\lambda_1}\right]_{\lambda=0} + \mu_2\left[\frac{\partial J}{\partial\lambda_1}\right]_{\lambda=0} = 0, \tag{12.4.13}$$

$$\mu_1\left[\frac{\partial I}{\partial\lambda_2}\right]_{\lambda=0} + \mu_2\left[\frac{\partial J}{\partial\lambda_2}\right]_{\lambda=0} = 0. \tag{12.4.14}$$

By writing explicitly the relations (12.4.13), (12.4.14), and recalling the linear combination introduced in (12.4.2), one finds

$$\int_a^b \left\{\mu_1\left[v_1\frac{\partial f}{\partial y}(\overline{y}) + v_1'\frac{\partial f}{\partial y'}(\overline{y})\right] + \mu_2\left[v_1\frac{\partial g}{\partial y}(\overline{y}) + v_1'\frac{\partial g}{\partial y'}(\overline{y})\right]\right\} \mathrm{d}x = 0, \tag{12.4.15}$$

$$\int_a^b \left\{\mu_1\left[v_2\frac{\partial f}{\partial y}(\overline{y}) + v_2'\frac{\partial f}{\partial y'}(\overline{y})\right] + \mu_2\left[v_2\frac{\partial g}{\partial y}(\overline{y}) + v_2'\frac{\partial g}{\partial y'}(\overline{y})\right]\right\} \mathrm{d}x = 0. \tag{12.4.16}$$

If the characteristic of the matrix in Eq. (12.4.12) is 0, then μ_1 and μ_2 can be chosen arbitrarily, and hence y must satisfy the Euler equations relative to the function f and to the function g, separately.

If instead the characteristic of (12.4.12) is 1, the pair (μ_1, μ_2) is then determined up to a multiplicative factor. One of the two Eqs. (12.4.15) and (12.4.16) is satisfied for all $v \in C_0^1(T)$; for example, if v_1 is chosen arbitrarily, v_2 is determined by Eq. (12.4.3). Thus, the necessary condition for the function y to be minimizing for the variational problem with constraints is that it satisfies the Euler integral equation (12.2.7) for the functional

$$H(y) \equiv \mu I(y) + J(y), \qquad (12.4.17)$$

where we have set in this case $\mu_2 = 1$ and $\mu_1 = \mu$. One can proceed along the same lines when several constraints are imposed.

Chapter 13
Classical Variational Problems

13.1 Isoperimetric Problems

The problems with constraints studied in Sect. 12.4 can be described as follows. One studies the functional $I(y)$ of Eq. (12.1.1), and one tries to determine the function y, hence the curve Γ joining A to B, in such a way that I achieves a minimum, but considering the curves Γ for which, $g(x, y, y')$ being another given function and C a constant, one has

$$\int_a^b g(x, y, y') = C. \tag{13.1.1}$$

The function g is taken to have the same properties of f. Let us assume that the problem has been solved, and consider the family of 1-parameter curves Γ' depending on a parameter λ, that reduces, for $\lambda = 0$, to Γ. The ordinate of a point of Γ' is $y = \varphi(x, \lambda)$, and the associated functional is

$$I(\lambda) = \int_{a'}^{b'} f(x, \varphi(x, \lambda), \varphi_x'(x, \lambda)) dx. \tag{13.1.2}$$

For the curves Γ', $I(\lambda)$ must have a vanishing derivative at $\lambda = 0$. On the other hand, for the curves we are looking for, the derivative of the left-hand side of (12.1.1), which is constant as a function of λ, must vanish. One has therefore, after integration by parts,

$$J(\delta y) \equiv \int_a^b \delta y \left[\frac{\partial f}{\partial y} - \frac{\mathrm{d}}{\mathrm{d}x} \left(\frac{\partial f}{\partial y'} \right) \right] \mathrm{d}x = 0, \tag{13.1.3}$$

for every function δy satisfying

$$\delta y \in C^1(a, b), \quad \delta y_a = \delta y(x = a) = 0, \quad \delta y_b = \delta y(x = b) = 0. \tag{13.1.4}$$

© Springer International Publishing AG 2017
G. Esposito, *From Ordinary to Partial Differential Equations*, UNITEXT 106,
DOI 10.1007/978-3-319-57544-5_13

183

We denote by y the ordinate of a point of Γ, and write $\delta y' \equiv \frac{d}{dx}\delta y$. The function f is still defined for the values x, $y + \lambda(\delta y)$, $y' + \lambda(\delta y')$, for all $\lambda \in (-1, 1)$, if Eq. (13.1.1) holds, therefore if

$$K(\delta y) \equiv \int_a^b \delta y \left[\frac{\partial g}{\partial y} - \frac{d}{dx}\left(\frac{\partial g}{\partial y'} \right) \right] dx = 0. \qquad (13.1.5)$$

Let us denote by δy_1 a first function δy for which $K(\delta y_1) \neq 0$ [134]. Such a function exists if Γ is not an extremal of the variational problem relative to g. The function δy being now arbitrary, but verifying the conditions (13.1.4), let us take

$$\varphi(x, \lambda) = y + \lambda(\delta y - K(\lambda)\delta y_1), \qquad (13.1.6)$$

where $K(\lambda)$ is obtained by the condition (13.1.1). The value $K(0)$ is given by the equality $K(\delta y) = \rho K(\delta y_1)$. One has therefore to find

$$J(\delta y) - \rho J(\delta y_1) = 0 \Longrightarrow J(\delta y) - \frac{J(\delta y_1)}{K(\delta y_1)} K(\delta y) = 0. \qquad (13.1.7)$$

As a consequence, for any value taken by δy, one finds

$$\int_a^b \delta y \left[\frac{\partial(f + \xi g)}{\partial y} - \frac{d}{dx}\frac{\partial(f + \xi g)}{\partial y'} \right] dx = 0, \ \xi = -\frac{J(\delta y_1)}{K(\delta y_1)}. \qquad (13.1.8)$$

One has to write the extremality conditions for the integral carrying over $f + \xi g$. This remains true when Γ is an extremal of the problem relative to g. For the problem to be solvable, the curve Γ must be an integral curve of the Euler equation

$$\frac{\partial(f + \xi g)}{\partial y} - \frac{d}{dx}\frac{\partial(f + \xi g)}{\partial y'} = 0, \qquad (13.1.9)$$

where ξ is now an unknown constant. If the curve Γ solves this equation, joins A to B and verifies the condition (13.1.1), the equality (13.1.3) holds for all functions δy for which $K(\delta y)$ vanishes. Since the integral curves of Eq. (13.1.9) depend on three constants, i.e. ξ and the two integration constants, the first-order problem can be treated in some cases.

The *isoperimetric problem* in the strict sense of the word, which has given its name to all other problems, consists in looking for a curve Γ joining the points A and B taken on Ox ($x = a$, $y = 0$; $x = b$, $y = 0$) having a given length L and such that the area of the domain bounded by Ox and Γ is maximal. We assume that Γ intersects a parallel to Oy only at a point, and that it has a tangent varying continuously. One has here

$$I = \int_a^b y \, dx, \ L = \int_a^b \sqrt{1 + (y')^2} \, dx. \qquad (13.1.10)$$

The equation determining the curves Γ is

$$1 + \xi \frac{d}{dx} \frac{y'}{\sqrt{1 + (y')^2}} = 0, \tag{13.1.11}$$

which admits a first integral

$$\frac{y'}{\sqrt{1 + (y')^2}} = -\frac{x}{\xi} + \lambda. \tag{13.1.12}$$

The result of the integration is

$$(y - \mu\xi)^2 + (x - \lambda\xi)^2 = \xi^2, \tag{13.1.13}$$

where λ, μ, ξ are some constants. Γ is therefore the arc of a circle. The problem here formulated is well posed provided that $L > (b - a)$ and $2L < \pi(b - a)$, because one has to join A and B by an arc of circle cut at only one point by a line parallel to Oy. But if one removes the condition that Γ should intersect a line parallel to Oy at only one point and if, assuming that the problem has been solved, one considers a line intersecting Γ at two consecutive points C and D, it is necessary that the arc CD solves the isoperimetric problem with the curves joining C to D and whose length is given. If the lines orthogonal to the string CD intersect the arc CD at only one point, this arc must be the arc of a circle. One can deduce from this that the whole arc AB is an arc of circle. Strictly, it remains to be seen whether the arc of circle provides truly the maximum.

Another example is obtained by assigning a curve Γ of length L joining A with B, and looking for the minimum of the functional

$$\int_a^b y \, ds,$$

ds being the arc element. This is equivalent to looking for the homogeneous curve having weight and of given length, whose centre of gravity is as low as possible, i.e. the equilibrium curve of a homogeneous heavy wire. One has here

$$\int_a^b ds = L. \tag{13.1.14}$$

The equation is obtained on writing that

$$\int_a^b (y + \xi) ds$$

is extremum. The resulting curve is a little chain whose base is parallel to Ox.

13.2   Double Integrals and Minimal Surfaces

Let us assume that we are given a function $f(x, y, z, p, q)$ of class C^2 in a certain domain, and consider the integral

$$I = \int \int_A f(x, y, z, p, q)\mathrm{d}x\,\mathrm{d}y, \qquad (13.2.1)$$

where A is a domain of the Oxy plane bounded by a simple closed curve C, z being the altitude $z = \varphi(x, y)$ of a point of a surface S, and p and q being the partial derivatives

$$p \equiv \frac{\partial \varphi}{\partial x}, \quad q \equiv \frac{\partial \varphi}{\partial y}. \qquad (13.2.2)$$

The surface S has to pass through a given curve C', whose projection on Oxy is C. One tries to determine the surface S in such a way that I is a maximum or a minimum. Continuity of p and q is assumed.

On assuming that a surface S exists which fulfills these requirements, a family of surfaces are introduced, defined by

$$Z = z + \lambda\,\delta z, \qquad (13.2.3)$$

z being the altitude on S, λ a parameter and δz an arbitrary function vanishing upon C. The derivative of the corresponding integral

$$I(\lambda) = \int \int_A f(x, y, z + \lambda\,\delta z, p + \lambda\,\delta p, q + \lambda\,\delta q)\mathrm{d}x\,\mathrm{d}y \qquad (13.2.4)$$

must vanish at $\lambda = 0$. The terms δp and δq are the partial derivatives

$$\delta p \equiv \frac{\partial}{\partial x}\delta z, \quad \delta q \equiv \frac{\partial}{\partial y}\delta z. \qquad (13.2.5)$$

One has therefore the condition

$$\int \int_A \left[\frac{\partial f}{\partial z}\delta z + \frac{\partial f}{\partial p}\delta p + \frac{\partial f}{\partial q}\delta q \right] \mathrm{d}x\,\mathrm{d}y = 0, \qquad (13.2.6)$$

the partial derivatives in the integrand being evaluated at the point x, y, z, p, q of S. The terms involving δp and δq are transformed by applying the Riemann-Green formula. For example, one has

$$\int_{C^+} \left(\frac{\partial f}{\partial p}\delta z \right) \mathrm{d}y = \int \int_A \left[\frac{\partial f}{\partial p}\delta p + \delta z \frac{\mathrm{d}}{\mathrm{d}x}\left(\frac{\partial f}{\partial p} \right) \right] \mathrm{d}x\,\mathrm{d}y. \qquad (13.2.7)$$

The left-hand side vanishes because $\delta z = 0$ on C, and hence

$$\int\int_A \left[\frac{\partial f}{\partial p}\delta p\right] dx\, dy = -\int\int_A \left[\delta z \frac{d}{dx}\left(\frac{\partial f}{\partial p}\right)\right] dx\, dy. \tag{13.2.8}$$

In analogous fashion, one finds

$$\int\int_A \left[\frac{\partial f}{\partial q}\delta q\right] dx\, dy = -\int\int_A \left[\delta z \frac{d}{dy}\left(\frac{\partial f}{\partial q}\right)\right] dx\, dy, \tag{13.2.9}$$

and therefore Eq. (13.2.6) takes the form

$$\int\int_A \left[\frac{\partial f}{\partial z} - \frac{d}{dx}\frac{\partial f}{\partial p} - \frac{d}{dy}\frac{\partial f}{\partial q}\right] \delta z\, dx\, dy = 0. \tag{13.2.10}$$

Since this equation is required to hold for all functions δz, it implies the local equation (cf. Chap. 12)

$$\frac{\partial f}{\partial z} - \frac{d}{dx}\frac{\partial f}{\partial p} - \frac{d}{dy}\frac{\partial f}{\partial q} = 0. \tag{13.2.11}$$

In other terms, if the problem we have formulated admits a solution, such a solution can only be found by means of the surfaces solving the partial differential equation (13.2.11). One has therefore to understand whether such a surface can intersect the given curve C', a problem much more difficult than the one of having to let an extremal curve pass through two points.

Equation (13.2.11) is the Euler-Lagrange equation of the problem. It is an equation of the Monge-Ampère type. The surfaces that solve this equation are said to be *extremal surfaces*. In particular, if one tries to minimize the area of the surface S passing through a closed curve C', one has to consider the integral

$$I = \int\int_A \sqrt{1 + p^2 + q^2}\, dx\, dy. \tag{13.2.12}$$

The associated Euler-Lagrange equation is in this case

$$\frac{d}{dx}\frac{p}{\sqrt{1 + p^2 + q^2}} + \frac{d}{dy}\frac{q}{\sqrt{1 + p^2 + q^2}} = 0, \tag{13.2.13}$$

and, upon expanding and using the Monge notation, according to which

$$r \equiv \frac{\partial^2 \varphi}{\partial x^2}, \quad s \equiv \frac{\partial^2 \varphi}{\partial x \partial y}, \quad t \equiv \frac{\partial^2 \varphi}{\partial y^2}, \tag{13.2.14}$$

one obtains the partial differential equation (e.g. [134])

$$r(1 + q^2) - 2pqs + t(1 + p^2) = 0. \tag{13.2.15}$$

13.3   Minimal Surfaces and Functions of a Complex Variable

Although the material in this section does not rely on variational methods and hence can be omitted in the lecture, we find it important to include it into the manuscript, in light of the topics covered in Chap. 4 and of our aim to encourage an interdisciplinary view of mathematics.

Minimal surfaces have the important property that the sum of their radii of curvature vanishes, and their characteristic property can be stated as follows [30]: *the two families of lines of vanishing length traced upon them must form a conjugate grid.* It is indeed well known that the equilateral hyperbola is the only conic admitting as conjugate diameters the two lines of angular coefficients $+i$ and $-i$, respectively. By taking for granted this property, one can choose as independent variables α and β the parameters of these two systems of lines of vanishing length. Since they form a conjugate grid, the Cartesian coordinates x, y, z are particular solutions of an equation of the form [30]

$$\left(\frac{\partial^2}{\partial\alpha\partial\beta} - A\frac{\partial}{\partial\alpha} - B\frac{\partial}{\partial\beta} \right) \theta = 0, \quad \theta = x, y, z. \tag{13.3.1}$$

Moreover, by virtue of the particular choice of α and β,

$$\left(\frac{\partial x}{\partial\alpha} \right)^2 + \left(\frac{\partial y}{\partial\alpha} \right)^2 + \left(\frac{\partial z}{\partial\alpha} \right)^2 = 0, \tag{13.3.2}$$

$$\left(\frac{\partial x}{\partial\beta} \right)^2 + \left(\frac{\partial y}{\partial\beta} \right)^2 + \left(\frac{\partial z}{\partial\beta} \right)^2 = 0. \tag{13.3.3}$$

Equations (13.3.1)–(13.3.3) express all conditions resulting from the definition of minimal surface and from the choice of coordinates.

If one differentiates Eq. (13.3.2) with respect to β, and Eq. (13.3.3) with respect to α, and if one replaces the second derivatives by their values obtained from the Eq. (13.3.1), one finds

$$\left(\frac{\partial x}{\partial\alpha}\frac{\partial x}{\partial\beta} + \frac{\partial y}{\partial\alpha}\frac{\partial y}{\partial\beta} + \frac{\partial z}{\partial\alpha}\frac{\partial z}{\partial\beta} \right) \chi = 0, \quad \chi = A, B. \tag{13.3.4}$$

Since one cannot have

$$\left(\frac{\partial x}{\partial \alpha}\frac{\partial x}{\partial \beta} + \frac{\partial y}{\partial \alpha}\frac{\partial y}{\partial \beta} + \frac{\partial z}{\partial \alpha}\frac{\partial z}{\partial \beta}\right) = 0 \tag{13.3.5}$$

without admitting that the arc of every curve traced upon the surface vanishes, we have to impose

$$A = 0, \ B = 0 \tag{13.3.6}$$

in order to satisfy Eq. (13.3.4). Equation (13.3.1) reduce therefore to the form

$$\frac{\partial^2 \theta}{\partial \alpha \partial \beta} = 0, \ \theta = x, y, z, \tag{13.3.7}$$

whose integration is immediate and yields

$$x = f_1(\alpha) + \varphi_1(\beta), \tag{13.3.8}$$

$$y = f_2(\alpha) + \varphi_2(\beta), \tag{13.3.9}$$

$$z = f_3(\alpha) + \varphi_3(\beta). \tag{13.3.10}$$

However, in order to satisfy Eqs. (13.3.2) and (13.3.3), one has to impose also the equations

$$\sum_{k=1}^{3}\left(\frac{\mathrm{d}f_k}{\mathrm{d}\alpha}\right)^2 = 0, \ \sum_{k=1}^{3}\left(\frac{\mathrm{d}\varphi_k}{\mathrm{d}\beta}\right)^2 = 0. \tag{13.3.11}$$

The most elegant solution of Eq. (13.3.11) was pointed out by Enneper [45], and obtained in a complete way for the first time by Weierstrass [139, 140]. They recognized that, upon defining

$$u(\alpha) \equiv \frac{\frac{\mathrm{d}}{\mathrm{d}\alpha}(f_1 + \mathrm{i}f_2)}{-\frac{\mathrm{d}}{\mathrm{d}\alpha}f_3}, \tag{13.3.12}$$

the first of Eq. (13.3.11) leads to

$$\frac{\mathrm{d}}{\mathrm{d}\alpha}(f_1 - \mathrm{i}f_2) = \frac{\frac{\mathrm{d}}{\mathrm{d}\alpha}f_3}{u(\alpha)}, \tag{13.3.13}$$

and hence the ratios of the three derivatives f_1', f_2', f_3' are determined by the formulae

$$\frac{\frac{d}{d\alpha} f_1}{(1-u^2)} = \frac{\frac{d}{d\alpha} f_2}{i(1+u^2)} = \frac{\frac{d}{d\alpha} f_3}{2u}. \tag{13.3.14}$$

By ruling out, for a moment, the exceptional case when u is a constant, we can therefore represent the common value of the previous ratios by

$$\frac{1}{2} \mathcal{F}(u) \frac{du}{d\alpha}.$$

One then finds

$$f_1(\alpha) = \frac{1}{2} \int (1-u^2) \mathcal{F}(u) du, \tag{13.3.15}$$

$$f_2(\alpha) = \frac{i}{2} \int (1+u^2) \mathcal{F}(u) du, \tag{13.3.16}$$

$$f_3(\alpha) = \int u \mathcal{F}(u) du. \tag{13.3.17}$$

In the same way one finds, by defining the function

$$u_1(\beta) \equiv \frac{\frac{d}{d\beta}(\varphi_1 - i\varphi_2)}{-\frac{d}{d\beta} \varphi_3}, \tag{13.3.18}$$

and following a method analogous to the one relying upon (13.3.12), the three functions of the variable β as given by

$$\varphi_1(\beta) = \frac{1}{2} \int \left[1 - (u_1)^2\right] \mathcal{F}_1(u_1) du_1, \tag{13.3.19}$$

$$\varphi_2(\beta) = -\frac{i}{2} \int \left[1 + (u_1)^2\right] \mathcal{F}_1(u_1) du_1, \tag{13.3.20}$$

$$\varphi_3(\beta) = \int u_1 \mathcal{F}_1(u_1) du_1. \tag{13.3.21}$$

The formulae that define a minimal surface in Cartesian coordinates admit therefore the remarkable integral representation

$$x = \frac{1}{2}\left[\int (1-u^2)\mathcal{F}(u)du + \int (1-(u_1)^2)\mathcal{F}_1(u_1)du_1\right],\qquad(13.3.22)$$

$$y = \frac{i}{2}\left[\int (1+u^2)\mathcal{F}(u)du - \int (1+(u_1)^2)\mathcal{F}_1(u_1)du_1\right],\qquad(13.3.23)$$

$$z = \int u\mathcal{F}(u)du + \int u_1\mathcal{F}_1(u_1)du_1.\qquad(13.3.24)$$

Interestingly, one can re-express these formulae in an equivalent way, not involving any quadrature, by replacing $\mathcal{F}(u)$ with the third derivative of a function $f(u)$, and similarly for $\mathcal{F}_1(u_1)$, i.e. the definitions

$$\mathcal{F}(u) \equiv \frac{d^3}{du^3}f(u),\ \ \mathcal{F}_1(u_1) \equiv \frac{d^3}{d(u_1)^3}f_1(u_1)\qquad(13.3.25)$$

lead to the Weierstrass formulae [30]

$$x = \left[\frac{1}{2}(1-u^2)\frac{d^2}{du^2} + u\frac{d}{du} - 1\right]f(u)$$
$$+ \left[\frac{1}{2}(1-(u_1)^2)\frac{d^2}{d(u_1)^2} + u_1\frac{d}{du_1} - 1\right]f_1(u_1),\qquad(13.3.26)$$

$$y = i\left[\frac{1}{2}(1+u^2)\frac{d^2}{du^2} - u\frac{d}{du} + 1\right]f(u)$$
$$+ i\left[-\frac{1}{2}(1+(u_1)^2)\frac{d^2}{d(u_1)^2} + u_1\frac{d}{du_1} - 1\right]f_1(u_1),\qquad(13.3.27)$$

$$z = \left(u\frac{d^2}{du^2} - \frac{d}{du}\right)f(u) + \left(u_1\frac{d^2}{d(u_1)^2} - \frac{d}{du_1}\right)f_1(u_1).\qquad(13.3.28)$$

Let us now try to understand under which conditions the surface represented by Eqs. (13.3.22)–(13.3.24) or (13.3.26)–(13.3.28) is real. Let us consider first Eqs. (13.3.22)–(13.3.24). If one takes therein for $\mathcal{F}(u)$, $\mathcal{F}_1(u)$ some complex conjugate functions and if, in addition, the integrals pertaining to u and u_1 are evaluated by taking complex conjugate contours, the expressions of the different coordinates are clearly real, because, to every element of the integral with respect to u, there corresponds, in each of the three formulae, a complex conjugate element of the other integral. We are going to show that these conditions, which are sufficient, are also necessary. For this purpose, let us consider the lines of vanishing length of the surface, for which

$$dx \otimes dx + dy \otimes dy + dz \otimes dz = 0, \tag{13.3.29}$$

where

$$dz = \frac{\partial z}{\partial x}dx + \frac{\partial z}{\partial y}dy = p\,dx + q\,dy. \tag{13.3.30}$$

These equations determine two different systems of values for the differentials dx, dy and dz. Let us denote them by [30]

$$\delta x, \delta y, \delta z \text{ and } \delta'x, \delta'y, \delta'z.$$

The functions u and u_1 are defined either by the formulae

$$\frac{\delta x + i\delta y}{\delta z} = -u, \quad \frac{\delta'x - i\delta'y}{\delta'z} = -u_1, \tag{13.3.31}$$

or by the formulae

$$\frac{\delta'x + i\delta'y}{\delta'z} = -u, \quad \frac{\delta x - i\delta y}{\delta z} = -u_1. \tag{13.3.32}$$

For every real point of a real sheet, one can assume that the two systems of ratios $\delta x, \delta y, \delta z$ and $\delta'x, \delta'y, \delta'z$, can be deduced the one from the other by replacing i with $-i$. In what follows, the values of u and u_1 obtained from one or the other of the systems (13.3.31) and (13.3.32) are complex conjugate. On the other hand, one derives from Eqs. (13.3.22)–(13.3.24) the following partial differential equations:

$$\frac{\partial x}{\partial u} - i\frac{\partial y}{\partial u} = \mathcal{F}(u), \quad \frac{\partial x}{\partial u_1} + i\frac{\partial y}{\partial u_1} = \mathcal{F}_1(u_1). \tag{13.3.33}$$

Since the first members of these relations are complex conjugate, the same holds for the second members. The functions $\mathcal{F}(u)$ and $\mathcal{F}_1(u_1)$ must therefore be complex conjugate, as it occurs for their arguments u and u_1.

This analysis shows that the real sheets of the minimal surface are represented by the equations

$$x = \text{Re} \int (1 - u^2)\mathcal{F}(u)du, \tag{13.3.34}$$

$$y = \text{Re} \int i(1 + u^2)\mathcal{F}(u)du, \tag{13.3.35}$$

$$z = \text{Re} \int 2u \mathcal{F}(u) du. \tag{13.3.36}$$

These formulae establish, as was first recognized by Weierstrass, the closest link between complex analysis and the theory of minimal surfaces. Indeed, to every function $\mathcal{F}(u)$, these formulae associate a real minimal surface whose different properties provide the most perfect and elegant geometric representation of all analytic relations to which the function $\mathcal{F}(u)$ gives rise.

This minimal surface is completely determined in its form and orientation, but it can be displaced parallel to itself, if one adds arbitrary constants to the integrals occurring in Eqs. (13.3.34)–(13.3.36). One can thus say that to a function $\mathcal{F}(u)$ of complex variable there corresponds a unique minimal surface, but the converse does not hold in general. In order to recognize it, it suffices to remark that Eqs. (13.3.31) and (13.3.32) lead to two different sets of values for u and u_1, and therefore to two expressions, different in general, for the function $\mathcal{F}(u)$.

At a deeper level, one proves the same property as follows. Let us revert to Eqs. (13.3.22)–(13.3.24), and let us try to replace them by similar formulae, where u and u_1 are replaced by v and v_1, and $\mathcal{F}(u)$, $\mathcal{F}_1(u_1)$ are replaced by other functions $\mathcal{G}(v)$, $\mathcal{G}_1(v_1)$. The parameters u and u_1 being, like v and v_1, those of lines of vanishing length of the surface, it is necessary to consider the functional relations

$$v = v(u), \quad v_1 = v_1(u_1), \tag{13.3.37}$$

or

$$v = v(u_1), \quad v_1 = v_1(u). \tag{13.3.38}$$

The first assumption yields of course

$$v = u, \quad v_1 = u_1, \quad \mathcal{G}(v) = \mathcal{F}(u), \quad \mathcal{G}_1(v_1) = \mathcal{F}_1(u_1), \tag{13.3.39}$$

while the second leads to the equations

$$(1 - u^2)\mathcal{F}(u)du = \left[1 - (v_1)^2\right]\mathcal{G}_1(v_1)dv_1, \tag{13.3.40}$$

$$\left[1 - (u_1)^2\right]\mathcal{F}_1(u_1)du_1 = (1 - v^2)\mathcal{G}(v)dv, \tag{13.3.41}$$

$$(1 + u^2)\mathcal{F}(u)du = -\left[1 + (v_1)^2\right]\mathcal{G}_1(v_1)dv_1, \tag{13.3.42}$$

$$\left[1 + (u_1)^2\right]\mathcal{F}_1(u_1)du_1 = -(1 + v^2)\mathcal{G}(v)dv, \tag{13.3.43}$$

$$u\mathcal{F}(u)du = v_1\mathcal{G}(v_1)dv_1, \tag{13.3.44}$$

$$u_1\mathcal{F}_1(u_1)du_1 = v\mathcal{G}(v)dv, \tag{13.3.45}$$

which are solved by

$$v_1 = -\frac{1}{u}, \quad v = -\frac{1}{u_1}, \tag{13.3.46}$$

$$\mathcal{G}_1(v_1) = -\frac{1}{(v_1)^4}\mathcal{F}\left(-\frac{1}{v_1}\right), \quad \mathcal{G}(v) = -\frac{1}{v^4}\mathcal{F}_1\left(-\frac{1}{v}\right). \tag{13.3.47}$$

Thus, if only real surfaces are studied, one finds that, to the same minimal surface, there correspond the two functions [30]

$$\mathcal{F}(u) \text{ and } -\frac{1}{u^4}\mathcal{F}_1\left(-\frac{1}{u}\right),$$

which are different, in general.

13.4   The Dirichlet Boundary-Value Problem

Now we revert to subjects where variational methods are applied. Let Ω be a bounded domain of n-dimensional Euclidean space $\mathbf{R}^n$ bounded by a surface S which is piecewise continuously differentiable or consists of a finite number of such surfaces.

The Dirichlet problem is the search of a harmonic function in the Sobolev space $W^{1,2}$, i.e. functions square integrable with their derivative, which assume on the boundary S the given permissible value φ:

$$u|_S = \varphi. \tag{13.4.1}$$

In order to solve this problem, we first solve a variational problem, and then show that the solution of such a variational problem is indeed a solution of the Dirichlet problem [131].

Let $W^{1,2}(\varphi)$ the Sobolev space of functions square-integrable with their first derivative, which take the value φ on S. Since φ is a permissible value, the space $W^{1,2}(\varphi)$ is not the empty set. On defining the functional

$$D(u) \equiv \int \cdots \int_\Omega \sum_{i=1}^n \left(\frac{\partial u}{\partial x^i}\right)^2 d\Omega < \infty. \tag{13.4.2}$$

one has, for each $v \in W^{1,2}(\varphi)$,

$$0 \le D(v) < \infty, \tag{13.4.3}$$

and hence there exists a greatest lower bound for the values of $D(v)$:

$$d = \inf_{v \in W^{1,2}(\varphi)} D(v), \ d \ge 0. \tag{13.4.4}$$

From the set $W^{1,2}(\varphi)$ one can choose a sequence $\{v_k\}$ for which

$$\lim_{k \to \infty} D(v_k) = d, \tag{13.4.5}$$

a property which follows from the definition of greatest lower bound. The sequence $\{v_k\}$ is said to be a *minimizing sequence*.

Theorem 13.1 *The minimizing sequence $\{v_k\}$ converges in the Sobolev space $W^{1,2}$. The limit function lies in $W^{1,2}(\varphi)$ and yields a proper minimum for the functional $D(v)$ among all such functions.*

Proof Following Sobolev [131], we define the norm in $W^{1,2}$ by the formula (see (13.4.2))

$$\|v\|_{W^{1,2}} \equiv \left\{ \left[\int \cdots \int_S v \, dS \right]^2 + D(v) \right\}^{\frac{1}{2}}. \tag{13.4.6}$$

From the equation

$$\int \cdots \int_S (v_k - v_m) dS = 0 \tag{13.4.7}$$

one obtains a formula for the Sobolev norm

$$\|v_k - v_m\|_{W^{1,2}} = \sqrt{D(v_k - v_m)}. \tag{13.4.8}$$

The convergence of the sequence $\{v_k\}$ in the Sobolev space $W^{1,2}$ is proved if one can show that

$$D(v_k - v_m) \to 0 \text{ for } k, m \to \infty. \tag{13.4.9}$$

Let ϵ positive be given. One can find a positive integer N such that $D(v_k) < d + \epsilon$ if $k > N$. Let k and m be bigger than N. Of course, the half-sum $\frac{(v_k + v_m)}{2}$ belongs to $W^{1,2}(\varphi)$, and hence

$$D\left(\frac{v_k + v_m}{2}\right) \ge d. \tag{13.4.10}$$

Now we can exploit the identity

$$D\left(\frac{v_k + v_m}{2}\right) + D\left(\frac{v_k - v_m}{2}\right) = \frac{1}{2}D(v_k) + \frac{1}{2}D(v_m) \tag{13.4.11}$$

to derive the inequality

$$d + D\left(\frac{v_k - v_m}{2}\right) < \frac{(d + \epsilon)}{2} + \frac{(d + \epsilon)}{2} = d + \epsilon, \tag{13.4.12}$$

which implies that

$$D\left(\frac{v_k - v_m}{2}\right) < \epsilon \implies D(v_k - v_m) < 4\epsilon, \tag{13.4.13}$$

and therefore

$$D(v_k - v_m) \to 0 \text{ for } k, m \to \infty. \tag{13.4.14}$$

Now from the completeness of the space $W^{1,2}$ it follows that the sequence $\{v_k\}$ converges to some function v_0 in $W^{1,2}$, so that one can write

$$\|v_0 - v_k\|_{W^{1,2}} \to 0, \ k \to \infty. \tag{13.4.15}$$

We now aim at showing that $D(v_0) = d$. Indeed, one has

$$
\begin{aligned}
|D(v_0) - D(v_k)| &= \left| \int \cdots \int_\Omega \sum_{i=1}^n \left[\left(\frac{\partial v_0}{\partial x^i}\right)^2 - \left(\frac{\partial v_k}{\partial x^i}\right)^2 \right] d\Omega \right| \\
&= \left| \int \cdots \int_\Omega \sum_{i=1}^n \left[\left(\frac{\partial v_0}{\partial x^i} - \frac{\partial v_k}{\partial x^i}\right)\left(\frac{\partial v_0}{\partial x^i} + \frac{\partial v_k}{\partial x^i}\right) \right] d\Omega \right| \\
&\leq \sum_{i=1}^n \left| \int \cdots \int_\Omega \left(\frac{\partial v_0}{\partial x^i} - \frac{\partial v_k}{\partial x^i}\right)\left(\frac{\partial v_0}{\partial x^i} + \frac{\partial v_k}{\partial x^i}\right) d\Omega \right| \\
&\leq \sum_{i=1}^n \sqrt{\int \cdots \int_\Omega \left(\frac{\partial v_0}{\partial x^i} - \frac{\partial v_k}{\partial x^i}\right)^2 d\Omega} \sqrt{\int \cdots \int_\Omega \left(\frac{\partial v_0}{\partial x^i} + \frac{\partial v_k}{\partial x^i}\right)^2 d\Omega} \\
&\leq \sum_{i=1}^n \|v_0 + v_k\|_{W^{1,2}} \sqrt{\int \cdots \int_\Omega \left(\frac{\partial v_0}{\partial x^i} - \frac{\partial v_k}{\partial x^i}\right)^2 d\Omega} \\
&\leq \|v_0 + v_k\|_{W^{1,2}} \|v_0 - v_k\|_{W^{1,2}}, \tag{13.4.16}
\end{aligned}
$$

from which there follows the limiting value

$$D(v_0) = \lim_{k \to \infty} D(v_k) = d. \tag{13.4.17}$$

The function v_0 belongs indeed to the Sobolev space $W^{1,2}$ and hence one can consider it on every $(n-1)$-dimensional manifold, where it is always square-integrable.

The value of the function v_0 on the surface S equals φ. Indeed, one has the majorization

$$\int \cdots \int_S (v_k - v_0)^2 dS \le (\|v_k - v_0\|_{W^{1,2}})^2 , \tag{13.4.18}$$

and therefore

$$\int \cdots \int_S (v_k - v_0)^2 dS \to 0. \tag{13.4.19}$$

Furthermore, we know that v_k equals φ on S, and hence Eq. (13.4.19) can be written as

$$\int \cdots \int_S (\varphi - v_0)^2 dS = 0. \tag{13.4.20}$$

In other words, the function v_0 takes on its boundary value φ by converging to it in the mean. Thus, $v_0 \in W^{1,2}$ has the properties

$$v_0|_S = \varphi(\vec{P}), \quad D(v_0) = d. \tag{13.4.21}$$

The variational problem is therefore solved [131]. Q.E.D.

Now we can go on and solve actually the Dirichlet problem, i.e.

Theorem 13.2 *The function providing a minimum to the functional $D(v)$ on the Sobolev space $W^{1,2}(\varphi)$ is the solution of the Dirichlet problem.*

Proof The method of proof consists in showing that, in the interior of Ω, the function v_0 has continuous derivatives of arbitrary order and satisfies the Laplace equation.

Let $\xi \in W^{1,2}$, $\xi|_S = 0$ and otherwise arbitrary. Consider [131]

$$D(v_0 + \lambda\xi) = D(v_0) + 2\lambda D(v_0, \xi) + \lambda^2 D(\xi), \tag{13.4.22}$$

where

$$D(v_0, \xi) = \int \cdots \int_\Omega \sum_{i=1}^n \frac{\partial v_0}{\partial x^i} \frac{\partial \xi}{\partial x^i} d\Omega. \tag{13.4.23}$$

Since also $v_0 + \lambda\xi \in W^{1,2}(\varphi)$, one finds

$$D(v_0 + \lambda \xi) \geq d = D(v_0), \tag{13.4.24}$$

and hence the functional (13.4.22) has a minimum for $\lambda = 0$. A theorem of Fermat ensures that

$$D(v_0, \xi) = 0. \tag{13.4.25}$$

The point $\xi(x^1, \ldots, x^n)$ is chosen of a special form. Let $\psi(\eta)$ be such that

$$\psi(\eta) = 1 \text{ if } \eta \in \left[0, \frac{1}{2}\right], \quad \psi(\eta) = 0 \text{ if } \eta \geq 1,$$

with $\psi(\eta)$ monotonic on the closed interval $\left[\frac{1}{2}, 1\right]$ and having continuous derivatives of arbitrary order for all $\eta \in [0, \infty]$. For example, one can define [131]

$$\psi(\eta) = \frac{1}{2}\left[1 + \text{th}\frac{\left(\eta - \frac{3}{4}\right)}{\left(\eta - \frac{1}{2}\right)(\eta - 1)}\right], \quad \eta \in \left]\frac{1}{2}, 1\right[. \tag{13.4.26}$$

For some interior point M_0 of the domain Ω, the distance from it to an arbitrary point is r, and the distance from M_0 to S is denoted by δ. After chhosing two numbers h_1 and h_2 in such a way that

$$0 < h_1 < h_2 < \delta, \tag{13.4.27}$$

one defines

$$\xi \equiv r^{2-n}\left[\psi\left(\frac{r}{h_1}\right) - \psi\left(\frac{r}{h_2}\right)\right]. \tag{13.4.28}$$

The choice of h_1 and h_2 enforces the vanishing of ξ on S. Moreover, ξ also vanishes for $r < \frac{h_1}{2}$, and hence the function ξ is continuous and has continuous derivatives of all orders, and belongs to the Sobolev space $W^{1,2}$. For such a function ξ, Eq. (13.4.25) holds. By virtue of the definition of generalized derivative $\frac{\partial v_0}{\partial x^i}$, one can write

$$\int \cdots \int_{\Omega} \frac{\partial v_0}{\partial x^i} \frac{\partial \xi}{\partial x^i} d\Omega = -\int \cdots \int_{\Omega} v_0 \frac{\partial^2 \xi}{\partial (x^i)^2} d\Omega. \tag{13.4.29}$$

Thus, Eq. (13.4.25) yields the integral condition

$$\int \cdots \int_{\Omega} v_0 \triangle \xi \, d\Omega = 0. \tag{13.4.30}$$

In the integrand, the action of the Laplacian on the function ξ can be expressed by means of the identity [131]

$$\Delta \xi = \Delta \left(r^{2-n} \psi \left(\frac{r}{h_1} \right) \right) - \Delta \left(r^{2-n} \psi \left(\frac{r}{h_2} \right) \right) = \frac{1}{(h_1)^n} \omega \left(\frac{r}{h_1} \right) - \frac{1}{(h_2)^n} \omega \left(\frac{r}{h_2} \right),$$

$$(13.4.31)$$

where ω has been defined according to

$$\omega \left(\frac{r}{h_i} \right) = (h_i)^n \Delta \left(r^{2-n} \psi \left(\frac{r}{h_i} \right) \right) = (h_i)^2 \Delta \left[\left(\frac{r}{h_i} \right)^{2-n} \psi \left(\frac{r}{h_i} \right) \right], \quad (13.4.32)$$

where the right-hand side is clearly a function only of the ratio $\frac{r}{h_i}$.

By exploiting the properties

$$\psi \left(\frac{r}{h_i} \right) = 1 \text{ for } r < \frac{h_i}{2}, \quad \Delta r^{2-n} = 0,$$

one finds

$$\omega \left(\frac{r}{h_i} \right) = 0 \text{ if } r < \frac{h_i}{2}. \qquad (13.4.33)$$

Thus, the action of the Laplacian on the ξ-function equals the difference of two differentiable functions of class C^k with k arbitrarily large on the whole Euclidean space, and Eq. (13.4.30) leads to the equality

$$\frac{1}{(h_1)^n} \int \cdots \int_\Omega v_0 \omega \left(\frac{r}{h_1} \right) d\Omega = \frac{1}{(h_2)^n} \int \cdots \int_\Omega v_0 \omega \left(\frac{r}{h_2} \right) d\Omega. \qquad (13.4.34)$$

It is now possible to multiply both sides of Eq. (13.4.34) by $\frac{1}{(n-2)\sigma_n}$, where σ_n is the surface area of the unit sphere in $\mathbf{R}^n$. Equation (13.4.34) becomes therefore

$$\frac{1}{(n-2)\sigma_n (h_1)^n} \int \cdots \int_\Omega v_0 \omega \left(\frac{r}{h_1} \right) d\Omega = \frac{1}{(n-2)\sigma_n (h_2)^n} \int \cdots \int_\Omega v_0 \omega \left(\frac{r}{h_2} \right) d\Omega.$$

$$(13.4.35)$$

Interestingly, one finds that the function

$$\frac{1}{(n-2)\sigma_n h^n} \omega \left(\frac{r}{h} \right)$$

may be considered as an *averaging kernel*, because its integral over the whole space equals 1, as can be seen below:

$$\frac{1}{(n-2)\sigma_n h^n} \int \cdots \int_\Omega \omega\left(\frac{r}{h}\right) d\Omega = \frac{1}{(n-2)\sigma_n} \int \cdots \int_\Omega \Delta\left(r^{2-n}\psi\left(\frac{r}{h}\right)\right) d\Omega$$

$$= \frac{1}{(n-2)\sigma_n} \int \cdots \int_{r=h} \frac{\partial}{\partial r}\left(r^{2-n}\psi\left(\frac{r}{h}\right)\right) dS$$

$$- \frac{1}{(n-2)\sigma_n} \int \cdots \int_{r=\frac{h}{2}} \frac{\partial}{\partial r}\left(r^{2-n}\psi\left(\frac{r}{h}\right)\right) dS = \frac{1}{\sigma_n} \int \cdots \int_{r=\frac{h}{2}} r^{1-n} dS$$

$$= \frac{1}{\sigma_n}\left(\frac{h}{2}\right)^{1-n}\left(\frac{h}{2}\right)^{n-1} \quad \sigma_n = 1.$$

$$\tag{13.4.36}$$

By virtue of this identity, Eq. (13.4.35) may be expressed in the form

$$(v_0)_{h_1} = (v_0)_{h_2}. \tag{13.4.37}$$

Thus, the averaged functions for v_0 do not change with a change in h (if $h < \delta$) at points lying at a distance greater than δ from the boundary, and hence the limit of the sequence $(v_0)_h$ coincides with v_0, i.e.

$$(v_0)_h = v_0. \tag{13.4.38}$$

Since $(v_0)_h$ has continuous derivatives of all orders, the same holds for v_0.

Now if ξ is an arbitrary function, continuous with its first derivatives in Ω and vanishing outside some interior sub-domain, an integrations by parts yields

$$D(v_0, \xi) = - \int \cdots \int_\Omega \xi \Delta v_0 \, d\Omega = 0, \tag{13.4.39}$$

and the arbitrariness of ξ implies therefore that v_0 is a harmonic function, which assumes on S the given value $\phi(\vec{P})$. Thus, v_0 is the desired solution of the Dirichlet problem [131].

Part IV
Linear and Non-linear Hyperbolic Equations

Chapter 14
Characteristics and Waves, 1

14.1 Systems of Partial Differential Equations

A system of m partial differential equations in the unknown functions $\varphi_1, \varphi_2, \ldots, \varphi_n$ of $n + 1$ independent variables $x^0, x^1, \ldots, x^n$ reads as [93]

$$E_\mu = 0, \qquad \mu = 1, 2, \ldots, m, \tag{14.1.1}$$

the E_μ being functions that depend on the x, on the φ and on the partial derivatives of the φ with respect to the x. Such a system is said to be *normal* with respect to the variable x^0 if it can be reduced to the form

$$\frac{\partial^{r_\nu} \varphi_\nu}{\partial (x^0)^{r_\nu}} = \Phi_\nu(x|\varphi|\psi|\chi), \qquad \nu = 1, 2, \ldots, m, \tag{14.1.2}$$

where the ψ occurring on the right-hand side are partial derivatives of each φ_ν with respect to x^0 only of order less than r_ν, and the χ are partial derivatives of the φ with respect to the x of arbitrary finite order, provided that, with respect to x^0, they are of order less than r_ν for the corresponding φ_ν.

The functions Φ_ν are taken to be real-analytic in the neighbourhood of a set of values of Cauchy data. Before stating the associated Cauchy–Kowalevsky theorem, it is appropriate to recall the existence theorem for integrals of a system of ordinary differential equations. Hence we consider momentarily the differential system (having set $x^0 = t$)

$$\frac{d^{r_\nu} \varphi_\nu}{dt^{r_\nu}} = \Phi_\nu(t|\varphi|\psi), \qquad \nu = 1, 2, \ldots, m. \tag{14.1.3}$$

This system can be re-expressed in canonical form, involving only first-order equations, by defining

© Springer International Publishing AG 2017
G. Esposito, *From Ordinary to Partial Differential Equations*, UNITEXT 106,
DOI 10.1007/978-3-319-57544-5_14

$$\frac{d}{dt}\varphi_\nu \equiv \varphi_\nu', \quad \frac{d}{dt}\varphi_\nu' \equiv \varphi_\nu'', \dots \frac{d}{dt}\varphi_\nu^{(r_\nu-2)} \equiv \varphi_\nu^{(r_\nu-1)}, \tag{14.1.4}$$

from which Eq. (14.1.3) read as

$$\frac{d}{dt}\varphi_\nu^{(r_\nu-1)} = \Phi_\nu(t|\varphi|\psi), \qquad \nu = 1, 2, \dots, m. \tag{14.1.5}$$

One can also denote by y_ρ the generic element of a table displaying φ_1 and its derivatives up to the order $(r_1 - 1)$ on the first column, φ_2 and its derivatives up to the order $(r_2 - 1)$ on the second column, ..., φ_m and its derivatives up to the order $(r_m - 1)$ on the last column. With such a notation, the canonical form (14.1.5) is further re-expressed as

$$\frac{d}{dt}y_\rho = Y_\rho(t|y), \quad (\rho = 1, 2, \dots, r; \ r \equiv \sum_{k=1}^m r_k). \tag{14.1.6}$$

If each Y_ρ is real-analytic in the neighbourhood of $t = t_0$, $y_\rho = b_\rho$, there exists a unique set of functions y_ρ, analytic in the t variable, which take the value b_ρ at $t = t_0$.

In order to prove such a theorem, one begins by remarking that the differential equations make it possible, by means of subsequent differentiations, to evaluate the derivatives of any order of an unknown function y_ρ at the point $t = t_0$ and hence to write, for each y_ρ, the Taylor expansion pertaining to such a point. The essential point of the proof consists in showing that such series converge in a suitable neighbourhood of $t = t_0$. For this purpose one considers some appropriate majorizing functions Y_ρ; the corresponding differential system (14.1.6), which can be integrated by elementary methods, defines some real-analytic functions in the neighbourhood of $t = t_0$, whose Taylor expansions majorize the Taylor expansions of the y_ρ functions.

The Cauchy theorem for the differential systems (14.1.6) holds also when the right-hand side depends on a certain number of parameters that can be denoted by $x_1, x_2, \dots, x_n$ provided that they vary in such a way that the functions Y_ρ are real-analytic. One can then state the following

Theorem 14.1 *Given the differential system*

$$\frac{d^{r_\nu}\varphi_\nu}{dt^{r_\nu}} = \Phi_\nu(t|x|\varphi|\psi), \qquad \nu = 1, 2, \dots, m, \tag{14.1.7}$$

by assigning at will, at $t = t_0$, the values of each φ_ν and of the subsequent derivatives up to the order $r_\nu - 1$ as functions of the parameters $x_1, x_2, \dots, x_n$, there exists a unique set of functions φ, real-analytic, of the variable t and of the parameters, satisfying Eq. (14.1.7), and equal to the assigned values at $t = t_0$.

This theorem admits an extension to systems of partial differential equations (14.1.2) in normal form, the new feature being that, on the right-hand side of Eq. (14.1.7), there occur also derivatives of the unknown functions with respect to the parameters, so that one deals with partial differential equations.

The Cauchy problem consists in finding the functions φ satisfying the system (14.1.2) in normal form, and the initial conditions given by the values of the unknown functions and their partial derivatives with respect to the variable x^0, of order less than the maximal order. Let S be the space of the variables, from now on denoted by $x^0, x^1, \ldots, x^n$. In order to fix the ideas, one can assume that S is endowed with an Euclidean metric, and interpret the x as Cartesian coordinates. Let ω be the hyperplane with equation

$$x^0 = a^0. \tag{14.1.8}$$

The Cauchy existence theorem states that, in the neighbourhood of the hyperplane ω, which is said to be the *carrier hyperplane*, one can find the values taken by the φ functions, once the initial values of φ and ψ functions are freely specified at each point of ω. An easy generalization of the theorem is obtained by replacing the hyperplane ω with an hypersurface σ of S. For example, if

$$z(x^0, x^1, \ldots, x^n) = z^0 = \text{constant} \tag{14.1.9}$$

is the equation of the hypersurface σ, it is enough to replace the x with $n+1$ independent combinations of the x, here denoted by $z, z^1, \ldots, z^n$, in such a way that one of them, i.e. z, is precisely the left-hand side of the Eq. (14.1.9) here written for σ.

14.2 Characteristic Manifolds for First- and Second-Order Systems

Hereafter we limit ourselves to consideration of differential systems for which the maximal order of derivation is $s = 1$ or $s = 2$. Such systems can be made explicit by writing them in the form [93]

$$E_\mu = \sum_{\nu=1}^{m} \sum_{i=0}^{n} E_{\mu\nu}^i \frac{\partial \varphi_\nu}{\partial x^i} + \Phi_\mu(x|\varphi) = 0, \quad \mu = 1, 2, \ldots, m, \tag{14.2.1}$$

and

$$E_\mu = \sum_{\nu=1}^{m} \sum_{i,k=0}^{n} E_{\mu\nu}^{ik} \frac{\partial^2 \varphi_\nu}{\partial x^i \partial x^k} + \Phi_\mu(x|\varphi|\chi) = 0, \quad \mu = 1, 2, \ldots, m, \tag{14.2.2}$$

respectively. In Eq. (14.2.1) the $E_{\mu\nu}^i$ and Φ_μ depend on the x and φ, whereas in Eq. (14.2.2) the $E_{\mu\nu}^{ik}$ and Φ_μ depend on the x, φ and on the first-order partial derivatives of the φ with respect to the x. Since the φ_ν are taken to fulfill the conditions under which one can exchange the order of derivatives, one can always assume that $E^{ik}_{\mu\nu}$ is symmetric in the lower case Latin indices.

In the particular case of a single unknown function φ, Eq. (14.2.2) reduce to the single equation

$$E = \sum_{i,k=0}^{n} E^{ik} \frac{\partial^2 \varphi}{\partial x^i \partial x^k} + \Phi(x|\varphi|\chi) = 0, \qquad (14.2.3)$$

where χ is a concise notation for the first-order partial derivatives of φ with respect to $x^0, x^1, \ldots, x^n$.

A remarkable equation of the type (14.2.3) is the scalar wave equation (having set $x^0 = t$ in $c = 1$ units)

$$\Box \varphi = \left(\frac{1}{V^2} \frac{\partial^2}{\partial t^2} - \Delta \right) \varphi = 0, \qquad (14.2.4)$$

where V is a constant and $\Delta = \sum_{i=1}^{3} \frac{\partial^2}{\partial (x^i)^2}$ the standard notation for minus the Laplacian in Euclidean three-dimensional space $\mathbf{R}^3$. The $\Box$ operator in Eq. (14.2.4) is the familiar D'Alembert operator for the wave equation in Minkowski space-time.

The systems (14.2.1) and (14.2.2) are not yet written in normal form, and we now aim at finding the conditions under which such systems are normal with respect to the variable x^0. For this purpose, we begin with the system (14.2.1) and point out that, since we are only interested in first-order partial derivatives with respect to x^0, we can re-express such equations in the form

$$\sum_{\nu=1}^{m} E^0_{\mu\nu} \frac{\partial \varphi_\nu}{\partial x^0} + \cdots = 0, \qquad \mu = 1, 2, \ldots, m. \qquad (14.2.5)$$

This system can be solved with respect to the derivatives $\frac{\partial \varphi_\nu}{\partial x^0}$ if the determinant of the matrix $E^0_{\mu\nu}$ does not vanish, i.e.

$$\Omega = \det E^0_{\mu\nu} \neq 0, \qquad \mu, \nu = 1, 2, \ldots, m. \qquad (14.2.6)$$

Such a determinant involves the independent variables $x^0, x^1, \ldots, x^n$ and also, in general, the unknown functions $\varphi_1, \varphi_2, \ldots, \varphi_m$.

Let us now consider the Eq. (14.2.2) of the second system, which are written more conveniently in the form

$$\sum_{\nu=1}^{m} E^{00}_{\mu\nu} \frac{\partial^2 \varphi_\nu}{\partial (x^0)^2} + \cdots = 0, \qquad \mu = 1, 2, \ldots, m, \qquad (14.2.7)$$

and are hence solvable with respect to $\frac{\partial^2 \varphi_\nu}{\partial (x^0)^2}$ if the determinant of the matrix $E_{\mu\nu}^{00}$ does not vanish, i.e.

$$\Omega = \det E_{\mu\nu}^{00} \neq 0, \qquad \mu, \nu = 1, 2, \ldots, m. \tag{14.2.8}$$

Furthermore, the single equation (14.2.3) can be put in normal form provided that

$$E^{00} \neq 0. \tag{14.2.9}$$

If the normality conditions (14.2.6), (14.2.8) and (14.2.9) are satisfied, for a given carrier hyperplane having equation $x^0 = a^0$, one can apply the Cauchy theorem, and the functions φ_ν, or the single function φ of Eq. (14.2.3), are uniquely determined in the neighbourhood of such hyperplane.

It is now necessary to investigate under which conditions the normal character is preserved, if the independent variables $x^0, x^1, \ldots, x^n$ are mapped into new variables $z, z^1, \ldots, z^n$, so that the hyperplane of equation $x^0 = a^0$ is turned into a hypersurface σ of the space S having equation

$$z(x^0, x^1, \ldots, x^n) = z^0, \tag{14.2.10}$$

starting from which one can determine (at least in a neighbourhood) the φ functions. For this purpose, one defines

$$p_i \equiv \frac{\partial z}{\partial x^i}, \qquad i = 0, 1, \ldots, n, \tag{14.2.11}$$

from which one obtains

$$\frac{\partial \varphi_\nu}{\partial x^i} = \frac{\partial \varphi_\nu}{\partial z} p_i + \sum_{j=1}^{3} \frac{\partial \varphi_\nu}{\partial z^j} \frac{\partial z_j}{\partial x^i}, \qquad \nu = 1, 2, \ldots, m, \tag{14.2.12}$$

where we need, on the right-hand side, only the first term, so that we write, following [93],

$$\frac{\partial \varphi_\nu}{\partial x^i} = \frac{\partial \varphi_\nu}{\partial z} p_i + \cdots, \qquad \nu = 1, 2, \ldots, m. \tag{14.2.13}$$

The insertion of (14.2.13) into the system (14.2.1) yields

$$\sum_{\nu=1}^{m} \frac{\partial \varphi_\nu}{\partial z} \sum_{i=0}^{n} E_{\mu\nu}^i p_i + \cdots = 0, \qquad \mu = 1, 2, \ldots, m. \tag{14.2.14}$$

If now one sets

$$\omega_{\mu\nu}^{(1)} \equiv \sum_{i=0}^{n} E_{\mu\nu}^{i} p_i, \tag{14.2.15}$$

the transformed system turns out to be normal provided that

$$\Omega^{(1)} \equiv \det\omega_{\mu\nu}^{(1)} \neq 0, \qquad \mu, \nu = 1, 2, \ldots, m. \tag{14.2.16}$$

As far as the system (14.2.2) is concerned, one finds in analogous way

$$\frac{\partial^2 \varphi_\nu}{\partial x^i \partial x^k} = \frac{\partial^2 \varphi_\nu}{\partial z^2} p_i p_k + \cdots, \tag{14.2.17}$$

and the Eq. (14.2.2) are turned into

$$\sum_{\nu=1}^{m} \frac{\partial^2 \varphi_\nu}{\partial z^2} \sum_{i,k=0}^{n} E_{\mu\nu}^{ik} p_i p_k + \cdots = 0, \qquad \mu = 1, 2, \ldots, m. \tag{14.2.18}$$

If one defines the matrix

$$\omega_{\mu\nu}^{(2)} \equiv \sum_{i,k=0}^{n} E_{\mu\nu}^{ik} p_i p_k, \tag{14.2.19}$$

the condition of normality of the system is expressed by non-singularity of this matrix, i.e.

$$\Omega^{(2)} \equiv \det\omega_{\mu\nu}^{(2)} \neq 0, \qquad \mu, \nu = 1, 2, \ldots, m. \tag{14.2.20}$$

Note that, in Eq. (14.2.16), the $\omega_{\mu\nu}^{(1)}$ are linear forms of the variables $p_0, p_1, \ldots, p_n$, and hence $\Omega^{(1)}$ is a form of degree m in such arguments, while in Eq. (14.2.20) the $\omega_{\mu\nu}^{(2)}$ are quadratic forms of the p, and hence $\Omega^{(2)}$ is a form of degree $2m$ of the arguments $p_0, p_1, \ldots, p_n$.

In the case of the unique Eq. (14.2.3), the determinant reduces to the only element

$$\Omega = \sum_{i,k=0}^{n} E^{ik} p_i p_k. \tag{14.2.21}$$

To sum up, to every function $z(x^0, x^1, \ldots, x^n)$ for which Ω does not vanish identically, there corresponds a family of hypersurfaces $z = z^0$, starting from each of which it is still possible to solve the Cauchy problem. This consists in determining the unknown functions when the initial data are relative to the hypersurface itself.

This holds by virtue of the normal character of the transformed system with respect to z.

When the function $z(x^0, x^1, \ldots, x^n)$ satisfies the equation

$$\Omega = 0, \tag{14.2.22}$$

it is no longer possible to apply (regardless of the value taken by the constant z^0) the Cauchy theorem starting from the carrier hypersurfaces $z = z^0$. In such a case, the carrier hypersurfaces are said to be *characteristic manifolds* [65, 93]. Equation (14.2.22) warns us that the system formed by the variables $z, z^1, \ldots, z^n$ is not normal with respect to z and makes it possible to assign the manifolds, in correspondence to which one cannot state that the unknown functions can be determined, once the values of unknown functions and their derivatives of order less than the maximum have been assigned on the manifold.

For the case of the single Eq. (14.2.3), the characteristic manifold is the one satisfying the equation

$$\sum_{i,k=0}^{n} E^{ik} p_i p_k = 0. \tag{14.2.23}$$

Such a manifold is necessarily complex if the quadratic form on the left-hand side of (14.2.23) is positive-definite. Otherwise the manifold is real, if the initial data, called Cauchy data, are real. Equation (14.2.23) can then be viewed as expressing the vanishing of the square of the pseudo-norm of the normal vector, and is therefore a null hypersurface (see details in [65]). In other words, with the nomenclature of relativity and pseudo-Riemannian geometry, *characteristic manifolds are null hypersurfaces*.

14.3 The Concept of Wavelike Propagation

The scalar wave equation (14.2.4) can be applied, in particular, to the air's acoustic vibrations, or to the vibrations of other gaseous masses, since one can neglect in a first approximation any dissipative effect and hence consider the motion as if it were irrotational, without heat exchange among particles (this behaviour is called *adiabatic*). If the velocity potential φ in Eq. (14.2.4) describes sound vibrations in the air, the three partial derivatives represent the speed of the air molecule located in (x^1, x^2, x^3) at time t. More precisely, what is vibrating at a generic time t is a certain air's stratum, placed in between the two surfaces

$$z(t|x) = c_1, \quad z(t|x) = c_2. \tag{14.3.1}$$

Outside this stratum there is rest, i.e. the solution of Eq. (14.2.4) vanishes, whereas within the stratum the acoustic phenomenon is characterized by a non-vanishing solution $\varphi(t|x)$.

From now on, without insisting on the acoustic interpretation of the solutions of Eq. (14.2.4), we assume that $\varphi(t|x)$ and $\varphi^*(t|x)$ are solutions of this equation within and outside of the stratum determined by the surfaces in Eq. (14.3.1), respectively. The phenomenon described by Eq. (14.2.4) is characterized by two distinct functions, depending on whether it is studied inside or outside the stratum. Throughout the surface of Eq. (14.3.1) the derivatives of various orders of φ will undergo, in general, sudden variations and, for this reason, one says we are dealing with *discontinuity surfaces*. Now it may happen that such discontinuities vary with time, in which case the discontinuity that undergoes propagation is said to be a *wave*.

Thus, if Eq. (14.2.4) is interpreted as characterizing a possible wavelike propagation, the discontinuity surface (or *wave surface*) bounds a stratum that undergoes displacement and, possibly, deformation with time. Following [93], we shall assume that, during the motion, no interpenetration or molecular cavitation occurs, so that, on passing through a wave surface, the normal component of the velocity of a generic particle does not undergo any discontinuity. We also rule out the possible occurrence of sliding phenomena of molecules on such surfaces, which would lead to tangential discontinuities of the velocity of particles. Moreover, in light of the postulates on the pressure that the mechanics of continua relies upon, the pressure cannot, under ordinary conditions, undergo sudden jumps, even if the state of motion were to change abruptly. The density μ is related to the pressure p by the equations of state $f(\mu, p) = 0$, which is the same on both sides of the discontinuity surface. The continuity of p implies therefore that also μ is continuous. On the other hand, by virtue of the equation [93]

$$\frac{\partial \varphi}{\partial t} + V^2 \sigma = 0, \tag{14.3.2}$$

the derivatives of φ and φ^* with respect to t represent a density up to a constant factor, hence also such derivatives must be continuous across the wave surface.

By virtue of all previous considerations one can say that, for the Eq. (14.2.4) to describe a wavelike propagation, one has to assume the existence of two different solutions, say φ and φ^*, taken to characterize the physical phenomenon, inside and outside of a stratum, that match each other, i.e. have equal first-order derivatives in time and space, through the wave surface which bounds the stratum at every instant of time. The second derivatives undergo instead sudden variations.

Let us now consider one of the wave surfaces σ which, at time t, bound the stratum where the perturbation is taking place, and let n be the outward-pointing normal to such a stratum at a generic point P. The surface undergoes motion and, at time $t + dt$, intersects the normal n at a point Q. The measure of the PQ segment, counted positively towards the exterior, can be denoted dn. The ratio

$$a \equiv \frac{dn}{dt} \tag{14.3.3}$$

is said to be the *progression velocity* of the wave surface at the point P at the instant of time under consideration. Under ordinary circumstances, at all points of one of the two limiting surfaces of the stratum, a is positive, while at all points of the other limiting surface a is negative. The former surface is said to be a *wave front* or a *bow*, while the latter is said to be a *poop*. The difference

$$v(P) \equiv a - \frac{d\varphi}{dn} \tag{14.3.4}$$

between the progression velocity and the component orthogonal to σ of the velocity of the fluid particle placed at the point P at the instant t is said to be the *normal propagation velocity* of the surface σ at the point P. This velocity measures the rate at which the surface is moving with respect to the medium (and not with respect to the fixed axes!).

If outside the stratum there is a rest condition, the solution φ^* vanishes and therefore, by virtue of the matching conditions at σ, one can write that

$$\frac{d\varphi}{dn} = 0 \Longrightarrow v(P) = a. \tag{14.3.5}$$

In this particular case the propagation velocity coincides with the progression velocity.

Note now that the surface σ is a characteristic manifold of Eq. (14.2.4), i.e. an integral of the equation

$$\frac{1}{V^2}(p_0)^2 - \sum_{i=1}^{3}(p_i)^2 = 0. \tag{14.3.6}$$

Indeed, if this were not true, a unique solution of Eq. (14.2.4) would be determined in the neighbourhood of σ by the mere knowledge of the values taken upon σ by φ and $\frac{\partial \varphi}{\partial t}$, in light of Cauchy's theorem. *The wavelike propagation is therefore possible because the wave surfaces are characteristic manifolds.*

In order to further appreciate how essential is the consideration of characteristic manifolds, let us study the following example [93]. Let us assume for simplicity that we study the wave equation (14.2.4) in two-dimensional Minkowski space-time, with x^1 denoted by x. Hence we write it in the form

$$\left(\frac{1}{V^2} \frac{\partial^2}{\partial t^2} - \frac{\partial^2}{\partial x^2} \right) \varphi = \left(\frac{1}{V} \frac{\partial}{\partial t} + \frac{\partial}{\partial x} \right) \left(\frac{1}{V} \frac{\partial}{\partial t} - \frac{\partial}{\partial x} \right) \varphi = 0. \tag{14.3.7}$$

The form of Eq. (14.3.7) suggests defining the new variables

$$z \equiv x - Vt, \ z_1 \equiv x + Vt, \tag{14.3.8}$$

from which the original variables are re-expressed as

$$x = \frac{1}{2}(z + z_1), \ t = \frac{1}{2}\frac{(z_1 - z)}{V}. \tag{14.3.9}$$

Moreover, the standard rules for differentiation of composite functions lead now to

$$\frac{\partial}{\partial z} = \frac{1}{2}\left(\frac{\partial}{\partial x} - \frac{1}{V}\frac{\partial}{\partial t}\right), \ \frac{\partial}{\partial z_1} = \frac{1}{2}\left(\frac{\partial}{\partial x} + \frac{1}{V}\frac{\partial}{\partial t}\right), \tag{14.3.10}$$

and hence Eq. (14.3.7) reads as

$$\frac{\partial^2 \varphi}{\partial z \partial z_1} = 0, \tag{14.3.11}$$

which is solved by a sum of arbitrary smooth functions

$$\varphi(z, z_1) = \alpha(z) + \beta(z_1) \tag{14.3.12}$$

depending only on z and on z_1, respectively. Thus, it is not possible in general to solve the Cauchy problem for a carrier line $z = c$, but it is necessary that the data satisfy a compatibility condition. In our case, from the solution (14.3.12) one finds

$$\varphi(c, z_1) = \alpha(c) + \beta(z_1), \ \left(\frac{\partial \varphi}{\partial z}\right)_{z=c} = \alpha'(c). \tag{14.3.13}$$

The functions $\varphi_0 = \varphi(z = c)$ and $\varphi_1 = \left(\frac{\partial \varphi}{\partial z}\right)_{z=c}$ of the variable z_1 cannot be therefore chosen at will, but the function $\varphi_1(z_1)$ must be a constant, in which case there exist infinitely many forms of the solution of the Cauchy problem for the scalar wave equation. In the following section we will discover also the advantage of characteristic manifolds, because we will see that they make it possible to solve the Cauchy problem pertaining to non-characteristic carrier manifolds.

14.4   The Concept of Hyperbolic Equation

The scalar wave equation (14.2.4) is a good example of hyperbolic equation, but before we go on it is appropriate to define what is an equation of hyperbolic type. Following [89], we first define this concept on a vector space and then on a manifold.

We consider a l-dimensional vector space X over the field of real numbers, whose dual vector space is denoted by Ξ. The point $x = (x^1, \dots, x^l) \in X$, and the point

$p = \left(\frac{\partial}{\partial x^1}, \ldots, \frac{\partial}{\partial x^l}\right) \in \Xi$. A differential equation of order m can be therefore written in the form

$$a(x, p)u(x) = v(x), \tag{14.4.1}$$

where $a(x, \xi)$ is a given real polynomial in ξ of degree m whose coefficients are functions defined on X, $u(x)$ is the unknown function and $v(x)$ a given function. Let $h(x, \xi)$ be the sum of the homogeneous terms of $a(x, \xi)$ of degree m (also called the *leading symbol* of the differential operator $a(x, p)$), and let $V_x(h)$ be the cone defined in Ξ by the equation

$$h(x, \xi) = 0. \tag{14.4.2}$$

The differential operator $a(x, p)$ is said to be *hyperbolic at the point* x if Ξ contains points ξ such that *any real line through ξ cuts the cone* $V_x(h)$ *at m real and distinct points*. These points ξ constitute the interior of two opposite convex and closed half-cones $\Gamma_x(a)$ and $-\Gamma_x(a)$, whose boundaries belong to $V_x(h)$.

Suppose that the following conditions hold:

(i) The operator $a(x, p)$ is hyperbolic at each point x of the vector space X.
(ii) The set

$$\Gamma_X \equiv \cap_{x \in X} \Gamma_x \tag{14.4.3}$$

has a non-empty interior.
(iii) No limit of $h(x, \xi)$ as the norm of x approaches 0 is vanishing.
(iv) No limit of the cones $V_x(h)$ as the norm of x approaches infinity has singular generator.

Under such circumstances, the operator $a(x, p)$ is said to be *regularly hyperbolic on X*.

When X is instead a l-dimensional $(m + M)$-smooth manifold, not necessarily complete, the operator $a(x, p)$ is said to be *hyperbolic on X* when the following conditions hold:

(1) $a(x, p)$ is hyperbolic at any point x of X, in the sense specified above.
(2) The set of timelike paths (i.e. with timelike tangent vector) from y to z is compact or empty for any y and $z \in X$.
(3) Either the coefficients of $a(x, p)$ have *locally bounded* (which means boundedness on any compact subset of X) derivatives of order M such that $1 \leq M \leq l$, or they have locally bounded derivatives of order $\leq l'$ and locally square integrable derivatives of order $> l'$ and $\leq M$, l' being the smallest integer $> \frac{l}{2}$.
(4) The total curvature of the interior of Γ_x is positive. If $M = 1$, then the first derivatives of the coefficients of $h(x, \xi)$ are continuous.

14.5   Riemann Kernel for a Hyperbolic Equation in 2 Variables

We here follow [28] in pointing out that the modern theory of hyperbolic equations was initiated by Riemann's representation of the solution of the initial-value problem for an equation of second order. Riemann was motivated by a very concrete problems in acoustics, but here we focus on the mathematical ingredients of his conceptual construction.

Indeed, any second-order linear hyperbolic equation in two variables can be written in one of the two equivalent forms

$$L[u] \equiv \left(\frac{\partial^2}{\partial x \partial y} + a(x,y) \frac{\partial}{\partial x} + b(x,y) \frac{\partial}{\partial y} + c(x,y) \right) u = f(x,y), \quad (14.5.1)$$

$$L[u] \equiv \left(\frac{\partial^2}{\partial y^2} - \frac{\partial^2}{\partial x^2} + a(x,y) \frac{\partial}{\partial x} + b(x,y) \frac{\partial}{\partial y} + c(x,y) \right) u = f(x,y), \quad (14.5.2)$$

where the functions a, b, c, f are of a suitable differentiability class. The initial curve C is taken to be nowhere tangent to a characteristic direction; the characteristics pertaining to Eq. (14.5.1) are straight lines parallel to the coordinate axes; the characteristics in Eq. (14.5.2) are the lines $x + y = $ const. and $x - y = $ const.

The aim is to represent a solution u at a point P in terms of f and of the initial data, i.e. the values taken by u and one derivative of u on C. If the initial curve degenerates into a right angle formed by the characteristics $x = \gamma, y = \delta$, one can no longer prescribe two conditions on the initial curve C, but one has to consider the *characteristic initial-value problem*, in which only the values of u on $x = \gamma$ and $y = \delta$ are prescribed.

The idea underlying Riemann's method is as follows. One multiplies the hyperbolic equation by a function v, integrates over a region $\mathcal{R}$, transforms the integral by Green's formula so that u appears as a factor of the integrand, then one tries to determine v in such a way that the required representation is obtained. This procedure is implemented by introducing the adjoint operator L^*, defined in such a way that $vL[u] - uL^*[v]$ is a divergence (here we are not dealing with linear operators on Hilbert spaces, although we will obtain formulae which are formally analogous to the formal adjoint of a linear operator on Hilbert space). For the hyperbolic equation in the form (14.5.1), the adjoint operator L^* turns out to be

$$L^*[v] = \left(\frac{\partial^2}{\partial x \partial y} - \frac{\partial a}{\partial x} - a(x,y) \frac{\partial}{\partial x} - \frac{\partial b}{\partial y} - b(x,y) \frac{\partial}{\partial y} + c(x,y) \right) v, \quad (14.5.3)$$

and one has

$$vL[u] - uL^*[v] = \frac{\partial}{\partial y}\left(v\frac{\partial u}{\partial x} + buv\right) - \frac{\partial}{\partial x}\left(u\frac{\partial v}{\partial y} - auv\right). \qquad (14.5.4)$$

At this stage, integration over a two-dimensional domain D with boundary Γ and Gauss' formula lead to

$$-\int\int_D \left(vL[u] - uL^*[v]\right)dx\,dy$$
$$= \int_\Gamma \left[\left(v\frac{\partial u}{\partial x} + buv\right)dx + \left(u\frac{\partial v}{\partial y} - auv\right)dy\right]. \qquad (14.5.5)$$

Now we have $L[u] = f$ from Eq. (14.5.1), and if we consider the case described by Fig. 14.1, for which the boundary Γ consists of the paths $AB + BP + PA$, we find

$$-\int\int_D \left(fv - uL^*[v]\right)dx\,dy$$
$$= \int_{AB+BP+PA}\left[v\left(\frac{\partial u}{\partial x} + bu\right)dx + u\left(\frac{\partial v}{\partial y} - av\right)dy\right]$$
$$= \int_{AB}\left[v\left(\frac{\partial u}{\partial x} + bu\right)dx + u\left(\frac{\partial v}{\partial y} - av\right)dy\right]$$
$$+ \int_{BP}u\left(\frac{\partial v}{\partial y} - av\right)dy - \int_{AP}v\left(\frac{\partial u}{\partial x} + bu\right)dx. \qquad (14.5.6)$$

For the last term on the right-hand side of this formula, we can use integration by parts to write

$$\int_{AP}v\left(\frac{\partial u}{\partial x} + bu\right)dx = v(P)u(P) - v(A)u(A) - \int_{AP}u\left(\frac{\partial v}{\partial x} - bv\right)dx, \qquad (14.5.7)$$

Fig. 14.1 Geometry of the characteristic initial-value problem

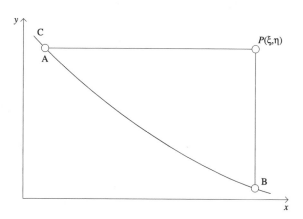

and hence we obtain, from (14.5.6) and (14.5.7),

$$
\begin{aligned}
u(P)v(P) = u(A)v(A) &+ \int_{AP} u \left(\frac{\partial v}{\partial x} - bv \right) dx + \int_{BP} u \left(\frac{\partial v}{\partial y} - av \right) dy \\
&+ \int_{AB} \left[v \left(\frac{\partial u}{\partial x} + bu \right) dx + u \left(\frac{\partial v}{\partial y} - av \right) dy \right] \\
&+ \int \int_{D} \left(fv - uL^*[v] \right) dx \, dy.
\end{aligned}
\tag{14.5.8}
$$

In order to represent $u(P) = u(\xi, \eta)$ we choose for v a two-point function or kernel $R(x, y; \xi, \eta)$ subject to the following conditions [28]:

(i) As a function of x and y, R satisfies the homogeneous equation

$$
L^*_{(x,y)}[R] = 0.
\tag{14.5.9}
$$

(ii) $\frac{\partial R}{\partial x} = bR$ on the segment AP parallel to the x-axis, and $\frac{\partial R}{\partial y} = aR$ on the segment BP parallel to the y-axis. More precisely, one has to write

$$
\frac{\partial}{\partial x} R(x, y; \xi, \eta) = b(x, \eta) R(x, y; \xi, \eta) \text{ on } y = \eta,
\tag{14.5.10}
$$

$$
\frac{\partial}{\partial y} R(x, y; \xi, \eta) = a(\xi, y) R(x, y; \xi, \eta) \text{ on } x = \xi.
\tag{14.5.11}
$$

(iii) The kernel R equals 1 at coinciding points, i.e.

$$
R(\xi, \eta; \xi, \eta) = 1.
\tag{14.5.12}
$$

Note that Eqs. (14.5.10) and (14.5.11) reduce to ordinary differential equations for the kernel R along the characteristics. Their integration, jointly with Eq. (14.5.12) to determine the constant of integration, yields

$$
R(x, \eta; \xi, \eta) = \exp \left[\int_{\xi}^{x} b(\lambda, \eta) d\lambda \right],
\tag{14.5.13}
$$

$$
R(\xi, y; \xi, \eta) = \exp \left[\int_{\eta}^{y} a(\lambda, \xi) d\lambda \right].
\tag{14.5.14}
$$

The formulae provide the value of R along the characteristics passing through the point $P(\xi, \eta)$. The problem of finding a solution R of Eq. (14.5.9) with data (14.5.13) and (14.5.14) is said to be a *characteristic initial-value problem*. Riemann did not actually prove the existence of such a solution, but brought this novel perspective in mathematics, i.e. solving hyperbolic equations by finding kernel functions that obey characteristic initial-value problems. In the case under examination, the desired Riemann's representation formula can be written as

$$u(P) = u(A)R(A; \xi, \eta) + \int_{AB} \left[R \left(\frac{\partial u}{\partial x} + bu \right) dx + u \left(\frac{\partial R}{\partial y} - aR \right) dy \right]$$
$$+ \int \int_D Rf \, dx \, dy.$$

$$(14.5.15)$$

Such an integral formula can be written in a more convenient, i.e. more symmetric expression. For this purpose, one adds the identity [28]

$$0 = \frac{1}{2} \left[u(B)R(B) - u(A)R(A) \right]$$
$$- \int_{AB} \left[\frac{1}{2} \left(\frac{\partial u}{\partial x} R + u \frac{\partial R}{\partial x} \right) dx + \frac{1}{2} \left(\frac{\partial u}{\partial y} R + u \frac{\partial R}{\partial y} \right) dy \right], \quad (14.5.16)$$

which is obtained by integrating the product uR by parts along the curve C between the points A and B of Fig. 14.1. Hence one obtains eventually the desired integral representation of the solution in the form

$$u(P) = \frac{1}{2} \left[u(A)R(A) + u(B)R(B) \right]$$
$$+ \int_{AB} \left(\left[\frac{1}{2} R \frac{\partial u}{\partial x} + \left(bR - \frac{1}{2} \frac{\partial R}{\partial x} \right) u \right] dx \right.$$
$$\left. - \left[\frac{1}{2} R \frac{\partial u}{\partial y} + \left(a \frac{\partial R}{\partial y} - \frac{1}{2} \frac{\partial R}{\partial y} \right) u \right] dy \right)$$
$$+ \int \int_D Rf \, dx \, dy.$$

$$(14.5.17)$$

Equation (14.5.17) represents the solution of Eq. (14.5.1) for arbitrary initial values assigned along an arbitrary non-characteristic curve C, by means of a solution R of the homogeneous adjoint equation which depends on x, y and on the parameters ξ, η.

The Riemann representation formula here obtained is a particular case of a property discovered by F. Riesz, according to which a continuous linear functional $u(P) = L[f]$ can be represented in the form

$$u(P) = \int K(Q, P) f(Q) dQ, \quad (14.5.18)$$

where the kernel K is a two-point function depending on Q and P, and the integration is performed over the range pertaining to the variable Q. Of course, suitable regularity conditions should be satisfied for this scheme to make sense. In general one can prove that, if the Riemann function exists and is of class C^k for a value of k sufficiently large, the function $u(P)$ written in the form (14.5.17) is indeed a solution of the equation $L[u] = f$ with the prescribed data, *provided that the curve is nowhere characteristic* [28].

14.6   Lack of Smooth Cauchy Problem for the Laplace Equation

At this stage, a naturally occurring question might be whether the Cauchy problem can be considered also for the linear elliptic equations of part II. A famous counterexample of Hadamard shows that the Cauchy problem is not the right question to pose for elliptic equations, at least not in the real domain [67, 74].

Following Hadamard, let us consider the Laplace equation in the two-dimensional Euclidean plane

$$\left(\frac{\partial^2}{\partial x^2} + \frac{\partial^2}{\partial y^2} \right) u = 0, \tag{14.6.1}$$

supplemented by the Cauchy data

$$u(0, y) = 0, \quad \frac{\partial u}{\partial x}(0, y) = \frac{1}{n} \sin(ny), \tag{14.6.2}$$

where n is any positive integer. The method of separation of variables suggests writing

$$u(x, y) = \frac{1}{n^2} \sinh(nx) \; \sin(ny) \tag{14.6.3}$$

for the solution of the Cauchy problem defined by Eqs. (14.6.1) and (14.6.2). Indeed, a factorized solution $u(x, y) = f(x)g(y)$ for the Laplace equation leads to

$$\frac{f''(x)}{f(x)} = k, \quad \frac{g''(y)}{g(y)} = -k, \tag{14.6.4}$$

where the first equation is solved by

$$f(x) = A_1 e^{\sqrt{k}x} + A_2 e^{-\sqrt{k}x}. \tag{14.6.5}$$

The Cauchy data lead to the algebraic system

$$A_1 + A_2 = 0, \tag{14.6.6}$$

$$\sqrt{k}(A_1 - A_2) = \frac{1}{n},\qquad (14.6.7)$$

and hence the solution (14.6.3) is obtained. Now when the parameter n becomes infinite, the Cauchy data approach zero uniformly, whereas for $x \neq 0$ the solution (14.6.3) oscillates between limits that increase indefinitely, by virtue of the exponentials e^{nx} and e^{-nx} in the hyperbolic sine function. On the other hand, a vanishing function is an obvious solution of Eq. (14.6.1) with vanishing Cauchy data, hence we have shown that, *for the Laplace equation, the dependence of the solution of Cauchy's problem on the data is not in general continuous.*

However, it is not equally widely known that, in [118], the author, without contradicting the Hadamard example, has shown that *positive solutions of Cauchy problems for the Laplace equation depend continuously on the data.*

Chapter 15
Characteristics and Waves, 2

15.1 Wavelike Propagation for a Generic Normal System

Following Levi-Civita [93], we here assume that, inside and outside the stratum determined by two hypersurfaces of equations

$$z = c_1, \ z = c_2, \tag{15.1.1}$$

one of the two systems

$$E_\mu = \sum_{\nu=1}^{m} \sum_{i=0}^{n} E_{\mu\nu}^i \frac{\partial \varphi_\nu}{\partial x^i} + \Phi_\mu(x|\varphi) = 0, \qquad \mu = 1, 2, ..., m, \tag{15.1.2}$$

$$E_\mu = \sum_{\nu=1}^{m} \sum_{i,k=0}^{n} E_{\mu\nu}^{ik} \frac{\partial^2 \varphi_\nu}{\partial x^i \partial x^k} + \Phi_\mu(x|\varphi|\chi) = 0, \qquad \mu = 1, 2, ..., m, \tag{15.1.3}$$

is satisfied by the m functions $\varphi_1, \varphi_2, ..., \varphi_m$ and $\varphi_1^*, \varphi_2^*, ..., \varphi_m^*$, respectively. We assume that the stratum determined by Eq. (15.1.1) undergoes motion and possibly also bendings, and that through the hypersurfaces (15.1.1) the partial derivatives of first order (for (15.1.2)) and of second order (for (15.1.3)) undergo sudden variations (or jumps) and are therefore discontinuous therein.

The solutions φ of Eq. (15.1.2) are taken to be continuous through the hypersurfaces (15.1.1), while the solutions φ^* of Eq. (15.1.3) are taken to be continuous together with their first derivatives through the confining hypersurfaces. This is what happens in a wavelike phenomenon, where the wave surfaces are those bounding the stratum.

For a system of maximal order s, the functions φ and φ^* should obey matching conditions through the wave surfaces of order less than s, whereas some discontinuities occur for the derivatives of order s. The wave surfaces turn out to be characteristic

© Springer International Publishing AG 2017
G. Esposito, *From Ordinary to Partial Differential Equations*, UNITEXT 106,
DOI 10.1007/978-3-319-57544-5_15

manifolds, because *out of them it is not possible to apply the theorem that guarantees uniqueness of the integrals.*

Hereafter we merely assume the existence of functions φ and φ^* with the associated wavelike propagation, and we describe some of its properties. If $z = c$ is a wave surface σ, the function z must satisfy the equation

$$\Omega(x|p) = 0, \tag{15.1.4}$$

where the p variables are given by

$$p_i = \frac{\partial z}{\partial x^i}, \qquad i = 0, 1, ..., n. \tag{15.1.5}$$

The validity of Eq. (15.1.4) is indeed established only on σ, i.e. for z equal to a particular value c. However, the limitation $z = c$ is inessential, because Ω certainly vanishes whenever the p_i are set equal to the derivatives of the function z. One therefore deals, with respect to z, with a partial differential equation. Such an equation can characterize z by itself provided that the E_μ functions occurring in such systems depend only on the x variables.

For the time being we have a simpler task, i.e. the study of the velocity of progression of the wave surface σ at a point P, by assuming that the space of variables $x^1, x^2, ..., x^n$ is endowed with an Euclidean metric, and that such variables are Cartesian coordinates. We suppose that

$$z(t|x) = c, \ z(t + dt|x) = c \tag{15.1.6}$$

are the equations of σ at the instants of time t and $t + dt$, respectively. The normal N to P at σ intersects the second of Eq. (15.1.6) at a point Q. If dN is the measure, with sign, of the segment PQ, counted positively towards the exterior of the stratum determined by σ and by the other wave surface pertaining to the instant $t + dt$, the ratio (Sect. 14.3)

$$a \equiv \frac{dN}{dt} \tag{15.1.7}$$

is said to be the *velocity of progression* of the wave surface at the point P at the instant of time under consideration. It can be evaluated in terms of the properties of the wave surface as follows [93].

The directional cosines of the normal N to σ at P are given by

$$\alpha_i = \frac{p_i}{|\rho|}, \qquad i = 1, 2, ..., n, \tag{15.1.8}$$

where

$$\rho^2 \equiv \sum_{i,j=1}^{n} \delta^{ij} p_i p_j = \sum_{i=1}^{n} p_i p^i. \tag{15.1.9}$$

If the points P and Q have coordinates x^i and $x^i + dx^i$, respectively, one has from (15.1.6)

$$z(t|x) = c, \quad z(t + dt|x + dx) = c, \tag{15.1.10}$$

and hence, by taking the difference,

$$dz = p_0 dt + \sum_{i=1}^{n} p_i dx^i = 0. \tag{15.1.11}$$

Since the dx^i are the components of the vector $Q - P$, one has also

$$dx^i = \pm \sum_{j=1}^{n} \delta^{ij} \alpha_j dN = \pm \alpha^i dN, \quad i = 1, 2, \dots, n. \tag{15.1.12}$$

The sign is $\pm$ depending on whether z is positive or negative outside of the stratum. We do not need to fix it. By virtue of (15.1.8) and (15.1.12), Eq. (15.1.11) leads to (we set $\varepsilon \equiv \pm 1$)

$$p_0 dt + \sum_{i=1}^{n} p_i \varepsilon \alpha^i dN = p_0 dt + \varepsilon \sum_{i=1}^{n} \frac{p_i p^i}{|\rho|} dN$$

$$= p_0 dt + \varepsilon |\rho| dN = 0, \tag{15.1.13}$$

from which

$$|a| = \left| \frac{dN}{dt} \right| = \left| \frac{p_0}{\rho} \right|. \tag{15.1.14}$$

This is the desired formula for the modulus of the velocity of progression, as the point P, the time parameter t and the wave surface are varying. It can be applied, for example, to acoustic vibrations. For perfect gases in adiabatic regime one has

$$a^2 = \gamma \frac{p_0}{\mu_0} \approx 331 \text{ m s}^{-1}, \tag{15.1.15}$$

which is in fairly good agreement with the experimental value. Historically, Newton regarded the propagation of sound as an isothermal phenomenon, hence he neglected the γ parameter in Eq. (15.1.15). Laplace was possibly the first who understood what was wrong with the Newton model: the compressions and attenuations of air's strata, resulting from wavelike propagation, engender variations of temperature, because the compressed strata get heated up, whereas rarefied strata cool down. The phenomenon is therefore not isothermal.

15.2   Cauchy's Method for Integrating a First-Order Equation

In Sect. 14.2 we have learned that the characteristic manifolds

$$z(x^0, x^1, ..., x^n) = \text{const.}$$

of a normal system of equations in the $n + 1$ independent variables $x^0, x^1, ..., x^n$ ensure the vanishing of a certain determinant

$$\Omega(x|p) = 0, \tag{15.2.1}$$

where the variables p_i are obtained as

$$p_i \equiv \frac{\partial z}{\partial x^i}, \qquad i = 0, 1, 2, ..., n. \tag{15.2.2}$$

In the most general case, Ω depends not only on the x and p, but also on the unknown functions φ of the normal system under consideration. There exists however a particular set of normal systems, of order $s = 1$ and $s = 2$, where Ω depends only on x and p variables, provided that the coefficients $E^i_{\mu\nu}$ in Eq. (15.1.2) and $E^{ik}_{\mu\nu}$ in Eq. (15.1.3) depend only on the x variables.

We are now going to describe the Cauchy method for integrating a first-order partial differential equation, considering, in particular, Eq. (15.2.1), where the unknown function z does not occur explicitly. We are therefore guaranteed that Ω contains at least one of the p functions, e.g. p_0. If Eq. (15.2.1) can be solved with respect to p_0, one can write

$$p_0 + H(t, x^1, ..., x^n|p_1, ..., p_n) = 0. \tag{15.2.3}$$

Let us study first the linear case, i.e. when H is a linear function of the p variables. We are going to show that the task of integrating Eq. (15.2.3) is turned into the integration of a system of ordinary differential equations. Indeed, Eq. (15.2.3) is then of the type

$$p_0 + A_0 + \sum_{i=1}^{n} A^i p_i = 0, \tag{15.2.4}$$

where the A's depend only on the variables $t, x^1, ..., x^n$. Let us consider the space S_{n+2} of the $(n + 2)$ variables $(t, x^1, ..., x^n, z)$ and an integral hypersurface

$$z = \varphi(t|x) \tag{15.2.5}$$

of Eq. (15.2.4), that we shall denote by σ. Let Γ be the section of σ with the hyperplane $t = 0$, i.e. the locus of points defined by the equation

$$\Gamma: \quad z = \varphi(0|x) = \varphi_0(x). \tag{15.2.6}$$

The fundamental guiding principle adopted at this stage consists in viewing σ as the locus of ∞^n curves obtainable by integration of a suitable system of ordinary differential equations of the kind

$$\frac{d}{dt} x^i = X^i(t|x), \qquad i = 1, 2, \ldots, n, \tag{15.2.7}$$

$$\frac{d}{dt} z = Z(t|x), \tag{15.2.8}$$

of rank $(n + 1)$ in the unknown functions $x^1, \ldots, x^n, z$ of the variable t. Such a system involves $(n + 1)$ arbitrary constants, but their number is reduced by 1 if one requires compatibility of the system with Eq. (15.2.5) for the integral hypersurface σ.

The basic assumption, which justifies the interest in the system (15.2.7) and (15.2.8), is that *it is independent of the preliminary integration of Eq. (15.2.4)*. Once we have made this statement, we must express the condition that any integral curve of Eqs. (15.2.7) and (15.2.8) belongs to σ.

Upon viewing z as a function of t and x, Eqs. (15.2.8) and (15.2.7) lead to

$$\frac{dz}{dt} = Z = p_0 + \sum_{i=1}^{n} p_i \frac{dx^i}{dt} = p_0 + \sum_{i=1}^{n} p_i X^i, \tag{15.2.9}$$

and, bearing in mind Eq. (15.2.4) to re-express p_0, one obtains

$$Z = -A_0 + \sum_{i=1}^{n} p_i (X^i - A^i). \tag{15.2.10}$$

Since we want to make sure that the differential system (15.2.7) and (15.2.8) is independent of the integration of Eq. (15.2.4), the coefficients of the p_i must vanish, and hence

$$X^i = A^i, \tag{15.2.11}$$

from which it follows that

$$Z = -A_0. \tag{15.2.12}$$

The desired differential system reads therefore

$$\frac{dx^i}{dt} = A^i, \qquad (i = 1, 2, \ldots, n), \tag{15.2.13}$$

$$\frac{dz}{dt} = -A_0, \tag{15.2.14}$$

or also, with the notation used until the end of nineteenth century,

$$\frac{dx^1}{A^1} = \frac{dx^2}{A^2} = \cdots = \frac{dx^n}{A^n} = -\frac{dz}{A_0} = dt, \qquad (15.2.15)$$

which is capable to determine the integral hypersurfaces σ of Eq. (15.2.4).

Indeed, in order to solve the Cauchy problem relative to a pre-assigned Γ of the hyperplane $t = 0$, it is enough to consider, in the first place, the whole set of integral curves, which are ∞^n, of the system of Eq. (15.2.13), in which the function z does not occur. The integration of the residual differential equation (15.2.14), which is performed by a simple quadrature, once the system (15.2.13) has been integrated, completes the determination of the curves of the space S_{n+2} (of the t, x and z variables) which are integral curves of the system (15.2.13) and (15.2.14). If one wants that these curves emanate from the points of Γ, it is necessary and sufficient that z takes the value $\varphi_0(x)$ at $t = 0$, the x referring to the same zero value of t and being therefore identifiable with the n arbitrary constants introduced from the integration of the system (15.2.13). Thus, the total number of arbitrary constants is n, and every integral hypersurface σ of Eq. (15.2.4) occurs as the locus of ∞^n integral curves of Eqs. (15.2.13) and (15.2.14), emanating from the points of Γ.

The concept of transforming the problem of the integration of a linear partial differential equation of first order into the problem of integrating a system of ordinary differential equations, originally developed by Lagrange, was generalized by Lagrange himself, Charpit, Cauchy and Jacobi to non-linear equations. Hereafter, following Levi-Civita [93], we describe the Cauchy method. For this purpose, let us revert to the general equation

$$p_0 + H(t, x^1, \ldots, x^n | p_1, \ldots, p_n) = 0, \qquad (15.2.16)$$

and let us try to understand whether it is possible to determine a generic integral hypersurface (the one whose existence is guaranteed by virtue of the Cauchy theorem for given initial data) as a locus of integral curves of a suitable differential system.

One can easily recognize that it is no longer possible, in general, to associate with Eq. (15.2.16) a congruence of curves of the space S_{n+2} that holds for whichever integral hypersurface, but it is necessary to pass to an auxiliary higher-dimensional space. It will be useful to regard as arguments, besides the x coordinates of a generic point P of the integral hypersurface σ, also the $p_0, p_1, \ldots, p_n$ which, geometrically, define a facet for P. In order to give a concrete metrical meaning to such p variables, we may regard t, the x and z as Cartesian coordinates of the space S_{n+2}. The variables $p_0, p_1, \ldots, p_n, -1$ are then proportional to the directional cosines of the normal to σ, with reference to the axes $t, x^1, \ldots, x^n, z$ respectively.

Having made this choice, let us try to associate with Eq. (15.2.16) a differential system of the kind

$$\frac{d}{dt}x^i = X^i(t, x | p), \quad \frac{d}{dt}p_i = P_i(t, x | p), \quad i = 1, 2, \ldots, n, \qquad (15.2.17)$$

$$\frac{d}{dt}z = Z(t, x|p). \tag{15.2.18}$$

Once the expressions of the X^i have been determined in terms of the t, x, p variables, one finds also the form of Z. Indeed, since z is a function of t by means of $x^0 = t$ and of $x^1, ..., x^n$, one has

$$\frac{dz}{dt} = p_0 + \sum_{i=1}^{n} p_i \frac{dx^i}{dt}, \tag{15.2.19}$$

and, by virtue of the first of Eq. (15.2.17),

$$\frac{dz}{dt} = Z(t, x|p) = p_0 + \sum_{i=1}^{n} p_i X^i. \tag{15.2.20}$$

Note that Eq. (15.2.18), with Z given by Eq. (15.2.20), should be associated after having integrated the system (15.2.17), because then z can be expressed in terms of t, by means of a quadrature.

Hereafter we denote by Γ a hypersurface in the hyperplane $t = 0$, M_0 a point of Γ, $\bar{\omega}_0$ the hyperplane tangent at M_0 to the integral hypersurface σ of Eq. (15.2.17) that is passing through Γ. We aim at expressing the condition for the integral curve C_0 of the system (15.2.17) and (15.2.18), that emanates from M_0 and is tangent to $\bar{\omega}_0$, to belong to the integral hypersurface σ, while still fulfilling the equations

$$p_i = \frac{\partial z}{\partial x^i}, \quad i = 0, 1, ..., n; x^0 = t, \tag{15.2.21}$$

and this for whatever hypersurface Γ passing through the point M_0.

On passing from t to $t + dt$, p_i undergoes an infinitesimal change

$$dp_i = P_i dt, \tag{15.2.22}$$

and on the other hand, for the Eq. (15.2.21) to remain valid, one requires that

$$dp_i = \sum_{j=0}^{n} p_{ij} dx^j, \quad i = 0, 1, ..., n, \tag{15.2.23}$$

having defined

$$p_{ij} \equiv \frac{\partial^2 z}{\partial x^i \partial x^j} = p_{ji}, \quad i, j = 0, 1, ..., n. \tag{15.2.24}$$

Of course, the formulae (15.2.22) and (15.2.23) for dp_i should agree. Note that the quantities p_{ij} when both indices are positive are arbitrary, because of the choice, arbitrary by hypothesis, of the hypersurface Γ. The components p_{0i} satisfy instead relations that can be obtained by differentiation of Eq. (15.2.16). In other words, one has the $(n + 1)$ equations

$$p_{0i} + \sum_{j=1}^{n} \frac{\partial H}{\partial p_j} p_{ji} + \frac{\partial H}{\partial x^i} = 0, \quad i = 0, 1, ..., n. \tag{15.2.25}$$

Since the full number of p_{ij} components is $\frac{1}{2}(n+1)(n+2)$, we are left with

$$\frac{1}{2}(n+1)(n+2) - (n+1) = \frac{1}{2}n(n+1)$$

free components, while we have at our disposal the $2n$ quantities

$$x^1, x^2, ..., x^n; \ p_1, p_2, ..., p_n.$$

It would therefore seem impossible to determine the P_i in such a way that

$$P_i dt = \sum_{j=0}^{n} p_{ij} dx^j, \tag{15.2.26}$$

independently of the p_{ij}.

However, Cauchy's idea works because, by virtue of $p_i = \frac{\partial z}{\partial x^i}$, one finds, by differentiation with respect to t,

$$\frac{dp_i}{dt} = P_i = p_{i0} + \sum_{j=1}^{n} p_{ij} \frac{dx^j}{dt} = p_{i0} + \sum_{j=1}^{n} p_{ij} X^j. \tag{15.2.27}$$

Now we eliminate the $p_{i0} = p_{0i}$ by means of Eq. (15.2.25) and we exploit the symmetry of the p_{ij}. Hence we find that Eq. (15.2.27) is equivalent to

$$P_i = -\frac{\partial H}{\partial x^i} + \sum_{j=1}^{n} \left(X^j - \frac{\partial H}{\partial p_j} \right) p_{ij}, \quad i = 1, ..., n. \tag{15.2.28}$$

Such equations are satisfied independently of the p_{ij} values provided that, for all i ranging from 1 through n, the following equations hold:

$$X^i = \frac{\partial H}{\partial p_i} \tag{15.2.29}$$

$$P_i = -\frac{\partial H}{\partial x^i}. \tag{15.2.30}$$

15.3 The Bicharacteristics

We have just shown that if, starting from a generic point M_0 of the integral hypersurface σ, one assigns to the t, x, p, z variables some increments which obey the differential system (15.2.17) and (15.2.18), which is uniquely characterized in the form

$$\frac{d}{dt}x^i = \frac{\partial H}{\partial p_i}, \quad \frac{d}{dt}p_i = -\frac{\partial H}{\partial x^i}, \quad i = 1, 2, ..., n, \tag{15.3.1}$$

$$\frac{d}{dt}z = \sum_{i=1}^{n} p_i \frac{\partial H}{\partial p_i} - H, \tag{15.3.2}$$

one reaches an infinitely close point M_1 which belongs again to σ and for which the $p_i + dp_i$ define the direction of the normal to the hypersurface σ itself.

The same considerations may be certainly repeated starting from the point M_1, and this for the essential reason that the system (15.2.17), (15.2.18) and hence (15.3.1), (15.3.2) has been built in such a way that it holds for all integral hypersurfaces σ passing through M_0 with given orientation of the normal, i.e. with given values of p. As far as the integral hypersurface σ is concerned, we are therefore at M_1 in the same conditions in which we found ourselves at M_0. Hence the whole curve C, defined uniquely from Eqs. (15.3.1) and (15.3.2) under the condition that the x, p, z take at $t = 0$ the values corresponding to M_0, belongs to the integral hypersurface under consideration, which (it should be stressed) is an integral hypersurface whatsoever among the many passing through M_0 and having therein the $\overline{w}_0$ as tangent hyperplane.

Thus, we discover the geometric corollary according to which, if two integral manifolds meet each other at a point, they meet each other along the whole curve C passing through that point.

The curves C were called characteristics by Cauchy, but, after Hadamard, it has become customary to call them *bicharacteristics*, whereas we call characteristics (of the space S of the t, x variables) the hypersurfaces having exceptional behaviour with respect to the Cauchy problem (Sect. 14.2).

15.4 Space-Time Manifold; Arc-Length; Geodesics

In the following chapters we shall use also the modern nomenclature for characteristics and bicharacteristics, therefore we find it appropriate to introduce some basic concepts of pseudo-Riemannian geometry [7, 48, 76].

A space-time (M, g) is the following collection of mathematical entities:
(i) A connected, four-dimensional, Hausdorff (distinct points belong always to disjoint open neighbourhoods) C^∞ manifold M;

(ii) A Lorentz metric g on M, i.e. the assignment of a non-degenerate bilinear form $g_{|p} : T_pM \times T_pM \rightarrow \mathbf{R}$ with diagonal form $(-, +, +, +)$ to each tangent space. Thus, g has signature $+2$ and is not positive-definite;

(iii) A time orientation, given by a globally defined timelike vector field $X : M \rightarrow TM$. A timelike or null tangent vector $v \in T_pM$ is said to be future-directed if $g(X(p), v) < 0$, or past-directed if $g(X(p), v) > 0$.

Some important remarks are now in order :

(a) The condition (i) can be formulated for each number of space-time dimensions ≥ 2 ;

(b) Also the convention $(+, -, -, -)$ for the diagonal form of the metric can be chosen. The definitions of timelike and spacelike will then become opposite to our definitions: X is timelike if $g(X(p), X(p)) > 0 \; \forall p \in M$, and X is spacelike if $g(X(p), X(p)) < 0 \; \forall p \in M$;

(c) The pair (M, g) is only defined up to equivalence. Two pairs (M, g) and (M', g') are said to be equivalent if there exists a diffeomorphism $\alpha : M \rightarrow M'$ such that $\alpha_* g = g'$. Thus, we are really dealing with *an equivalence class of pairs* (M, g).

The distance function in Riemannian geometry

Let Ω_{pq} be the set of piecewise smooth curves in M from p to q. Given the curve $c : [0, 1] \rightarrow M$ and belonging to Ω_{pq}, there is a finite partition of $[0, 1]$ such that c restricted to the sub-interval $[t_i, t_{i+1}]$ is smooth $\forall i$. The Riemannian arc-length of c with respect to g_0 is defined by

$$L_0(c) \equiv \sum_{i=1}^{k-1} \int_{t_i}^{t_{i+1}} \sqrt{g_0(c'(t), c'(t))} \; \mathrm{d}t. \qquad (15.4.1)$$

The Riemannian distance function $d_0 : M \times M \rightarrow [0, \infty)$ is then defined by

$$d_0(p, q) \equiv \inf \left\{ L_0(c) : c \in \Omega_{pq} \right\}. \qquad (15.4.2)$$

Thus, d_0 has the following properties:

(1) $d_0(p, q) = d_0(q, p) \quad \forall p, q \in M$;

(2) $d_0(p, q) \leq d_0(p, r) + d_0(r, q) \quad \forall p, q, r \in M$;

(3) $d_0(p, q) = 0$ if and only if $p = q$;

(4) d_0 is continuous and, $\forall p \in M$ and $\epsilon > 0$, the family of metric balls $B(p, \epsilon) = \{q \in M : d_0(p, q) < \epsilon\}$ is a basis for the manifold topology.

The distance function in Lorentzian geometry

Let Ω_{pq} be the space of all future-directed non-spacelike curves $\gamma : [0, 1] \rightarrow M$ with $\gamma(0) = p$ and $\gamma(1) = q$. Given $\gamma \in \Omega_{pq}$ we choose a partition of the closed interval $[0, 1]$ such that γ, when restricted to $[t_i, t_{i+1}]$, is smooth $\forall i = 0, 1, ..., n - 1$. The Lorentzian arc-length is then defined as [7]

$$L(\gamma) \equiv \sum_{i=0}^{n-1} \int_{t_i}^{t_{i+1}} \sqrt{-g(\gamma'(t), \gamma'(t))} \, dt. \tag{15.4.3}$$

The Lorentzian distance function $d : M \times M \to \mathbf{R} \cup \{\infty\}$ is thus defined as follows. Given $p \in M$, if q does not belong to the causal future of p, i.e. the set $J^+(p)$ of all points $q \in M$ such that there exists a future-directed timelike or null curve from p to q, one sets $d(p, q) = 0$. Otherwise, if $q \in J^+(p)$, one sets

$$d(p, q) \equiv \sup \left\{ L_g(\gamma) : \gamma \in \Omega_{pq} \right\}. \tag{15.4.4}$$

Thus such $d(p, q)$ may not be finite, if timelike curves from p to q attain arbitrarily large arc-lengths. It also fails to be symmetric in general, and one has : $d(p, q) \geq d(p, r) + d(r, q)$ if there exist future-directed non-spacelike curves from p to r and from r to q.

Geodesics

If T is a tensor field defined along a curve λ of class C^r, and if $\overline{T}$ is an arbitrary tensor field of class C^r which extends T in an open neighbourhood of λ, the covariant derivative of T along $\lambda(t)$ can be denoted by $\frac{DT}{\partial t}$ and is equal to

$$\frac{DT}{\partial t} \equiv \nabla_{\frac{\partial}{\partial t}} \overline{T}, \tag{15.4.5}$$

where ∇ is the connection of the Riemannian or pseudo-Riemannian manifold we are considering. The formula (15.4.5) describes a tensor field of class C^{r-1}, defined along the curve λ, and independent of the extension $\overline{T}$ [76]. In particular, if λ has local coordinates $x^a(t)$, and $X^a = \frac{dx^a}{dt}$ are the components of its tangent vector, the expression in local coordinates of the covariant derivative of a vector Y along a curve is

$$\frac{DY^a}{\partial t} = \frac{\partial Y^a}{\partial t} + \sum_{b,c=1}^{n} \Gamma\{a, [b, c]\} \frac{dx^b}{dt} Y^c, \tag{15.4.6}$$

where the Γ's are the Christoffel symbols of second kind of Chap. 4.

The tensor field T (and also, in particular, the vector field corresponding to Y) is said to undergo *parallel transport along λ* if

$$\frac{DT}{\partial t} = 0. \tag{15.4.7}$$

In particular, one may consider the covariant derivative of the tangent vector itself along λ. The curve λ is said to be a *geodesic* if

$$\nabla_X X = \frac{D}{\partial t} \left(\frac{\partial}{\partial t} \right)_\lambda,$$

is parallel to the tangent vector $\left(\frac{\partial}{\partial t}\right)_\lambda$. This implies that there exists a smooth function on the manifold M for which (the semicolon being used to denote the covariant derivative ∇_b)

$$\sum_{b=1}^{n} X^a{}_{;b} X^b = f X^a. \tag{15.4.8}$$

The parameter $v(t)$ along the curve λ such that

$$\frac{D}{\partial v}\left(\frac{\partial}{\partial v}\right)_\lambda = 0 \tag{15.4.9}$$

is said to be an *affine parameter*. The corresponding tangent vector

$$U \equiv \left(\frac{\partial}{\partial v}\right)_\lambda \tag{15.4.10}$$

obeys the equation

$$\sum_{b=1}^{n} U^a{}_{;b} U^b = 0, \tag{15.4.11}$$

i.e., by virtue of (15.4.6),

$$\frac{d^2 x^a}{dv^2} + \sum_{b,c=1}^{n} \Gamma\{a, [b, c]\} \frac{dx^b}{dv} \frac{dx^c}{dv} = 0. \tag{15.4.12}$$

The affine parameter is determined up to a linear transformation

$$v' = av + b. \tag{15.4.13}$$

We stress that our geodesics are *auto-parallel* curves [76], whereas other authors define geodesics as extremal curves which solve a variational problem in pseudo-Riemannian geometry [92]. We prefer auto-parallel curves because they involve the full connection, whereas extremal curves for the Lorentzian arc-length involve only the Christoffel symbols of the Levi-Civita connection (which is the unique torsion-free, symmetric connection such that $\nabla g = 0$).

The mathematical theory of parallel transport is a great merit of Gregorio Ricci-Curbastro and his student, Tullio Levi-Civita. They first studied in detail the theory of surfaces [121], and they defined parallelism first with respect to surfaces, and then for a generic Riemannian manifold [92]. The emphasis on coordinate-independent formulations of calculus is one of the great merits of Ricci-Curbastro, who did much more than simply developing the tools that Einstein needed for general relativity. The introduction of the work in [119] makes it quite clear in the first paragraph therein (see also the fundamental and the intrinsic equations of surfaces in [120]).

Chapter 16
Fundamental Solution and Characteristic Conoid

16.1 Relation Between Fundamental Solution and Riemann's Kernel

In Chap. 3 of his famous book [74], Hadamard studies the fundamental solutions of partial differential operators, starting from the familiar form of the fundamental solution

$$\mathcal{U} \log \frac{1}{r} + w, \ r \equiv \sqrt{(x - x_0)^2 + (y - y_0)^2} \tag{16.1.1}$$

for the equation

$$\left(\frac{\partial^2}{\partial x^2} + \frac{\partial^2}{\partial y^2} + C(x, y) \right) u = 0. \tag{16.1.2}$$

In the formula (16.1.1), $\mathcal{U}$ and w are properly chosen functions of $(x, y; x_0, y_0)$ which are regular in the neighbourhood of $x = x_0, y = y_0$. The function w remains arbitrary to some extent, because any regular solution of Eq. (16.1.2) might be added to it.

As a next step, Hadamard went on to consider the more general equation

$$\left(\frac{\partial^2}{\partial x^2} + \frac{\partial^2}{\partial y^2} + A(x, y)\frac{\partial}{\partial x} + B(x, y)\frac{\partial}{\partial y} + C(x, y) \right) u = 0, \tag{16.1.3}$$

where A, B, C are taken to be analytic functions. In this analytic case, there is no essential distinction between Eq. (16.1.3) and the equation

$$\mathcal{F}(u) = \left(\frac{\partial^2}{\partial x \partial y} + A(x, y)\frac{\partial}{\partial x} + B(x, y)\frac{\partial}{\partial y} + C(x, y) \right) u = 0, \tag{16.1.4}$$

© Springer International Publishing AG 2017
G. Esposito, *From Ordinary to Partial Differential Equations*, UNITEXT 106,
DOI 10.1007/978-3-319-57544-5_16

which can be obtained from Eq. (16.1.3) by changing

$$(x + iy) \rightarrow x, \ (x - iy) \rightarrow y.$$

This map has the effect of changing r^2 in Eq. (16.1.1) into $(x - x_0)(y - y_0)$. Thus, Hadamard wrote the fundamental solution of Eq. (16.1.4) in the form

$$u = \mathcal{U} \log\left[(x - x_0)(y - y_0)\right] + w. \tag{16.1.5}$$

For this to be a solution of Eq. (16.1.4) at all $x \neq x_0$, $y \neq y_0$, we have to require that

$$\mathcal{F}\left[\mathcal{U} \log(x - x_0)(y - y_0)\right] = \mathcal{M}, \tag{16.1.6}$$

where $\mathcal{M}$ is a regular function, while for w we can take any regular solution of the equation

$$\mathcal{F}(w) = -\mathcal{M}. \tag{16.1.7}$$

Indeed, by virtue of the definition (16.1.4) of the operator $\mathcal{F}$, one finds

$$\mathcal{F}\left\{\mathcal{U} \log[(x - x_0)(y - y_0)]\right\} = \mathcal{F}(\mathcal{U}) \log[(x - x_0)(y - y_0)]$$
$$+ \frac{1}{(x - x_0)}\left(\frac{\partial}{\partial y} + A(x, y)\right)\mathcal{U} + \frac{1}{(y - y_0)}\left(\frac{\partial}{\partial x} + B(x, y)\right)\mathcal{U}. \tag{16.1.8}$$

This is found to be a regular function of x, y near each of the lines $x = x_0$, $y = y_0$ if and only if the following conditions hold [74]:
(i) The logarithmic term vanishes, so that $\mathcal{U}$ itself is a solution of Eq. (16.1.4).
(ii) The numerators of the two fractions on the second line of (16.1.8) vanish at the same time as the denominators, i.e.

$$\left(\frac{\partial}{\partial y} + A\right)\mathcal{U} = 0 \quad \text{(for } x = x_0\text{)}, \tag{16.1.9}$$

$$\left(\frac{\partial}{\partial x} + B\right)\mathcal{U} = 0 \quad \text{(for } y = y_0\text{)}. \tag{16.1.10}$$

Note now that these conditions, together with

$$\mathcal{U} = 1 \text{ at } x = x_0, \ y = y_0, \tag{16.1.11}$$

are precisely the conditions of Sect. 14.5 for the Riemann kernel. Thus, we have just proved that *Riemann's kernel coincides with the coefficient of the logarithmic term in the fundamental solution of Eq. (16.1.4)*.

16.2 The Concept of Characteristic Conoid

In general, the fundamental solution is singular not only at a point, e.g. the pole, but along a certain surface. What this surface must be was the content of an important theorem of Le Roux [91] and Delassus [38, 39], who proved that *any singular surface of a solution of a linear differential equation must be characteristic*. Such a singular surface must therefore satisfy the first-order differential equation (see Sect. 14.2)

$$\Omega\left(\frac{\partial z}{\partial x^1}, \frac{\partial z}{\partial x^2}, \dots, \frac{\partial z}{\partial x^m}; x^1, x^2, \dots, x^m\right) = 0. \tag{16.2.1}$$

Among the solutions of Eq. (16.2.1), one which was especially considered by Darboux [29] is the one which has a given point $a(a^1, a^2, \dots, a^m)$ as a conic point, which is called, since Hadamard, the *characteristic conoid*. It coincides with the characteristic cone itself when the coefficients of the equation, or at least the coefficients of the terms of second order, are constants. In general, however, it is a kind of cone with curved generatrices. More precisely, one deals with the following framework [24].

If M is a connected, four-dimensional, Hausdorff four-manifold of class C^∞, a linear partial differential operator is a linear map

$$L : u \in C^\infty(M) \to (Lu) \in C^k(M), \tag{16.2.2}$$

with coefficients $a^{i_1 \cdots i_m}$ given by functions of class C^k. The *characteristic polynomial* of the operator L at a point $x \in M$ is

$$H(x, \xi) = \sum_{i_1, \dots, i_m} a^{i_1 \cdots i_m}(x)\xi_{i_1} \cdots \xi_{i_m}, \tag{16.2.3}$$

where ξ_i is a cotangent vector at x. The cone in the cotangent space T_x^* at x defined by

$$H(x, \xi) = 0 \tag{16.2.4}$$

is called the *characteristic conoid*. By construction, such a cone is independent of the choice of coordinates, because the terms of maximal order (also called leading or principal symbol) of L transform into terms of the same order by a change of coordinates.

The concept of *hyperbolicity* at x of the operator L, already encountered in Sect. 14.4, requires the existence of a domain Γ_x, a convex open cone in T_x^*, such that every line through $\lambda \in \Gamma_x$ cuts the characteristic conoid in m real distinct points. In particular, second-order differential operators with higher-order terms

$$(g^{-1})^{\alpha\beta}(x)\frac{\partial}{\partial x^\alpha}\frac{\partial}{\partial x^\beta}$$

are hyperbolic at x if and only if the cone defined by

$$H_2(x, \xi) \equiv \sum_{\alpha,\beta=1}^{n} (g^{-1})^{\alpha\beta}(x)\xi_\alpha\xi_\beta = 0 \tag{16.2.5}$$

is convex, i.e. if the quadratic form $H_2(x, \xi)$ has signature $(1, n - 1)$.

16.3 Fundamental Solutions with an Algebraic Singularity

Following [74], we study in the first place the case of a surface without a singular point. We look for fundamental solutions of Eq. like (16.1.3) (but in m variables) having the form

$$u = UG^p + w, \tag{16.3.1}$$

where $G = 0$ is the equation of the desired regular surface, p a given constant, U and w regular functions. Since we assume for u the homogeneous equation

$$\mathcal{F}(u) = \left(\sum_{i,k=1}^{m} A^{ik} \frac{\partial^2}{\partial x^i \partial x^k} + \sum_{i=1}^{m} B^i \frac{\partial}{\partial x^i} + C \right) u = 0, \tag{16.3.2}$$

the insertion of the factorized ansatz $u = U F(G)$ into Eq. (16.3.2) yields, upon defining

$$p_i \equiv \frac{\partial G}{\partial x^i}, \tag{16.3.3}$$

terms involving the first derivatives

$$\frac{\partial u}{\partial x^i} = U p_i F'(G) + \frac{\partial U}{\partial x^i} F(G), \tag{16.3.4}$$

and terms involving the second derivatives

$$\frac{\partial^2 u}{\partial x^i \partial x^k} = U p_i p_k F''(G) + \left(p_i \frac{\partial U}{\partial x^k} + p_k \frac{\partial U}{\partial x^i} + U \frac{\partial^2 G}{\partial x^i \partial x^k} \right) F'(G)$$

$$+ \frac{\partial^2 U}{\partial x^i \partial x^k} F(G). \tag{16.3.5}$$

Now we have to multiply Eq. (16.3.4) for every i by B^i, and Eq. (16.3.5) for every i, k by A^{ik}, and add to $Cu = CUF$. In this combination one finds that [74]:

(i) The coefficient of $F''(G)$ is $A(p_1, \ldots, p_m)$;
(ii) In the coefficient of $F'(G)$, the terms in $\frac{\partial U}{\partial x^i}$ are

$$\frac{\partial U}{\partial x^i} \sum_{k=1}^{m} 2A^{ik} p_k = \frac{\partial U}{\partial x^i} \frac{\partial A}{\partial p_i}.$$

Thus, Eq. (16.3.2) becomes

$$U F''(G) A(p_1, \ldots, p_m) + F'(G)\left(\sum_{i=1}^{m} \frac{\partial U}{\partial x^i} \frac{\partial A}{\partial p_i} + MU\right) + F(G)\mathcal{F}(U) = 0,$$

$$(16.3.6)$$

where M denotes

$$M \equiv \mathcal{F}(G) - CG. \tag{16.3.7}$$

In particular, if $F(G)$ reduces to the p-th power of G, i.e. $F(G) = G^p$, one gets from Eq. (16.3.6) the equation

$$p(p-1)G^{p-2}U A(p_1, \ldots, p_m) + pG^{p-1}\left(\sum_{i=1}^{m} \frac{\partial U}{\partial x^i} \frac{\partial A}{\partial p_i} + MU\right) + G^p\mathcal{F}(U) = 0.$$

$$(16.3.8)$$

If the cases $p = 0, 1$ are ruled out, the left-hand side of Eq. (16.3.8) cannot vanish identically or even be a regular function if the coefficient $A(p_1, \ldots, p_m)$ does not vanish. In other words, $G = 0$ is not a characteristic. The equation

$$A\left(\frac{\partial G}{\partial x^1}, \ldots, \frac{\partial G}{\partial x^m}\right) = 0$$

must be either an identity or a consequence of $G = 0$, so that there exists a function A_1, regular also for $G = 0$, such that

$$A\left(\frac{\partial G}{\partial x^1}, \ldots, \frac{\partial G}{\partial x^m}\right) = A_1 G. \tag{16.3.9}$$

The Delassus theorem mentioned earlier is therefore proved. Hereafter we assume that Eq. (16.3.9) is satisfied, so that the term involving G^{p-2} disappears from Eq. (16.3.8) [74]. More precisely, one finds, by virtue of Eq. (16.3.9), that Eq. (16.3.8) reads as

$$pG^{p-1}\left[(p-1)A_1 U + MU + \sum_{i=1}^{m} \frac{\partial U}{\partial x^i} \frac{\partial A}{\partial p_i}\right] + G^p\mathcal{F}(U) = 0. \tag{16.3.10}$$

At this stage, multiplication by G^{1-p} and subsequent restriction to the surface $G = 0$ imply that Eq. (16.3.10) leads to

$$\sum_{i=1}^{m} \frac{\partial U}{\partial x^i} \frac{\partial A}{\partial p_i} + \left[M + (p-1)A_1 \right] U = 0. \tag{16.3.11}$$

This is a linear partial differential equation of first order in U, whose integration would lead to the introduction of the lines defined by the ordinary differential equations (Sects. 15.2 and 15.3)

$$\frac{dx^1}{\frac{1}{2}\frac{\partial A}{\partial p_1}} = \cdots = \frac{dx^m}{\frac{1}{2}\frac{\partial A}{\partial p_m}} = ds. \tag{16.3.12}$$

In the denominators one can recognize the direction cosines of the transversal to $G = 0$; this is, in the case here considered, tangent to that surface (since the latter is a characteristic; the transversal is the direction of the generatrix of contact between the plane $(p_1, \ldots, p_m)$ and the characteristic cone). Thus, a line satisfying Eq. (16.3.12) and issuing from a point of $G = 0$ is lying entirely on that surface. These lines are in fact the bicharacteristics of Eq. (16.2.1), with $\Omega = A$ and $z = G$. If the function A_1 in Eq. (16.3.9) vanishes, so that the function G satisfies identically the equation $A = 0$, the theory of partial differential equations of first order shows that, besides Eq. (16.3.12), the bicharacteristics satisfy also the equations

$$\frac{dp_1}{-\frac{1}{2}\frac{\partial A}{\partial x^1}} = \cdots = \frac{dp_m}{-\frac{1}{2}\frac{\partial A}{\partial x^m}} = ds, \tag{16.3.13}$$

so that they can be determined without knowing the equation $G = 0$ by integrating the system of ordinary differential equations (16.3.12) and (16.3.13).

16.4   Geodesic Equations with and Without Reparametrization Invariance

The characteristic conoid with any point $a(a^1, \ldots, a^m)$ as its vertex has that point for a singular point, and to study this new case one has to first form the equation of the characteristic conoid. That is *the locus of all bicharacteristics issuing from a.* One has to take any set of quantities $p_1, \ldots, p_m$ fulfilling the equation

$$A(p_1, \ldots, p_m; x^1, \ldots, x^m) = 0 \tag{16.4.1}$$

and, with the initial conditions

$$x^i(s = 0) = a^i, \quad p_i(s = 0) = p_{0i}, \tag{16.4.2}$$

integrate the Eqs. (16.3.12) and (16.3.13), here written concisely in Hamilton form

$$\frac{dx^i}{ds} = \frac{1}{2}\frac{\partial A}{\partial p_i}, \quad \frac{dp_i}{ds} = -\frac{1}{2}\frac{\partial A}{\partial x^i}. \tag{16.4.3}$$

Since the ratios of the quantities $p_{01}, \ldots, p_{0m}$ under condition (16.4.1) depend on $(m-2)$ parameters, the locus of the line generated in such a way is a surface. Our task is now to obtain a precise form for the equation of this surface. According to Hadamard [74], it was Coulon who first suggested the desired form of the equation of this surface.

For this purpose, we construct every line issuing from the point $(a^1, \ldots, a^m)$ and satisfying the differential system (16.4.3), whether or not the initial values $p_{01}, \ldots, p_{0m}$ of the variables p_i satisfy Eq.(16.4.1). Such lines are indeed the geodesics of a suitably chosen line element. Within this framework we here recall that, if

$$\mathbf{H}(dx^1, \ldots, dx^m; x^1, \ldots, x^m) = \sum_{i,k=1}^{m} H_{ik} dx^i \otimes dx^k \tag{16.4.4}$$

is any non-singular quadratic form, the coefficients H_{ik} being given functions of $x^1, \ldots, x^m$, then if the differentials dx^l are viewed as differentials of the corresponding x^l, the $\mathbf{H}$ can be taken as the squared line element in a m-dimensional manifold. The integral

$$\int \sqrt{\mathbf{H}(dx^1, \ldots, dx^m)} = \int \sqrt{\mathbf{H}(x'^1, \ldots, x'^m)} \, dt, \tag{16.4.5}$$

where $x'^i \equiv \frac{dx^i}{dt}$, is therefore the arc-length (see Sect. 15.4) of a smooth curve. The corresponding geodesics are the lines which make the variation of this functional vanish. Their differential equations are

$$\frac{d}{dt}\left(\frac{\partial}{\partial x'^i}\sqrt{\mathbf{H}}\right) - \frac{\partial}{\partial x^i}\sqrt{\mathbf{H}} = 0. \quad (i = 1, \ldots, m) \tag{16.4.6}$$

On the other hand, Lagrangian dynamics leads to writing these differential equations in a different form, i.e.

$$\frac{d}{ds}\left(\frac{\partial}{\partial x'^i}\mathbf{H}\right) - \frac{\partial}{\partial x^i}\mathbf{H} = 0, \quad (i = 1, \ldots, m) \tag{16.4.7}$$

this being the law governing the motion of a system whose *vis viva* is $\mathbf{H}(x', x)$, and on which no forces act. The equations (16.4.6) and (16.4.7) are not exactly equivalent, but are *conditionally equivalent* [74]. The former determines the required lines but not t, the time remaining an arbitrary parameter whose choice is immaterial. In other words, Eq. (16.4.6) are reparametrization-invariant, because they remain unchanged if t gets replaced by any smooth function $\phi(t)$.

However the latter equations, i.e. (16.4.7), define not only a line, but a motion on that line, and this motion is no longer arbitrary in time: it must satisfy the vis viva integral

$$\mathbf{H} = \text{constant}, \tag{16.4.8}$$

hence the representative point $(x^1, \ldots, x^m)$ must move on the curve with constant kinetic energy. But on taking into account Eq. (16.4.8), the systems (16.4.6) and (16.4.7) become in general equivalent. A simple way to see this is to point out that, if we choose t in Eq. (16.4.6) in such a way that $\mathbf{H}$ is constant in time, then the denominator $2\sqrt{\mathbf{H}}$ in the identity

$$\frac{\partial}{\partial x'^i}\sqrt{\mathbf{H}} = \frac{1}{2\sqrt{\mathbf{H}}}\frac{\partial}{\partial x'^i}\mathbf{H} \tag{16.4.9}$$

is not affected by the time derivative, and we obtain eventually Eq. (16.4.7).

Conversely, if one wants to write Eq. (16.4.7) in such a way that the independent variable t may become arbitrary, one has to note that, as a function of t, the variable s can be easily evaluated from the vis-viva integral (16.4.8) according to

$$ds = \sqrt{\mathbf{H}}\,dt. \tag{16.4.10}$$

On replacing ds by this value, and accordingly

$$x'^i \text{ by } \frac{x'^i}{\sqrt{\mathbf{H}}},$$

one recovers Eq. (16.4.6) [31, 74].

All these recipes no longer hold for bicharacteristics, for which $A = \mathbf{H} = 0$. For them the system (16.4.6) becomes meaningless, whereas Eq. (16.4.7) remain valid.

Chapter 17
How to Build the Fundamental Solution

17.1 Hamiltonian Form of Geodesic Equations

Let us now try to describe in a more systematic way the latest findings of Chap. 16. For this purpose, we here consider the fundamental form of n-dimensional Euclidean space [67]

$$g_E = \sum_{i,j=1}^{n} A_{ij} dx^i \otimes dx^j, \qquad (17.1.1)$$

where $A_{ij} = A_{ji} = A_{ij}(x^1, ..., x^n)$. Assuming that we deal with smooth curves for simplicity, the associated arc-length interval reads as

$$I \equiv \int_{t_0}^{t_1} \sqrt{\sum_{i,j=1}^{n} A_{ij} \frac{dx^i}{dt} \frac{dx^j}{dt}} \, dt, \qquad (17.1.2)$$

and at the ends of the integration interval we define

$$x^i(t_0) \equiv y^i, \quad x^i(t_1) \equiv z^i, \quad i = 1, \ldots, n. \qquad (17.1.3)$$

Let a^{ij} be the contravariant components of the inverse metric, for which

$$\sum_{k=1}^{n} a^{ik} A_{kj} = \delta^i{}_j. \qquad (17.1.4)$$

If we denote by Q the argument of the square root in Eq. (17.1.2), i.e.

$$Q \equiv \sum_{i,j=1}^{n} A_{ij} \frac{dx^i}{dt} \frac{dx^j}{dt}, \qquad (17.1.5)$$

© Springer International Publishing AG 2017
G. Esposito, *From Ordinary to Partial Differential Equations*, UNITEXT 106,
DOI 10.1007/978-3-319-57544-5_17

the Lagrangian for the variational problem defining the geodesic distance is

$$L = \sqrt{Q}, \tag{17.1.6}$$

and the associated Hessian matrix is singular, i.e.

$$\det \frac{\partial^2 L}{\partial \dot{x}^i \partial \dot{x}^j} = 0, \tag{17.1.7}$$

because L is a function homogeneous of degree 1 in the $\dot{x}^j = \frac{dx^j}{dt}$. It is indeed well-known that non-singularity of the Hessian matrix is necessary to define the Legendre transform. However, one can overcome this difficulty by writing the Euler–Lagrange equations (see Sect. 16.4)

$$\frac{d}{dt} \frac{\partial L}{\partial \dot{x}^i} - \frac{\partial L}{\partial x^i} = 0 \tag{17.1.8}$$

in terms of Q, to obtain

$$\frac{d}{dt} \frac{1}{\sqrt{Q}} \frac{\partial Q}{\partial \dot{x}^i} - \frac{1}{\sqrt{Q}} \frac{\partial Q}{\partial x^i} = 0, \quad i = 1, \ldots, n. \tag{17.1.9}$$

This suggests taking t, the parameter along the geodesics, as the arc-length measured from the point $y^1, \ldots, y^n$. From the constancy in time of $\sqrt{Q}$, Eq. (17.1.9) becomes formally analogous to Eq. (17.1.8).

The integral

$$J \equiv \int_0^s Q \, dt \tag{17.1.10}$$

is stationary. Now fix the terminal values $y^1, \ldots, y^n$ and $z^1, \ldots, z^n$ of $x^1, \ldots, x^n$, but allow the upper limit of integration s to vary in Eq. (17.1.10). This implies that the extremals of J come to depend on s and on the variable of integration t according to [67]

$$x^i = x^i \left(\frac{I}{s} t \right) \tag{17.1.11}$$

for a change of scale, since Q is a function homogeneous of degree 2 in the velocity variables $\dot{x}^1, \ldots, \dot{x}^n$. Thus, the constant value of Q becomes $\frac{I^2}{s^2}$ along each such extremal curve, and

$$J = \frac{I^2(z^1, \ldots, z^n)}{s^2}(s - 0) = \frac{I^2(z^1, \ldots, z^n)}{s}. \tag{17.1.12}$$

Now we can apply the Hamilton-Jacobi theory to the equations of motion that we are studying. Our Hamiltonian reads as

$$H = Q = \frac{1}{4} \sum_{i,j=1}^{n} a^{ij} p_i p_j. \tag{17.1.13}$$

Since the momenta are given by

$$p_i = \frac{\partial L}{\partial \dot{x}^i} = \frac{\partial Q}{\partial \dot{x}^i} = 2 \sum_{j=1}^{n} A_{ij} \frac{dx^j}{dt}, \tag{17.1.14}$$

we can re-express the velocity variables in the form

$$\frac{dx^j}{dt} = \frac{1}{2} \sum_{i=1}^{n} a^{ji} p_i, \tag{17.1.15}$$

and the $\mathbf{p} \cdot \dot{\mathbf{x}}$ term in the Legendre transform becomes

$$\sum_{i=1}^{n} p_i \frac{dx^i}{dt} = 2 \sum_{i,j=1}^{n} A_{ij} \frac{dx^i}{dt} \frac{dx^j}{dt} = \frac{1}{2} \sum_{i,j=1}^{n} a^{ij} p_i p_j. \tag{17.1.16}$$

The functional J defined in Eq. (17.1.10) satisfies the Hamilton-Jacobi equation

$$\frac{\partial J}{\partial s} + \frac{1}{4} \sum_{i,j=1}^{n} a^{ij}(z^1, \dots, z^n) \frac{\partial J}{\partial z^i} \frac{\partial J}{\partial z^j} = 0. \tag{17.1.17}$$

In this equation we can insert the form (17.1.12) of J and set eventually $s = 1$. The non-vanishing factor I^2, common to both terms, drops therefore out, and our Hamilton-Jacobi equation (17.1.17) reduces to

$$\sum_{i,j=1}^{n} a^{ij} \frac{\partial I}{\partial z^i} \frac{\partial I}{\partial z^j} = 1. \tag{17.1.18}$$

If we now define

$$\Gamma \equiv I^2, \tag{17.1.19}$$

we obtain $I = \sqrt{\Gamma}$, and Eq. (17.1.18) takes the remarkable form

$$\sum_{i,j=1}^{n} a^{ij} \frac{\partial \Gamma}{\partial z^i} \frac{\partial \Gamma}{\partial z^j} = 4\Gamma. \tag{17.1.20}$$

This equation coincides with Eq. (16.3.9) upon setting therein

$$G = \Gamma, \quad A_1 = 4, \quad A = a(\text{grad}\Gamma, \text{grad}\Gamma). \tag{17.1.21}$$

The function Γ is a conoidal solution of Eq. (17.1.20), generated by all bicharacteristics of this equation passing through $(y^1, \ldots, y^n)$ which are, of course, geodesics of the metric (17.1.1).

The geodesics satisfy the equations of motion in Hamiltonian form

$$\frac{dx^i}{ds} = \frac{1}{2} \sum_{j=1}^{n} a^{ij} p_j, \tag{17.1.22}$$

$$\frac{dp_i}{ds} = -\frac{1}{4} \sum_{j,k=1}^{n} \frac{\partial a^{jk}}{\partial x^i} p_j p_k, \tag{17.1.23}$$

together with the initial conditions

$$x^i(0) = y^i, \quad p_i(0) = \gamma_i. \tag{17.1.24}$$

In a generic space-time manifold (see Sect. 15.4), the a^{ij} of Eq. (17.1.20) will denote the contravariant components $(g^{-1})^{ij}$ of the space-time metric, i.e.

$$g^{-1} = \sum_{i,j=1}^{n} a^{ij} \frac{\partial}{\partial x^i} \otimes \frac{\partial}{\partial x^j}, \tag{17.1.25}$$

the signature of g being $(n-2)$. Equation (17.1.19) will then be interpreted by stating that Γ is a two-point function, called the *world function* and equal to the square of the geodesic distance between the space-time points $x = (x^1, \ldots, x^n)$ and $y = (y^1, \ldots, y^n)$. This means that such a formalism can only be used *locally*, when there exists a unique geodesic from x to y [65]. Such a space-time is said to be geodesically convex.

17.2 The Unique Real-Analytic World Function

Equation (17.1.20), or (16.3.9), is the fundamental equation in the theory of the characteristic conoid, in that any function real-analytic in the neighbourhood of the desired vertex a, vanishing on the conoid and satisfying Eq. (17.1.20), can only be the world function Γ itself (besides this, there exist infinitely many non-analytic solutions of Eq. (17.1.20), as pointed out by Hadamard [74]).

Proof The desired function should be of the form $\Gamma\Pi$, where Π is a real-analytic function. By insertion into Eq. (17.1.20), this yields

$$4\Gamma\Pi = \sum_{i,j=1}^{n} a^{ij} \left(\Pi \frac{\partial\Gamma}{\partial x^i} + \Gamma \frac{\partial\Pi}{\partial x^i} \right) \left(\Pi \frac{\partial\Gamma}{\partial x^j} + \Gamma \frac{\partial\Pi}{\partial x^j} \right)$$
$$= \Pi^2 \nabla_1 \Gamma + 2\Pi\Gamma\nabla_1(\Pi, \Gamma) + \Gamma^2 \nabla_1 \Pi, \tag{17.2.1}$$

where we have used the differential parameters used in Sect. 4.3. On the right-hand side of Eq. (17.2.1), the term involving the mixed differential parameter can be re-expressed because the derivative of a generic function U along a geodesic reads as

$$\frac{dU}{ds} = \sum_{i=1}^{n} \frac{\partial U}{\partial x^i} \frac{dx^i}{ds} = \frac{1}{2} \sum_{i=1}^{n} \frac{\partial U}{\partial x^i} \frac{\partial A}{\partial p_i} = \sum_{i=1}^{n} a^{ij} \frac{\partial U}{\partial x^i} p_j, \tag{17.2.2}$$

while, on the other hand,

$$\nabla_1(\Gamma, U) = \sum_{i,j=1}^{n} a^{ij} \frac{\partial\Gamma}{\partial x^i} \frac{\partial U}{\partial x^j} = 2s \sum_{i,j=1}^{n} a^{ij} p_i \frac{\partial U}{\partial x^j} = 2s \frac{dU}{ds}, \tag{17.2.3}$$

where the last equality follows from Eq. (17.2.2) and from the symmetry of a^{ij}. Thus, Eq. (17.2.1) becomes

$$\Pi^2 \nabla_1 \Gamma + 4s\Gamma\Pi \frac{d\Pi}{ds} + \Gamma^2 \nabla_1 \Pi = 4\Gamma\Pi. \tag{17.2.4}$$

In this equation, we can divide both sides by $4\Gamma\Pi$, finding therefore

$$\Pi \frac{\nabla_1\Gamma}{4\Gamma} + s \frac{d\Pi}{ds} + \frac{\Gamma}{4\Pi}\nabla_1\Pi - 1 = \left(\Pi + s\frac{d\Pi}{ds} - 1 \right) + \frac{\Gamma}{4\Pi}\nabla_1\Pi$$
$$= \frac{d}{ds}[s(\Pi - 1)] + \frac{\Gamma}{4\Pi}\nabla_1\Pi = 0. \tag{17.2.5}$$

This equation shows that the function Π equals 1 over the whole conoid, and hence we can write the general formula

$$\Pi = 1 + \Gamma^l R, \tag{17.2.6}$$

where l is a positive exponent, and R is yet another real-analytic function, not vanishing over the whole surface of the conoid. But this leads to a contradiction, because the insertion of Eq. (17.2.6) for Π into Eq. (17.2.5) yields

$$\frac{d}{ds}(s\Gamma' R) + \frac{\Gamma}{4\Pi}\nabla_1(\Gamma' R)$$

$$= \Gamma' R + s\frac{d\Gamma'}{ds}R + s\Gamma'\frac{dR}{ds}$$

$$+ \frac{\Gamma}{4\Pi}\sum_{i,j=1}^{n} a^{ij}\left(l\Gamma^{l-1}\frac{\partial\Gamma}{\partial x^i}R + \Gamma'\frac{\partial R}{\partial x^i}\right)\left(l\Gamma^{l-1}\frac{\partial\Gamma}{\partial x^j}R + \Gamma'\frac{\partial R}{\partial x^j}\right)$$

$$= \Gamma'\left[s\frac{dR}{ds} + (2l+1)R\right]$$

$$+ \frac{\Gamma}{4\Pi}\left[l^2\Gamma^{2l-2}R^2(\Delta_1\Gamma) + 2l\Gamma^{2l-1}R\,\Delta_1\,(\Gamma, R) + \Gamma^{2l}\,\Delta_1\,R\right]. \qquad (17.2.7)$$

This equation, when restricted to the characteristic conoid, reduces to

$$s\frac{dR}{ds} + (2l+1)R = 0, \qquad (17.2.8)$$

which is solved by

$$R = R_0\left(\frac{s}{s_0}\right)^{-(2l+1)}, \qquad (17.2.9)$$

which can only be regular if identically vanishing, i.e.

$$R_0 = 0 \Longrightarrow R = 0. \qquad (17.2.10)$$

$\square$

17.3 Fundamental Solution with Odd Number of Variables

We here study the linear partial differential equation (16.3.2), with associated world function Γ, the square of the geodesic distance between two points, obeying Eq. (17.1.20), with coefficients a^{ij} equal to the contravariant components (of the inverse metric) A^{ij} in Eq. (16.3.2). A fundamental solution of $\mathcal{F}(u) = 0$ is a two-point function $S(x, \xi)$, with $x = (x^1, \ldots, x^n)$ and $\xi = (\xi^1, \ldots, \xi^n)$, which solves Eq. (16.3.2) in its dependence on x and possesses, at the parameter point ξ, a singularity characterized by the split reading as (Garabedian [67])

$$S = \frac{U}{\Gamma^m} + V\log(\Gamma) + W, \qquad (17.3.1)$$

where U, V and W are taken to be smooth functions of x in a neighbourhood of ξ, with $U \neq 0$ at ξ, and where the exponent m is given by

$$m = \frac{n}{2} - 1. \tag{17.3.2}$$

The simplest example is possibly provided by the fundamental solution of an elliptic equation in n-dimensional Euclidean space $\mathbf{R}^n$, i.e. the Laplace equation, for which

$$S = \frac{1}{r^{n-2}}, \quad r \equiv \sqrt{\sum_{i=1}^{n}(x^i - \xi^i)^2}. \tag{17.3.3}$$

We will prove in due course that, when n is even, the term W is redundant, whereas the coefficient V of the logarithm vanishes for n odd. Thus, the dimension of Euclidean space affects in a non-trivial way the conceivable form of the fundamental solution.

Following Garabedian [67], we consider first the odd values of n. We then put $V = W = 0$ in Eq. (17.3.1), and look for a convergent series expressing the unknown function U, in the form

$$U = U_0 + U_1\Gamma + U_2\Gamma^2 + O(\Gamma^3), \tag{17.3.4}$$

with regular coefficients U_l. A patient calculation, where the symmetry of the inverse metric a^{ij} is exploited, shows that

$$
\begin{aligned}
\mathcal{F}[U_l\Gamma^{l-m}] &= (l-m)(l-m-1)U_l\Gamma^{l-m-2}\sum_{i,j=1}^{n}a^{ij}\frac{\partial\Gamma}{\partial x^i}\frac{\partial\Gamma}{\partial x^j} \\
&\quad + (l-m)\left[2\sum_{i,j=1}^{n}a^{ij}\frac{\partial U}{\partial x^i}\frac{\partial\Gamma}{\partial x^j} + 4DU_l\right]\Gamma^{l-m-1} \\
&\quad + \mathcal{F}[U_l]\Gamma^{l-m},
\end{aligned}
\tag{17.3.5}
$$

where D is the term

$$D \equiv \frac{1}{4}\sum_{i,j=1}^{n}a^{ij}\frac{\partial^2\Gamma}{\partial x^i\partial x^j} + \frac{1}{4}\sum_{i=1}^{n}B^i\frac{\partial\Gamma}{\partial x^i}. \tag{17.3.6}$$

One should stress that the possibility of eliminating the lowest power of Γ on the right-hand side of Eq. (17.3.5) by means of the first-order partial differential equation (17.1.20) now shows why the fundamental solution S should be expanded in terms of this particular function, i.e. the world function Γ.

It is now convenient to introduce again a parameter s which is measured along the geodesics that generate Γ. We can then exploit Eq. (17.2.3) and write that

$$\sum_{i,j=1}^{n}a^{ij}\frac{\partial U_l}{\partial x^i}\frac{\partial\Gamma}{\partial x^j} = 2s\frac{dU_l}{ds}. \tag{17.3.7}$$

Hence we arrive at a simplified form of Eq. (17.3.5), i.e. [67].

$$\mathcal{F}[U_l \Gamma^{l-m}] = 4(l-m)\left\{s\frac{dU_l}{ds} + (D+l-m-1)U_l\right\}\Gamma^{l-m-1} + \mathcal{F}[U_l]\Gamma^{l-m}. \qquad (17.3.8)$$

At this stage, in order to solve, $\forall x \neq \xi$, the equation

$$\mathcal{F}[S] = \mathcal{F}\left[\sum_{l=0}^{\infty} U_l \Gamma^{l-m}\right] = 0, \qquad (17.3.9)$$

we set to zero all coefficients of the various powers of Γ. This leads to the fundamental recursion formulae

$$\left[s\frac{d}{ds} + (D-m-1)\right]U_0 = 0, \qquad (17.3.10)$$

$$\left[s\frac{d}{ds} + (D+l-m-1)\right]U_l = -\frac{1}{4(l-m)}\mathcal{F}[U_{l-1}], \quad l \geq 1, \quad (17.3.11)$$

for the evaluation of $U_0, U_1, U_2, \ldots$. The crucial detail is that, for odd values of n, the division by $(l-m)$ on the right-hand side of (17.3.11) is always legitimate by virtue of (17.3.2), because $(l-m)$ never vanishes.

Note that, when Eq. (16.3.2) is hyperbolic, the fundamental solution (17.3.1) becomes infinite along a two-sheeted conoid $\Gamma = 0$ separating n-dimensional space into three parts. This conoid is indeed a characteristic surface for the second-order equation (16.3.2), since Eq. (17.1.20) reduces on the level surface $\Gamma = 0$ to the first-order partial differential equation

$$\sum_{i,j=1}^{n} a^{ij} \frac{\partial \Gamma}{\partial x^i} \frac{\partial \Gamma}{\partial x^j} = 0$$

for such a characteristic. The basic property involved is that *any locus of singularities of a solution of a linear hyperbolic equation can be expected to form a characteristic surface* [67].

The geodesics that lie on the conoid $\Gamma = 0$ are the bicharacteristics of the original equation (16.3.2). We have found that, along the characteristic conoid $\Gamma = 0$, the ordinary differential operators occurring on the left in the transport equations (17.3.10) and (17.3.11) apply in the directions of the bicharacteristics. This happens because, within any of its characteristic surfaces, Eq. (16.3.2) reduces to an ordinary differential equation imposed on the Cauchy data along each bicharacteristic [67].

17.4 Convergence of the Power Series for U

In the course of evaluating the functions $U_0, U_1, U_2, \ldots$ it is convenient to work in a new space with coordinates $\theta_1, \ldots, \theta_n$ defined by [67, 74]

$$\theta_i = s p_i(0). \tag{17.4.1}$$

It is possible to do so in a sufficiently small neighbourhood of the parameter point $\xi = (\xi^1, .., \xi^n)$ because the relevant Jacobian does not vanish. In this new space the geodesics become rays emanating from the origin, while the parameter s can be chosen to coincide with the distance from the origin along each such ray. Each coefficient U_l in the expansion (17.3.4) can be written in the form of a series

$$U_l = \sum_{j=0}^{\infty} P_{lj} \tag{17.4.2}$$

of polynomials P_{lj} homogeneous in the coordinates $\theta_1, \ldots, \theta_n$ of a degree equal to the index j.

Note that the differential operator $s\frac{d}{ds}$ in Eqs. (17.3.10) and (17.3.11) does not alter the degree of any of the polynomials P_{lj}, with the exception that it reduces a polynomial of degree zero, i.e. a constant, to zero. Thus, unless the coefficient $(D - m - 1)$ vanishes for $\theta_1 = \ldots = \theta_n = 0$, there does not exist a solution U_0 of Eq. (17.3.10) satisfying the requirement $P_{00} \neq 0$. However, we have chosen the exponent as in (17.3.2) precisely so that this will be the case, because our $D = \frac{n}{2}$ at the parameter point $x = \xi$. Thus, we can integrate Eq. (17.3.10) to find

$$U_0 = P_{00} e^{-\int_0^s (D-m-1)\frac{d\tau}{\tau}}, \tag{17.4.3}$$

where P_{00} is a constant as a function of x that might vary with ξ.

Similarly, Eq. (17.3.11) may be solved by the recursion formula

$$U_l = -\frac{U_0}{4(l-m)s^l} \int_0^s \frac{\mathcal{F}[U_{l-1}]\tau^{l-1}}{U_0} d\tau, \ l \geq 1. \tag{17.4.4}$$

The linear operator on the right turns any convergent series U_{l-1} of the type (17.4.2) into another series of the same kind for U_l. At this stage, one has still to prove uniform convergence of the expansion of U in powers of Γ, for sufficiently small values of s. This can be obtained by using the method of majorants.

For the purpose of proving convergence, it is sufficient to treat only the particular case $U_0 = $ constant, because the substitution

$$u_1 \equiv \frac{u}{U_0}, \tag{17.4.5}$$

with U_0 given by Eq. (17.4.3), reduces Eq. (16.3.2) to a new partial differential equation reading as

$$\mathcal{F}_1[u_1] = \mathcal{F}[U_0 u_1] = 0, \tag{17.4.6}$$

for which such an assumption is verified.

Let K and ϵ be positive numbers such that the geometric series

$$\sum_{j=0}^{\infty} \frac{K}{\epsilon^j} \left(|\theta_1| + \ldots + |\theta_n| \right)^j = \frac{K\epsilon}{\epsilon - |\theta_1| - \ldots - |\theta_n|} \tag{17.4.7}$$

is a majorant for the Taylor expansions in powers of $\theta_1, \ldots, \theta_n$ of all the coefficients of $\mathcal{F}$, which is now a differential operator expressed in these new coordinates. Hence one finds that, if

$$M\{U_l\} = \frac{M_l}{\left(1 - \frac{|\theta_1| + \ldots + |\theta_n|}{\epsilon} \right)^{2l}} \tag{17.4.8}$$

denotes a majorant for U_l, with M_l taken as a suitably large constant, then

$$M\{\mathcal{F}[U_l]\} = \frac{2l(2l+1)\left[1 + \frac{n}{\epsilon} + \frac{n^2}{\epsilon^2} \right] K M_l}{\left(1 - \frac{|\theta_1| + \ldots + |\theta_n|}{\epsilon} \right)^{2l+3}} \tag{17.4.9}$$

is a majorant for $\mathcal{F}[U_l]$. We now apply the recursion formula (17.4.4) to (17.4.9) in order to establish that when l is replaced by $(l+1)$, and with

$$M_{l+1} = \frac{l(2l+1)}{2(l+1)(l-m+1)} \left[1 + \frac{n}{\epsilon} + \frac{n^2}{\epsilon^2} \right] K M_l \tag{17.4.10}$$

the rule (17.4.8) also defines a majorant for U_{l+1}.

Since we have recognized that it is enough to consider the case $U_0 = $ constant, the proof that we are interested in reduces to a verification that

$$M\left\{ s^{-l-1} \int_0^s \frac{\tau^l}{(1 - \gamma\tau)^{2l+3}} d\tau \right\} = \frac{1}{(l+1)} (1 - \gamma s)^{-2l-2} \tag{17.4.11}$$

is a majorant for the integral inside curly brackets on the left. This can be proved with the help of the convenient choice

$$M\left\{\frac{s^l}{(1-\gamma s)^{2l+3}}\right\} = [1+\gamma s]\frac{s^l}{(1-\gamma s)^{2l+3}}$$

$$= \frac{1}{(l+1)}\frac{d}{ds}\frac{s^{l+1}}{(1-\gamma s)^{2l+2}} \qquad (17.4.12)$$

of a majorant for the integrand. With this notation (Garabedian [67]), γ is specified by

$$|\theta_1| + \ldots + |\theta_n| = \epsilon\gamma s. \qquad (17.4.13)$$

By induction, we conclude that the majorants (17.4.8) are valid for all $l \geq 1$, provided that M_1 is sufficiently large and that $M_2, M_3, \ldots$ are given by (17.4.10). Thus, the series for U in powers of the world function Γ converges in a neighbourhood specified by the upper bound

$$|\Gamma| < \frac{\left(1 - \frac{|\theta_1|+\ldots+|\theta_n|}{\epsilon}\right)^2}{\left[1 + \frac{n}{\epsilon} + \frac{n^2}{\epsilon^2}\right]K} \qquad (17.4.14)$$

of the parameter point $\xi = (\xi^1, \ldots, \xi^n)$.

To sum up, we obtain *locally* a fundamental solution S of Eq. (16.3.2) having the special form

$$S = \frac{U}{\Gamma^m} = \sum_{l=0}^{\infty} U_l \Gamma^{l-m}, \qquad (17.4.15)$$

when the number n of independent variables is odd. The addition of a regular term W to the right-hand side of (17.4.15) is not mandatory.

Chapter 18
Examples of Fundamental Solutions

18.1 Even Number of Variables and Logarithmic Term

When the number $n \geq 4$ of independent variables is even, the exponent m defined in Eq. (17.3.2) is a positive integer and the construction of U in Sect. 17.4 no longer holds, because the whole algorithm (see Eq. (17.4.4)) involves division by $(l - m)$, which vanishes when $l = m$. Only the functions $U_0, U_1, \ldots, U_{m-1}$ can then be obtained as in Sect. 17.4. This is why a logarithmic term is needed in the formula (17.3.1) for the fundamental solution S in a space with even number of dimensions. Hence we look for S in the form

$$S = \sum_{l=0}^{m-1} U_l \Gamma^{l-m} + V \log(\Gamma) + W. \tag{18.1.1}$$

If the formula (18.1.1) is inserted into the homogeneous equation (16.3.2), one finds

$$\sum_{l=0}^{m-1} \mathcal{F}[U_l \Gamma^{l-m}] + \mathcal{F}[V \log(\Gamma)] + \mathcal{F}[W] = \mathcal{F}[U_{m-1}] \frac{1}{\Gamma} - \frac{V}{\Gamma^2} \sum_{i,j=1}^{n} a^{ij} \frac{\partial \Gamma}{\partial x^i} \frac{\partial \Gamma}{\partial x^j}$$

$$+ \left[2 \sum_{i,j=1}^{n} a^{ij} \frac{\partial V}{\partial x^i} \frac{\partial \Gamma}{\partial x^j} + 4DV \right] \frac{1}{\Gamma} + \mathcal{F}[V] \log(\Gamma) + \mathcal{F}[W] = 0, \tag{18.1.2}$$

by virtue of equation (17.3.8) and of the transport equations (17.3.10) and (17.3.11). In Eq. (18.1.2) the term which is non-linear in the derivatives of Γ is re-expressed from Eq. (17.1.20) (with z^i therein written as x^i), and we arrive at

$$\left\{ \mathcal{F}[U_{m-1}] + 4 \left[s \frac{dV}{ds} + (D - 1)V \right] \right\} \frac{1}{\Gamma} + \mathcal{F}[V] \log(\Gamma) + \mathcal{F}[W] = 0. \tag{18.1.3}$$

© Springer International Publishing AG 2017

G. Esposito, *From Ordinary to Partial Differential Equations*, UNITEXT 106,

DOI 10.1007/978-3-319-57544-5_18

We are now going to prove that this equation determines V uniquely, whereas W can be selected in a number of ways, in order to satisfy the requirements imposed on it.

We note first that the $\log(\Gamma)$ is not balanced by other terms in Eq. (18.1.3), hence the function V must solve the homogeneous equation

$$\mathcal{F}[V] = 0. \tag{18.1.4}$$

Second, the coefficient of $\frac{1}{\Gamma}$ in Eq. (18.1.3) must vanish along the whole characteristic conoid $\Gamma = 0$, since the remaining regular term $\mathcal{F}[W]$ cannot balance the effect of $\frac{1}{\Gamma}$. Thus, the function V has to solve also the ordinary differential equation.

$$\left[s\frac{d}{ds} + (D-1) \right] V = -\frac{1}{4}\mathcal{F}[U_{m-1}] \tag{18.1.5}$$

on each bicharacteristic that generates the conoid. From our study of the transport equations (17.3.10) and (17.3.11) we know that Eq. (18.1.5) determines the function V uniquely on the characteristic surface $\Gamma = 0$, and that V must indeed coincide there with the function V_0 defined in a neighbourhood of the parameter point ξ by the integral

$$V_0 = -\frac{U_0}{4s^m} \int_0^s \frac{\mathcal{F}[U_{m-1}]\tau^{m-1}}{U_0} d\tau. \tag{18.1.6}$$

We have therefore formulated a characteristic initial-value problem for the partial differential equation (18.1.4), in which the unknown function V is prescribed on the conoid $\Gamma = 0$. This result agrees with our previous findings in Sect. 16.1, according to which the coefficient of the logarithm is a Riemann kernel satisfying a characteristic initial-value problem.

From another point of view [67], one can think of Eq. (18.1.6) as a substitute for the recursion formula (17.4.4) in the case $l = m$. This property suggests trying to find V as a convergent power series

$$V = \sum_{l=0}^{\infty} V_l \Gamma^l. \tag{18.1.7}$$

Insertion of Eq. (18.1.7) into Eq. (18.1.4) leads to infinitely many powers of Γ whose coefficients should all be set to zero. We do so, and integrate the resulting ordinary differential equations, which are the counterpart of Eq. (17.3.11), finding therefore

$$V_l = -\frac{U_0}{4ls^{l+m}} \int_0^s \frac{\mathcal{F}[V_{l-1}]\tau^{l+m-1}}{U_0} d\tau, \quad \forall l \geq 1. \tag{18.1.8}$$

The first term V_0 is instead obtained from Eq. (18.1.6). The method of majorants can be used to deduce estimates like (17.4.8) for the functions V_l provided by

Eq. (18.1.8). Thus, the series (18.1.7) converges uniformly in a region like (17.4.14) surrounding the parameter point ξ. The proof of this property follows the same path as in Sect. 17.4.

18.2 Smooth Part of the Fundamental Solution

We have just shown how V can be constructed in such a way that Eq. (18.1.3) becomes a partial differential equation for W with a inhomogeneous term that is regular in the neighbourhood of ξ. This determines W only *up to the addition of an arbitrary solution of the homogeneous equation* (16.3.2). A particular choice for W that agrees with the method used so far demands that

$$W = \sum_{l=1}^{\infty} W_l \Gamma^l, \qquad (18.2.1)$$

where the coefficient functions $W_1, W_2, \ldots$ will be found from a recursive scheme like (18.1.8). By requiring that the series for W should not include the term W_0 corresponding to the value $l = 0$, one obtains a unique determination of the fundamental solution (18.1.1) such that the functions $U_0, \ldots, U_{m-1}, V$ and W are all regular as functions of the parameter point ξ.

Garabedian [67] has stressed that, for a *hyperbolic equation*, the *bicharacteristics and the characteristic conoid* $\Gamma = 0$ along which the fundamental solution becomes infinite appear to play the key role, whereas for an *elliptic equation* the *geodesics* seem a more natural concept (but our reader knows that bicharacteristics are null geodesics anyway). In both the hyperbolic and the elliptic case, the limitation of the Hadamard approach described so far is that it yields the fundamental solution only locally, i.e. in a sufficiently small neighbourhood of ξ. Furthermore, when the inverse metric components a^{ij} in Eq. (17.1.20) are varying, also the world function Γ is defined only in the small.

18.3 Parametrix of Scalar Wave Equation in Curved Space-Time

When we study the wave equation

$$\Box \phi = 0, \qquad (18.3.1)$$

in Minkowski space-time, we discover that the solution formula involves amplitude and phase functions, as well as the Cauchy data

$$\phi(t = 0, x) \equiv u_0(x), \tag{18.3.2}$$

$$\frac{\partial \phi}{\partial t}(t = 0, x) \equiv u_1(x), \tag{18.3.3}$$

which are assumed to be Fourier transformable. However, the variable nature of the coefficients of the d'Alembert wave operator in curved space-time demands for a non-trivial generalization of the integral representation

$$\phi(t, x, y, z) = \int_{-\infty}^{\infty} d\xi_1 \int_{-\infty}^{\infty} d\xi_2 \int_{-\infty}^{\infty} d\xi_3 A(\xi_1, \xi_2, \xi_3, t) e^{i(\xi_1 x + \xi_2 y + \xi_3 z)}. \tag{18.3.4}$$

This is indeed available, since a theorem guarantees that the solution of the Cauchy problem (18.3.1)–(18.3.3) can be expressed in the form [54]

$$\phi(x, t) = \sum_{j=0}^{1} E_j(t) u_j(x), \tag{18.3.5}$$

where, on denoting by $\hat{u}_j$ the Fourier transform of the Cauchy data, the operators $E_j(t)$ act according to [hereafter, $(x) \equiv (x^1, x^2, x^3)$, with covariable $(\xi) \equiv (\xi_1, \xi_2, \xi_3)$]

$$E_j(t) u_j(x) = \sum_{k=1}^{2} (2\pi)^{-3} \int e^{i \varphi_k(x, t, \xi)} \alpha_{jk}(x, t, \xi) \hat{u}_j(\xi) d^3\xi + R_j(t) u_j(x),$$
$$\tag{18.3.6}$$

where the φ_k are real-valued phase functions which satisfy the initial condition

$$\varphi_k(t = 0, x, \xi) = x \cdot \xi = \sum_{s=1}^{3} x^s \xi_s, \tag{18.3.7}$$

and $R_j(t)$ is a regularizing operator which smoothes out the singularities acted upon by it. In other words, the Cauchy problem is here solved by a pair of Fourier-Maslov integral operators [133] of the form (18.3.6), and such a construction (leaving aside, for the moment, its global version, which can be built as shown in Chap. 7 of [133]) generalizes the monochromatic plane waves for the d'Alembert operator from Minkowski space-time to curved space-time. Strictly, we are dealing with the parametrix for the wave equation. In our case, since we know a priori that (18.3.5) and (18.3.6) yield an exact solution of the Cauchy problem, we can insert them into Eq. (18.3.1) with $P = \square$, finding that, for all $j = 0, 1$,

$$P[E_j(t)u_j(x)] \sim \sum_{k=1}^{2}(2\pi)^{-3} \int P[e^{\mathrm{i}\varphi_k}\alpha_{jk}]\hat{u}_j(\xi)\mathrm{d}^3\xi, \tag{18.3.8}$$

where $PR_j(t)u_j(x)$ can be neglected with respect to the integral on the right-hand side of Eq. (18.3.6), because $R_j(t)$ is a regularizing operator. Next, we find from Eq. (18.3.1) that

$$P[e^{\mathrm{i}\varphi_k}\alpha_{jk}] = e^{\mathrm{i}\varphi_k}(\mathrm{i}A_{jk} + B_{jk}), \tag{18.3.9}$$

where, on considering the form of P in Kasner space-time [54], i.e.

$$P = -\frac{\partial^2}{\partial t^2} - \frac{1}{t}\frac{\partial}{\partial t}\sum_{l=1}^{3}t^{-2p_l}\frac{\partial^2}{\partial x^{l2}}, \quad \sum_{k=1}^{3}p_k = \sum_{k=1}^{3}(p_k)^2 = 1, \tag{18.3.10}$$

one finds

$$A_{jk} \equiv \frac{\partial^2\varphi_k}{\partial t^2}\alpha_{jk} + 2\frac{\partial\varphi_k}{\partial t}\frac{\partial\alpha_{jk}}{\partial t} + \frac{1}{t}\frac{\partial\varphi_k}{\partial t}\alpha_{jk} - \sum_{l=1}^{3}t^{-2p_l}\left(\frac{\partial^2\varphi_k}{\partial x_l^2}\alpha_{jk} + 2\frac{\partial\varphi_k}{\partial x_l}\frac{\partial\alpha_{jk}}{\partial x_l}\right), \tag{18.3.11}$$

$$B_{jk} \equiv \frac{\partial^2\alpha_{jk}}{\partial t^2} - \left(\frac{\partial\varphi_k}{\partial t}\right)^2\alpha_{jk} + \frac{1}{t}\frac{\partial\alpha_{jk}}{\partial t} - \sum_{l=1}^{3}t^{-2p_l}\left(\frac{\partial^2\alpha_{jk}}{\partial x_l^2} - \left(\frac{\partial\varphi_k}{\partial x_l}\right)^2\alpha_{jk}\right). \tag{18.3.12}$$

If the phase functions φ_k are real-valued, since the exponentials $e^{\mathrm{i}\varphi_k}$ can be taken to be linearly independent, we can fulfill Eq. (18.3.1), up to the negligible contributions resulting from $PR_j(t)u_j(x)$, by setting to zero in the integrand (18.3.8) both A_{jk} and B_{jk}. This leads to a coupled system of partial differential equations. Our Cauchy problem (18.3.1)–(18.3.3) is therefore equivalent to solving the equations

$$A_{jk} = 0, \quad B_{jk} = 0. \tag{18.3.13}$$

18.4 Non-linear Equations for Amplitude and Phase Functions

Equation (18.3.13) is the *dispersion relation* for the scalar wave equation in Kasner spacetime. Such a dispersion relation takes a neater geometric form upon bearing in mind the form (18.3.10) of the wave (or d'Alembert) operator $P = \square$ in Kasner coordinates, i.e.

$$A_{jk} = 0 \implies \left[-\alpha_{jk}(\Box\varphi_k) - 2\sum_{\beta,\gamma=1}^{4}(g^{-1})^{\beta\gamma}(\varphi_k)_{,\beta}(\alpha_{jk})_{,\gamma} \right] = 0, \qquad (18.4.1)$$

$$B_{jk} = 0 \implies \left[-\Box + \sum_{\beta,\gamma=1}^{4}(g^{-1})^{\beta\gamma}(\varphi_k)_{,\beta}(\varphi_k)_{,\gamma} \right]\alpha_{jk} = 0. \qquad (18.4.2)$$

Let us bear in mind that the indices j and k are not tensorial, but they merely count the number of functions contributing to the Fourier-Maslov integral operator. We can therefore exploit the four-dimensional concept of gradient of a function as the four-dimensional covariant vector defined by the differential of the function, i.e.

$$\mathrm{d}f = \sum_{\alpha=1}^{4}\frac{\partial f}{\partial x^{\alpha}}\mathrm{d}x^{\alpha} = \sum_{\alpha=1}^{4}(\nabla_{\alpha}f)\mathrm{d}x^{\alpha} = \sum_{\alpha=1}^{4}(\mathrm{grad}\,f)_{\alpha}\mathrm{d}x^{\alpha}, \qquad (18.4.3)$$

where ∇ is the Levi-Civita connection on four-dimensional space-time, and we exploit the identity $f_{,\alpha} = \nabla_{\alpha}f$, $\forall f \in C^{\infty}(M)$. The consideration of $\nabla_{\alpha}f$ is not mandatory at this stage, but it will be helpful in a moment, when we write (see below) in tensor language the equations expressing the dispersion relation.

We arrive therefore, upon multiplying Eq. (18.4.1) by α_{jk}, while dividing Eq. (18.4.2) by α_{jk}, at the following geometric form of dispersion relation in Kasner space-time (with our notation we actually write it in the same way in any Lorentzian space-time):

$$\sum_{\beta,\gamma=1}^{4}(g^{-1})^{\beta\gamma}\nabla_{\beta}\left[(\alpha_{jk})^2\nabla_{\gamma}\varphi_k\right] = \mathrm{div}\left[(\alpha_{jk})^2\mathrm{grad}\varphi_k\right] = 0, \qquad (18.4.4)$$

$$\sum_{\beta,\gamma=1}^{4}(g^{-1})^{\beta\gamma}(\nabla_{\beta}\varphi_k)(\nabla_{\gamma}\varphi_k) = \langle\mathrm{grad}\varphi_k, \mathrm{grad}\varphi_k\rangle = \frac{(\Box\alpha_{jk})}{\alpha_{jk}}, \qquad (18.4.5)$$

where the four-dimensional divergence operator acts according to

$$\mathrm{div}\,F = \sum_{\beta=1}^{4}\nabla^{\beta}F_{\beta} = \sum_{\alpha,\beta=1}^{4}(g^{-1})^{\alpha\beta}\nabla_{\beta}F_{\alpha}. \qquad (18.4.6)$$

18.5 Tensor Generalization of the Ermakov-Pinney Equation

Note that, if the ratio $\frac{(\Box \alpha_{jk})}{\alpha_{jk}}$ is much smaller than a suitable parameter having dimension length^{-2}, Eq. (18.4.5) reduces to the eikonal equation and hence the phase function reduces to the Hadamard-Ruse-Synge world function that we have defined in the course of studying the characteristic conoid. Interestingly, it is possible to devise a strategy to solve exactly Eqs. (18.4.4) and (18.4.5). For this purpose we remark that, upon defining the covariant vector

$$\psi_\gamma \equiv (\alpha_{jk})^2 \nabla_\gamma \varphi_k, \tag{18.5.1}$$

Equation (18.4.4) is equivalent to solving the first-order partial differential equation expressing the vanishing divergence condition for ψ_γ, i.e.

$$\sum_{\gamma=1}^4 \nabla^\gamma \psi_\gamma = \mathrm{div}\psi = 0. \tag{18.5.2}$$

Of course, this equation is not enough to determine the four components of ψ_γ, but there are cases where further progress can be made (see below). After doing that, we can express the (covariant) derivative of the phase function from the definition (18.5.1), i.e.

$$\nabla_\gamma \varphi_k = \partial_\gamma \varphi_k = (\alpha_{jk})^{-2} \psi_\gamma, \tag{18.5.3}$$

and the insertion of Eq. (18.5.3) into Eq. (18.4.5) yields

$$(\alpha_{jk})^3 \Box \alpha_{jk} = g(\psi, \psi) = \sum_{\beta,\gamma=1}^4 (g^{-1})^{\beta\gamma} \psi_\beta \psi_\gamma = \sum_{\gamma=1}^4 \psi_\gamma \psi^\gamma. \tag{18.5.4}$$

Interestingly, this is a tensorial generalization of a famous non-linear ordinary differential equation, i.e. the Ermakov-Pinney equation [47, 110]

$$y'' + py = qy^{-3}. \tag{18.5.5}$$

If y'' is replaced by $\Box y$, p is set to zero and q is promoted to a function of space-time location, Eq. (18.5.5) is mapped into Eq. (18.5.4). After solving this nonlinear equation for $\alpha_{jk} = \alpha_{jk}[g(\psi, \psi)]$, the task remains of finding the phase function φ_k by writing and solving the four components of Eq. (18.5.3). To sum up, we have proved the following original result.

Theorem 18.1 *For any Lorentzian space-time manifold (M, g), the amplitude functions $\alpha_{jk} \in C^2(T^*M)$ and phase functions $\varphi_k \in C^1(T^*M)$ in the parametrix (18.3.6)* *for the scalar wave equation can be obtained by solving, first, the linear condition (18.5.2) of vanishing divergence for a covariant vector ψ_γ. All non-linearities of the coupled system are then mapped into solving the non-linear equation (18.5.4) for the amplitude function α_{jk}. Eventually, the phase function φ_k is found by solving the first-order linear equation (18.5.3).*

18.6 Damped Waves

Among the other examples of fundamental solutions in section 69 of [74], we here select the equation for damped waves

$$\left(\frac{\partial^2}{\partial(x^1)^2} + \cdots + \frac{\partial^2}{\partial(x^{m-1})^2} - \frac{1}{\omega^2}\frac{\partial^2}{\partial t^2} + K\right) u = 0. \tag{18.6.1}$$

On defining the new variables

$$i\omega t = x^m, \quad i\omega t^0 = a^m, \tag{18.6.2}$$

Equation (18.6.1) is turned into

$$\left(\frac{\partial^2}{\partial(x^1)^2} + \cdots + \frac{\partial^2}{\partial(x^{m-1})^2} + \frac{\partial^2}{\partial(x^m)^2} + K\right) u = 0. \tag{18.6.3}$$

One can take for u a function of the variable ρ whose square is given by

$$\rho^2 = \sum_{h=1}^{m}(x^h - a^h)^2, \tag{18.6.4}$$

i.e. one takes for u a solution of

$$\left[\frac{d^2}{d\rho^2} + \frac{(m-1)}{\rho}\frac{d}{d\rho} + K\right] u = 0. \tag{18.6.5}$$

If one knows a solution u of this equation for $m = m_0$, one has also a solution u_1 corresponding to $m = m_0 + 2$, by considering

$$u_1 = -\frac{1}{\rho}\frac{du}{d\rho}, \tag{18.6.6}$$

so that only the integration for the values $m = 1, 2$ is strictly necessary. Indeed, for $m = 1$, one finds the linearly independent integrals $u = \cos(\sqrt{K}\rho)$ and $u = \sin(\sqrt{K}\rho)$, so that for $m = 3$ one gets, according to (18.6.6), the integrals $u = \frac{1}{\rho}\sin(\sqrt{K}\rho)$ and $u = \frac{1}{\rho}\cos(\sqrt{K}\rho)$.

The equation of damped cylindrical waves in two spatial dimensions can be written as

$$\left[\frac{1}{\omega^2}\frac{\partial^2}{\partial t^2} - \left(\frac{\partial^2}{\partial x^2} + \frac{\partial^2}{\partial y^2}\right) - K\right]u = 0,$$ (18.6.7)

with associated fundamental solution

$$u = \omega\frac{\text{Ch}\sqrt{K[\omega^2(t - t_0)^2 - (x - x_0)^2 - (y - y_0)^2}}{\sqrt{\pm[\omega^2(t - t_0)^2 - (x - x_0)^2 - (y - y_0)^2]}}.$$ (18.6.8)

When $m = 2$, Eq. (18.6.5) has a real-analytic solution and a logarithmic one, the latter providing our fundamental solution. In particular, the logarithmic fundamental solution reads as

$$u = J_0(\rho\sqrt{-K})\log(\rho) + w,$$ (18.6.9)

where J_0 is the Bessel function of order 0, the variable

$$\rho^2 \equiv \omega^2(t - t_0)^2 - (x - x_0)^2,$$ (18.6.10)

and w is a real-analytic function.

If $m = 4$, a solution of Eq. (18.6.5) is obtained by applying the operator (18.6.6) to the previous equation, i.e.

$$\frac{1}{\rho^2}J_0(\rho\sqrt{-K}) + \frac{\sqrt{-K}}{\rho}J_0'(\rho\sqrt{-K})\log(\rho) + w$$
$$= \frac{1}{\rho^2}j\left(\frac{K\rho^2}{4}\right) + \frac{K}{4}j'\left(\frac{K\rho^2}{4}\right)\log(\rho) + w,$$ (18.6.11)

where we set [74] $j(\lambda) \equiv J_0(2\sqrt{-\lambda})$, and we allow for three spatial dimensions in

$$\rho^2 = \omega^2(t - t_0)^2 - (x - x_0)^2 - (y - y_0)^2 - (z - z_0)^2 = \Gamma.$$ (18.6.12)

Chapter 19
Linear Systems of Normal Hyperbolic Form

19.1 Einstein Equations and Non-linear Theory

The development of classical field theory led first to the Maxwell equations, which are a linear system of hyperbolic equations for the electromagnetic potential, and then to the Einstein equations, which are instead a non-linear system of hyperbolic equations. The Cauchy problem associated to the latter is technically harder than in the linear theory, but in the early fifties the case of analytic Cauchy data for the Einstein gravitational equations was well understood, although the hypothesis of analyticity for such data was unsatisfactory, because in Einstein's theory the coordinate changes are only restricted to be sufficiently differentiable.

The great achievement of the work by Fourès-Bruhat [64] was a rigorous and constructive proof that the Cauchy problem for Einstein's theory is well posed and admits a unique solution also with non-analytic Cauchy data. From now on we shall therefore devote our efforts to a systematic presentation of what Fourès-Bruhat did. More precisely, she studied the Cauchy problem, in the non-analytic case, for a system of non-linear second-order hyperbolic partial differential equations with n unknown functions W_s and four independent variables x^α, having the form

$$E_s = \sum_{\lambda,\mu=1}^{4} A^{\lambda\mu} \frac{\partial^2 W_s}{\partial x^\lambda \partial x^\mu} + f_s = 0, \quad s = 1, 2, \ldots, n, \tag{E}$$

where $A^{\lambda\mu}$ and f_s are given functions of the unknown W_s and their first derivatives.

She used, for this solution, a system of integral equations fulfilled by the seven-times differentiable solutions of equations (E). This system is obtained for some linear equations by integrating over the characteristic conoid Σ with vertex M some linear combinations of the equations (the coefficients of these combinations are some auxiliary functions which possess at M the parametrix properties, an approximation of the fundamental solution of Hadamard) and by adjoining to the formulae of Kirchhoff type obtained in such a way the equations determining the characteristic conoid

© Springer International Publishing AG 2017
G. Esposito, *From Ordinary to Partial Differential Equations*, UNITEXT 106,
DOI 10.1007/978-3-319-57544-5_19

and the auxiliary functions. The results admit of an easy extension to non-linear equations, subject to the condition of integrating over Σ not the (E) equations themselves but the equations derived from (E) by five derivatives, and provided one supplements the previous integral equations with the equations relating among themselves the derivatives of the unknown functions up to the fifth order. Such a system had been formed by Sobolev [129] for an hyperbolic linear second-order partial differential equation (with analytic coefficients) and by Christianovich [25] for a non-linear equation in four variables.

By extending these methods, Fourès-Bruhat wrote in its complete form the system of integral equations satisfied from a system whatsoever of type (E) and studied in detail the various quantities occurring in these integral equations in light of the goal of solving them. She discovered that the kernel, occurring in the Kirchhoff formula, is only bounded under differentiability assumptions made on the unknown functions. Some difficulties occur therefore in the process of solving directly the system of integral equations obtained, and of using it to solve the Cauchy problem relative to (E).

She then considered the Cauchy problem for system (E) by using the system of integral equations fulfilled by the solutions of partial differential equations E_1 that approximate (E). The proof is performed in detail in the case, a bit simpler, that involves derivatives of a lower order, which is the one of equations of general relativity, where the coefficients of second derivatives depend on the unknown functions but not on their first derivatives. She proved that, to Cauchy data five times differentiable, assigned on a compact domain d of the initial surface $x^4 = 0$, there corresponds a unique solution, four times differentiable, of equations (E) in a domain D, section of a cone having as basis the domain d, if the coefficients of these equations are four times differentiable.

The solution of the Cauchy problem for a system (E) whatsoever can be obtained in a completely analogous way: it is enough to consider equations approaching not (E) itself but some equations previously derived.

Eventually, she applied the previous results to the equations of gravitation.

The vacuum Einstein equations $R_{\alpha\beta} = 0$ get reduced, in isothermal coordinates, to equations of the type (E), $G_{\alpha\beta} = 0$. She proved, by using the conservation equations, that the solution of the Cauchy problem, pertaining to the equations $G_{\alpha\beta} = 0$, satisfies over the whole of its existence domain the isothermal conditions if the same is true for the initial data. This solution satisfies therefore the equations of gravitation. She proved that it is unique up to a coordinate change. She also built an Einstein space-time corresponding to non-analytic initial data, assigned on a spacelike domain, and in such a way that it highlights the propagation character which is peculiar of relativistic gravitation.

The Fourès-Bruhat work begins by considering a system $[E]$ of n second-order partial differential equations, with n unknown functions u_s and four variables x, hyperbolic and *linear*, of the following type:

$$E_r = \sum_{\lambda,\mu=1}^{4} A^{\lambda\mu} \frac{\partial^2 u_r}{\partial x^\lambda \partial x^\mu} + \sum_{s=1}^{n} \sum_{\mu=1}^{4} B_r^{s\,\mu} \frac{\partial u_s}{\partial x^\mu} + f_r = 0, \quad r = 1, 2, \ldots, n. \quad [E]$$

The coefficients $A^{\lambda\mu}$ (which are the same for all n equations), $B_r^{s\,\mu}$ and f_r are given functions of the four variables x^α. They are taken to satisfy, within a domain defined by

$$\left| x^i - \overline{x}^i \right| \le d, \ \left| x^4 \right| \le \varepsilon \ (i = 1, 2, 3)$$

(where $\overline{x}^i$, d and ε are some given numbers) the following assumptions:

Assumptions on the Coefficients

(1) The coefficients $A^{\lambda\mu}$ and $B_s^{r\,\lambda}$ possess continuous and bounded derivatives up to the orders *four* and *two*, respectively. The coefficients f_r are continuous and bounded.

(2) The quadratic form $\sum_{\lambda,\mu=1}^{4} A^{\lambda\mu} x_\lambda x_\mu$ is of the normal hyperbolic type, has one positive square and three negative squares. We will assume in addition that the variable x^4 is a temporal variable, the three variables x^i being spatial, i.e.

$$A^{44} > 0 \text{ and the quadratic form } \sum_{i,j=1}^{3} A^{ij} x_i x_j < 0 \text{ (negative } - \text{ definite).}$$

(3) Partial derivatives of the $A^{\lambda\mu}$ and $B_s^{r\,\lambda}$ of order four and two, respectively, satisfy Lipschitz conditions with respect to all their arguments.

Summary of Results in Linear Theory

Fourès-Bruhat proved, in light of her aim to solve the Cauchy problem, that every system of n functions (continuous and bounded within D with their first partial derivatives), satisfying the $[E]$ equations and taking at $x^4 = 0$, as for their first partial derivatives, some given values, is a solution of a system of integral equations (I). These equations (I) express the values, at a point $M_0(x_0)$ belonging to D, of the unknown u_s in terms of their values on the characteristic conoid (Σ_0) of vertex M_0 and in terms of the initial data.

These equations are obtained by integrating over Σ_0 some linear combinations of equations $[E]$, the coefficients of these combinations being n^2 auxiliary functions which exhibit a singularity at M_0.

The coefficients $A^{\lambda\mu}$ are initially assumed to take at M_0 some particular values $(1, 0 \text{ and } -1)$. Later on, this restriction will be suppressed.

19.2 Equations Defining the Characteristic Conoid

The characteristic surfaces of system $[E]$ are three-dimensional manifolds of the space of four variables x^α, solutions of the differential system

$$F = \sum_{\lambda,\mu=1}^{4} A^{\lambda\mu} y_\lambda y_\mu = 0$$

with

$$\sum_{\lambda=1}^{4} y_\lambda dx^\lambda = 0.$$

The four quantities y_λ denote a system of directional parameters of the normal to the contact element, having support x^α. Let us take this system, which is only defined up to a proportionality factor, in such a way that $y_4 = 1$ and let us set $y_i = p_i$. The desired surfaces are solution of

$$F = A^{44} + 2\sum_{i=1}^{3} A^{i4} p_i + \sum_{i,j=1}^{3} A^{ij} p_i p_j = 0,$$

$$dx^4 + \sum_{i=1}^{3} p_i dx^i = 0. \qquad (19.2.1)$$

The characteristics of this differential system, bicharacteristics of equations $[E]$, satisfy the following differential equations:

$$\frac{dx^i}{A^{i4} + \sum_{j=1}^{3} A^{ij} p_j} = \frac{dx^4}{A^{44} + \sum_{i=1}^{3} A^{i4} p_i} = \frac{-dp_i}{\frac{1}{2}\left(\frac{\partial F}{\partial x^i} - p_i \frac{\partial F}{\partial x^4}\right)} = d\lambda_1,$$

λ_1 being an auxiliary parameter.

The characteristic conoid Σ_0 with vertex $M_0(x_0^\alpha)$ is the characteristic surface generated from the bicharacteristics passing through M_0. Any such bicharacteristic satisfies the system of integral equations

$$x^i = x_0^i + \int_0^{\lambda_1} T^i d\lambda, \quad T^i \equiv A^{i4} + \sum_{j=1}^{3} A^{ij} p_j, \qquad (19.2.2a)$$

$$x^4 = x_0^4 + \int_0^{\lambda_1} T^4 d\lambda, \quad T^4 \equiv A^{44} + \sum_{i=1}^{3} A^{i4} p_i, \qquad (19.2.2b)$$

$$p_i = p_i^0 + \int_0^{\lambda_1} R_i d\lambda, \quad R_i \equiv -\frac{1}{2}\left(\frac{\partial F}{\partial x^i} - p_i \frac{\partial F}{\partial x^4}\right), \qquad (19.2.2c)$$

where the p_i^0 verify the relation

$$A_0^{44} + 2 \sum_{i=1}^{3} A_0^{i4} p_i^0 + \sum_{i,j=1}^{3} A_0^{ij} p_i^0 p_j^0 = 0, \tag{19.2.3}$$

where $A_0^{\lambda\mu}$ denotes the value of the coefficient $A^{\lambda\mu}$ at the vertex M_0 of the conoid Σ_0. We note that we are dealing with non-linear integral equations, the integrand R_i being a polynomial of given functions of the unknown functions.

We will assume that at the point M_0 the coefficients $A^{\lambda\mu}$ take the following values:

$$A_0^{44} = 1, \ A_0^{i4} = 0, \ A_0^{ij} = -\delta^{ij}. \tag{19.2.4}$$

The relation (19.2.3) takes therefore the simple form

$$\sum_{i=1}^{3} (p_i^0)^2 = 1.$$

We will introduce to define the points of the surface Σ_0, besides the parameter λ_1 which defines the position of a point on a given bicharacteristic, two new parameters λ_2 and λ_3 that vary with the bicharacteristic under consideration, by setting

$$p_1^0 = \sin \lambda_2 \cdot \cos \lambda_3, \ p_2^0 = \sin \lambda_2 \cdot \sin \lambda_3, \ p_3^0 = \cos \lambda_2.$$

19.3 A Domain of the Characteristic Conoid

The assumptions made on the coefficients $A^{\lambda\mu}$ make it possible to prove that there exists a number ε_1 defining a variation domain Λ of the parameters λ_i by means of

$$|\lambda_1| \le \varepsilon_1, \ 0 \le \lambda_2 \le \pi, \ 0 \le \lambda_3 \le 2\pi, \tag{Λ}$$

such that the integral equations (19.2.2) possess within (Λ) a unique solution, continuous and bounded

$$x^\alpha = x^\alpha(x_0^\alpha, \lambda_1, \lambda_2, \lambda_3), \ p_i = p_i(x_0^\alpha, \lambda_1, \lambda_2, \lambda_3), \tag{19.3.1}$$

satisfying the inequalities

$$|x^i - \overline{x}^i| \le d, \ |x^4| \le \varepsilon$$

and possessing partial derivatives, continuous and bounded, of the first three orders with respect to the overabundant variables λ_1, p_i^0 (hence with respect to the three variables λ_i).

The first four equations (19.3.1) define, as a function of the three parameters λ_i, varying within the domain Λ, a point of a domain V of the characteristic conoid Σ_0.

We shall be led, in the following part of this work, to consider other parametric representations of the domain V:

(1) We shall take as independent parameters the three quantities $x^4, \lambda_2, \lambda_3$. The function $x^4(\lambda_1, \lambda_2, \lambda_3)$ satisfies the equation

$$x^4 = \int_0^{\lambda_1} T^4 d\lambda + x_0^4 \text{ where } T^4 \equiv A^{44} + \sum_{i=1}^{3} A^{i4}p_i. \tag{19.3.2}$$

Or it turns out from (19.2.3) that, on Σ_0, one has

$$2\sum_{i=1}^{3} A^{i4}p_i = -\sum_{i,j=1}^{3} A^{ij}p_ip_j - A^{44} \geq -A^{44},$$

from which

$$T^4 \geq \frac{A^{44}}{2} > 0;$$

x^4 is thus a monotonic increasing function of λ_1, the correspondence between $(x^4, \lambda_2, \lambda_3)$ and $(\lambda_1, \lambda_2, \lambda_3)$ is bijective.

(2) We shall take as representative parameters of a point of Σ_0 his three spatial coordinates x^i. The elimination of $\lambda_1, \lambda_2, \lambda_3$ among the four equations yields x^4 as a function of the x^i.

From the relation

$$dx^4 + \sum_{i=1}^{3} p_i dx^i = 0,$$

identically verified from the solutions of equations (19.2.2) on the characteristic surface Σ_0, one infers that the partial derivatives of this function x^4 with respect to the x^i verify the relation

$$\frac{\partial x^4}{\partial x^i} = -p_i.$$

If we denote by $[\varphi]$ the value of a function φ of four coordinates x^α on Σ_0 and if we express $[\varphi]$ as a function of the three parameters x^i representatives of Σ_0, the partial derivatives of this function with respect to the x^i fulfill therefore:

$$\frac{\partial[\varphi]}{\partial x^i} = \left[\frac{\partial\varphi}{\partial x^i}\right] - \left[\frac{\partial\varphi}{\partial x^4}\right]p_i. \tag{19.3.3}$$

19.4   Integral Equations for Derivatives of x^i and p_i

We shall set

$$\frac{\partial x^i}{\partial p_j^0} \equiv y_j^i, \quad \frac{\partial^2 x^i}{\partial p_j^0 \partial p_h^0} \equiv y_{jh}^i, \quad \frac{\partial^3 x^i}{\partial p_j^0 \partial p_h^0 \partial p_k^0} \equiv y_{jhk}^i,$$

$$\frac{\partial p_i}{\partial p_j^0} \equiv z_j^i, \quad \frac{\partial^2 p_i}{\partial p_j^0 \partial p_h^0} \equiv z_{jh}^i, \quad \frac{\partial^3 p_i}{\partial p_j^0 \partial p_h^0 \partial p_k^0} \equiv z_{jhk}^i.$$

These functions satisfy the integral equations obtained by derivation under the summation sign with respect to the p_i^0 of Eq. (19.2.2) (the quantities obtained under the integration signs being continuous and bounded). Formula (19.3.3) shows that these equations can be written (the derivatives $\frac{\partial x^4}{\partial p_i^0}$ being useless)

$$y_j^i = \int_0^{\lambda_1} T_j^i \mathrm{d}\lambda, \quad T_j^i = \frac{\partial T^i}{\partial p_j^0} = \sum_{h=1}^3 \left\{ \frac{\partial}{\partial x^h} \sum_{k=1}^3 [A^{ik}] p_k + \frac{\partial}{\partial x^h} [A^{i4}] \right\} y_j^h + [A^{ih}] z_j^h,$$

$$z_j^i = \int_0^{\lambda_1} R_j^i \mathrm{d}\lambda, \quad R_j^i = \frac{\partial R_i}{\partial p_j^0} = \sum_{k=1}^3 \left(\frac{\partial R_i}{\partial x_k} y_j^k + \frac{\partial R_i}{\partial p_k} z_j^k \right),$$

$$y_{jk}^i = \int_0^{\lambda_1} T_{jk}^i \mathrm{d}\lambda, \quad T_{jk}^i = \frac{\partial T_j^i}{\partial p_k^0} = \sum_{h=1}^3 \left(\frac{\partial T^i}{\partial x^h} y_{jk}^h + \frac{\partial T^i}{\partial p_h} z_{jk}^h \right) + \phi_{jk}^i,$$

$$z_{jk}^i = \int_0^{\lambda_1} R_{jk}^i \mathrm{d}\lambda, \quad R_{jk}^i = \frac{\partial R_j^i}{\partial p_k^0} = \sum_{h=1}^3 \left(\frac{\partial R_i}{\partial x^h} y_{jk}^h + \frac{\partial R_i}{\partial p_h} z_{jk}^h \right) + \psi_{jk}^i,$$

where ϕ_{jk}^i and ψ_{jk}^i are polynomials of the functions $p_i(\lambda)$, $y_j^i(\lambda)$, $z_j^i(\lambda)$, of the coefficients $A^{\lambda\mu}(x^\alpha)$ and of their partial derivatives with respect to the x^α up to the third order included. In these functions the x^α are replaced from the $x^\alpha(\lambda)$ given by the formulae (19.3.1).

We would find by analogous fashion

$$y_{jhk}^i = \int_0^{\lambda_1} T_{jhk}^i \mathrm{d}\lambda, \quad T_{jhk}^i = \frac{\partial T_{jh}^i}{\partial p_k^0} = \sum_{l=1}^3 \left(\frac{\partial T^i}{\partial x^l} y_{jhk}^l + \frac{\partial T^i}{\partial p_l} z_{jhk}^l \right) + \phi_{jhk}^i,$$

$$z_{jhk}^i = \int_0^{\lambda_1} R_{jhk}^i \mathrm{d}\lambda, \quad R_{jhk}^i = \frac{\partial R_{jh}^i}{\partial p_k^0} = \sum_{l=1}^3 \left(\frac{\partial R^i}{\partial x^l} y_{jhk}^l + \frac{\partial R^i}{\partial p_l^0} z_{jhk}^l \right) + \psi_{jhk}^i,$$

where ϕ^i_{jhk} and ψ^i_{jhk} are polynomials of the functions p_i, y^i_j, z^i_j, y^i_{jh}, z^i_{jh} as well as of the coefficients $A^{\lambda\mu}$ and of their partial derivatives up to the fourth order included (functions of the functions x^α).

19.5   Relations on the Conoid Satisfied by the Unknown Functions

We will denote by $[\varphi]$ the value of a function φ of the four coordinates x^α on the surface of the characteristic conoid Σ_0; $[\varphi]$ can be expressed as a function of three variables of a parametric representation of Σ_0, in particular of the three coordinates x^i. In light of the equality (19.3.3) the partial derivatives of this function with respect to the x^i verify the relation

$$\left[\frac{\partial[\varphi]}{\partial x^i}\right] = \left[\frac{\partial\varphi}{\partial x^i}\right] - \left[\frac{\partial\varphi}{\partial x^4}\right]p_i.$$

One applies again this rule to the evaluation of the derivatives

$$\frac{\partial}{\partial x^j}\left[\frac{\partial\varphi}{\partial x^i}\right] \text{ and } \frac{\partial}{\partial x^j}\left[\frac{\partial\varphi}{\partial x^4}\right],$$

from which it follows easily

$$\left[\frac{\partial^2\varphi}{\partial x^i\partial x^4}\right] = \frac{\partial}{\partial x^i}\left[\frac{\partial\varphi}{\partial x^4}\right] + \left[\frac{\partial^2\varphi}{\partial(x^4)^2}\right]p_i,$$

$$\left[\frac{\partial^2\varphi}{\partial x^i\partial x^j}\right] = \frac{\partial^2[\varphi]}{\partial x^i\partial x^j} + \frac{\partial}{\partial x^i}\left[\frac{\partial\varphi}{\partial x^4}\right]p_j + \frac{\partial}{\partial x^j}\left[\frac{\partial\varphi}{\partial x^4}\right]p_i + \left[\frac{\partial\varphi}{\partial x^4}\right]\frac{\partial p_i}{\partial x^j}$$

$$+ \left[\frac{\partial^2\varphi}{\partial(x^4)^2}\right]p_i p_j.$$

These identities make it possible to write the following relations satisfied by the unknown functions u_s on the characteristic conoid:

$$[E_r] = \sum_{i,j=1}^{3}[A^{ij}]\frac{\partial^2[u_r]}{\partial x^i\partial x^j} + \left\{\sum_{i,j=1}^{3}[A^{ij}]p_i p_j + 2\sum_{i=1}^{3}[A^{i4}]p_i + [A^{44}]\right\}\frac{\partial^2 u_r}{\partial(x^4)^2}$$

$$+ 2\sum_{i=1}^{3}\left\{\sum_{j=1}^{3}[A^{ij}]p_j + [A^{i4}]\right\}\frac{\partial}{\partial x^i}\left[\frac{\partial u_r}{\partial x^4}\right] + \left[\frac{\partial u_r}{\partial x^4}\right]\sum_{i,j=1}^{3}[A^{ij}]\frac{\partial p_i}{\partial x^j}$$

$$+ \sum_{s=1}^{n}\sum_{\mu=1}^{4}B^{s\mu}_r\left[\frac{\partial u_s}{\partial x^\mu}\right] + [f_r] = 0. \tag{19.5.1}$$

The coefficient of the term $\left[\frac{\partial^2 u_r}{\partial (x^4)^2}\right]$ is the value on the characteristic conoid of the first member of Eq. (19.2.1); it therefore vanishes. We might have expected on the other hand that the equations $[E_r] = 0$ would not contain second derivatives of the functions u_r but those obtained by derivation on the surface Σ_0, the assignment on a characteristic surface of the unknown functions $[u_r]$ and of their first derivatives $\left[\frac{\partial u_r}{\partial x^\alpha}\right]$ not being able to determine the set of second derivatives.

19.6 The Auxiliary Functions σ_s^r

We form n^2 linear combinations $\sum_{r=1}^{n} \sigma_s^r [E_r]$ of the Eq. (19.5.1) verified by the unknown functions within the domain V of Σ_0, the σ_s^r denoting n^2 auxiliary functions which possess at M_0 a singularity.

We set

$$M(\varphi) = \sum_{i,j=1}^{3} [A^{ij}] \frac{\partial^2 \varphi}{\partial x^i \partial x^j},$$

φ denoting a function whatsoever of the three variables x^i, and we write

$$\sum_{r=1}^{n} \sigma_s^r E_r = \sum_{r=1}^{n} \left\{ M([u_r]) + 2 \sum_{i=1}^{3} \left(\sum_{j=1}^{3} [A^{ij}] p_j + [A^{i4}] \right) \frac{\partial}{\partial x^i} \left[\frac{\partial u_r}{\partial x^4}\right] \right.$$

$$\left. + \left[\frac{\partial u_r}{\partial x^4}\right] \sum_{i,j=1}^{3} [A^{ij}] \frac{\partial p_i}{\partial x^j} + \sum_{t,\mu} [B_r^{t\mu}] \left[\frac{\partial u_t}{\partial x^\mu}\right] + [f_r] \right\} \sigma_s^r = 0. \qquad (19.6.1)$$

We will transform these equations in such a way that a divergence occurs therein, whose volume integral will get transformed into a surface integral, while the remaining terms will contain only $[u_r]$ and $\left[\frac{\partial u_r}{\partial x^4}\right]$. We will use for that purpose the following identity, verified by two functions whatsoever φ and ψ of the three variables x^i:

$$\psi M(\varphi) = \sum_{i,j=1}^{3} \frac{\partial}{\partial x^i} \left([A^{ij}] \psi \frac{\partial \varphi}{\partial x^j}\right) - \sum_{i,j=1}^{3} \frac{\partial \varphi}{\partial x^j} \frac{\partial}{\partial x^i} ([A^{ij}] \psi)$$

or

$$\psi M(\varphi) = \sum_{i,j=1}^{3} \frac{\partial}{\partial x^i} \left([A^{ij}] \psi \frac{\partial \varphi}{\partial x^j} - \varphi \frac{\partial}{\partial x^j} ([A^{ij}] \varphi)\right) + \varphi \overline{M}(\psi),$$

where $\overline{M}$ is the adjoint operator of M, i.e.

$$\overline{M}(\psi) = \sum_{i,j=1}^{3} \frac{\partial^2 ([A^{ij}]\psi)}{\partial x^i \partial x^j},$$

and the identity (19.3.3), previously written, which yields here

$$\left[\frac{\partial u_r}{\partial x^i}\right] = \frac{\partial [u_r]}{\partial x^i} + p_i \left[\frac{\partial u_r}{\partial x^4}\right].$$

We see without difficulty that the expressions $\sum_{r=1}^{n} \sigma_s^r [E_r]$ take the form

$$\sum_{r=1}^{n} \sigma_s^r [E_r] = \sum_{i=1}^{3} \frac{\partial}{\partial x^i} E_s^i + \sum_{r=1}^{n} [u_r] L_s^r + \sum_{r=1}^{n} \sigma_s^r f_r - \sum_{r=1}^{n} \left[\frac{\partial u_r}{\partial x^4}\right] D_s^r,$$

where one has defined

$$E_s^i = \sum_{j=1}^{3} \sum_{r=1}^{n} \left([A^{ij}]\sigma_s^r \frac{\partial [u_r]}{\partial x^j} - [u_r] \frac{\partial}{\partial x^j}([A^{ij}]\sigma_s^r)\right)$$

$$+ 2 \sum_{r=1}^{n} \sigma_s^r \left\{\sum_{j=1}^{3} [A^{ij}] p_j + [A^{i4}]\right\} \left[\frac{\partial u_r}{\partial x^4}\right]$$

$$+ \sum_{r,t=1}^{n} [B_{ri}^t][u_t]\sigma_s^r, \tag{19.6.2a}$$

$$L_s^r = \overline{M}(\sigma_s^r) - \sum_{i=1}^{3} \sum_{t=1}^{n} \frac{\partial}{\partial x^i}\left([B_t^{ri}]\sigma_s^t\right), \tag{19.6.2b}$$

$$D_s^r = \sigma_s^r \left\{2 \sum_{i=1}^{3} \frac{\partial}{\partial x^i}\left(\sum_{j=1}^{3} [A^{ij}] p_j + [A^{i4}]\right) - \sum_{i,j=1}^{3} [A^{ij}] \frac{\partial p_j}{\partial x^i}\right\}$$

$$+ 2 \sum_{i=1}^{3} \left(\sum_{j=1}^{3} [A^{ij}] p_j + [A^{i4}]\right) \frac{\partial \sigma_s^r}{\partial x^i}$$

$$-\sum_{t=1}^{n}([B_t^{r4}]+\sum_{i=1}^{3}[B_t^{ri}]p_i)\sigma_s^t. \tag{19.6.2c}$$

We will choose the auxiliary functions σ_s^r in such a way that, in every equation, the coefficient of $\left[\frac{\partial u_r}{\partial x^4}\right]$ vanishes. These functions will therefore have to fulfill n^2 partial differential equations of first order

$$D_s^r = 0. \tag{19.6.3}$$

We will see that these equations possess a solution having at M_0 the desired singularity. If the auxiliary functions σ_s^r verify these n^2 relations, the equations, verified by the unknown functions u_r on the characteristic conoid Σ_0, take the simple form

$$\sum_{r=1}^{n}\left([u_r]L_s^r+\sigma_s^r[f_r]\right)+\sum_{i=1}^{3}\frac{\partial}{\partial x^i}E_s^i = 0. \tag{19.6.4}$$

19.7 Integrating Linear Combinations of the Equations

We will integrate the equations so obtained with respect to the three variables x^i on a portion V_η of hypersurface of the characteristic conoid Σ_0, limited by the hypersurfaces $x^4 = 0$ and $x^4 = x_0^4 - \eta$. This domain V_η is defined to be simply connected and internal to the domain V if the coordinate x_0^4 is sufficiently small. As a matter of fact:

$$|x_0^4| < \varepsilon_0 \text{ implies within } V_\eta \ |x^4 - x_0^4| < \varepsilon_0.$$

The formula (19.3.2) shows in such a case that, for a suitable choice of ε_0, we will have

$$\lambda_1 \leq \varepsilon_1.$$

Since the boundary of V_η consists of two-dimensional domains S_0 and S_η cut over Σ_0 from the hypersurfaces $x^4 = 0, x^4 = x_0^4 - \eta$ we will have, upon integrating the Eq. (19.6.4) within V_η, the following fundamental relations:

$$\int\int_{V_\eta}\int \sum_{r=1}^{n}\{[u_r]L_s^r+\sigma_s^r[f_r]\}\,dV + \int\int_{S_\eta}\sum_{i=1}^{3}E_s^i\cos(n,x^i)dS$$

$$-\int\int_{S_0}\sum_{i=1}^{3}E_s^i\cos(n,x^i)dS = 0, \tag{19.7.1}$$

where dV, dS and $\cos(n, x^i)$ denote, in the space of three variables x^i, the volume element, the area element of a surface $x^4 = C^{te}$ and the directional cosines of the outward-pointing normal to one of such surfaces, respectively.

The limit of these equations, when η approaches zero, will provide us with Kirchhoff formulae that we will build in the last part of this chapter.

19.8   Determination of the Auxiliary Functions σ_s^r

We will look for a solution of Eq. (19.6.3) in the form

$$\sigma_s^r = \sigma \omega_s^r,$$

where σ is infinite at the point M_0 and the ω_s^r are bounded.

The Eq. (19.6.3) read as

$$\sigma_s^r \left\{ \sum_{i=1}^{3} \frac{\partial}{\partial x^i} \left(\sum_{j=1}^{3} [A^{ij}] p_j + [A^{i4}] \right) + \sum_{i,j=1}^{3} p_j \frac{\partial}{\partial x^i} [A^{ij}] + \sum_{i=1}^{3} \frac{\partial}{\partial x^i} [A^{i4}] \right\}$$

$$- \sum_{t=1}^{n} \left([B_t^{r4}] + \sum_{i=1}^{3} [B_t^{ri}] p_i \right) \sigma_s^t + 2 \sum_{i=1}^{3} \left(\sum_{j=1}^{3} [A^{ij}] p_j + [A^{i4}] \right) \frac{\partial \sigma_s^r}{\partial x^i} = 0.$$

The coefficients $A^{\lambda\mu}$, $B_s^{t\lambda}$, the first derivatives of the $A^{\lambda\mu}$ and the functions p_i are bounded within the domain V, the coefficients of the linear first-order partial differential equations are therefore a sum of bounded terms, perhaps with exception of the terms

$$\sum_{i=1}^{3} \frac{\partial}{\partial x^i} \left\{ \sum_{j=1}^{3} [A^{ij}] p_j + [A^{i4}] \right\}.$$

We will therefore choose the ω_s^r, that we want to be bounded, as satisfying the equation

$$\omega_s^r \left(\sum_{i,j=1}^{3} p_j \frac{\partial}{\partial x^i} [A^{ij}] + \sum_{i=1}^{3} \frac{\partial}{\partial x^i} [A^{i4}] \right) - \sum_{t=1}^{n} \omega_s^t \left\{ [B_t^{r4}] + \sum_{i=1}^{3} [B_t^{ri}] p_i \right\}$$

$$+ 2 \sum_{i=1}^{3} \left\{ \sum_{j=1}^{3} [A^{ij}] p_j + [A^{i4}] \right\} \frac{\partial \omega_s^r}{\partial x^i} = 0, \tag{19.8.1}$$

fulfilling in turn

$$\sigma \sum_{i=1}^{3} \frac{\partial}{\partial x^i} \left(\sum_{j=1}^{3} [A^{ij}]p_j + [A^{i4}] \right) + 2 \sum_{i=1}^{3} \left(\sum_{j=1}^{3} [A^{ij}]p_j + [A^{i4}] \right) \frac{\partial \sigma}{\partial x^i} = 0. \quad (19.8.2)$$

19.9 Evaluation of the ω_s^r

We see easily that Eq. (19.8.1) can be written in form of integral equations analogous to the Eq. (19.2.2) obtained in the search for the conoid Σ_0. We have indeed, on Σ_0:

$$\sum_{j=1}^{3} [A^{ij}]p_j + [A^{i4}] = T^i = \frac{\partial x^i}{\partial \lambda_1},$$

from which, for an arbitrary function φ defined on Σ_0,

$$\sum_{i=1}^{3} T^i \frac{\partial \varphi}{\partial x^i} = \frac{\partial \varphi}{\partial \lambda_1}.$$

Let us impose upon the ω_s^r the limiting conditions

$$\omega_s^r = \delta_s^r \text{ for } \lambda_1 = 0.$$

These quantities satisfy therefore the integral equations

$$\omega_s^r = \int_0^{\lambda_1} \left(\sum_{t=1}^{n} Q_t^r \omega_s^t + Q \omega_s^r \right) d\lambda + \delta_s^r \quad (19.9.1)$$

with

$$Q_t^r = \frac{1}{2} ([B_t^{r4}] + \sum_{i=1}^{3} [B_t^{ri}]p_i) \text{ and } Q = -\frac{1}{2} \left(\sum_{i,j=1}^{3} p_j \frac{\partial}{\partial x^i} [A^{ij}] + \sum_{i=1}^{3} \frac{\partial}{\partial x^i} [A^{i4}] \right),$$

the assumptions made upon the coefficients $A^{\lambda\mu}$ and $B_s^{r\lambda}$ and the results obtained on the functions x^i, p_i enabling moreover to prove that, for a convenient choice of ε_1, these equations have a unique, continuous, bounded solution which has partial derivatives of the first two orders with respect to the p_i^0, continuous and bounded within the domain Λ. We will denote these derivatives by ω_{si}^r and ω_{sij}^r.

19.10   Calculation of σ

Let us consider Eq. (19.8.2) verified by σ. We know that

$$\sum_{i=1}^{3}(\sum_{j=1}^{3}[A^{ij}]p_j + [A^{i4}])\frac{\partial\sigma}{\partial x^i} = \frac{\partial\sigma}{\partial\lambda_1},$$

and we are going to evaluate the coefficient of σ,

$$\sum_{i=1}^{3}\frac{\partial}{\partial x^i}\left(\sum_{j=1}^{3}[A^{ij}]p_j + [A^{i4}]\right),$$

by relating it very simply to the determinant

$$\frac{D(x^1,x^2,x^3)}{D(\lambda_1,\lambda_2,\lambda_3)} \equiv J_{x\lambda}.$$

This Jacobian $J_{x\lambda}$ of the change of variables $x^i = x^i(\lambda_j)$ on the conoid Σ_0, has for elements

$$\frac{\partial x^i}{\partial\lambda_1} = T^i, \quad \frac{\partial x^i}{\partial\lambda_2} = \sum_{j=1}^{3}y_j^i\frac{\partial p_j^0}{\partial\lambda_2}, \quad \frac{\partial x^i}{\partial\lambda_3} = \sum_{j=1}^{3}y_j^i\frac{\partial p_j^0}{\partial\lambda_3}. \tag{19.10.1}$$

Let us denote by J_j^i the minor relative to the element $\frac{\partial x^i}{\partial\lambda_j}$ of the determinant $J_{x\lambda}$.
A function whatsoever φ, defined on Σ_0, verifies the identities

$$\frac{\partial\varphi}{\partial x^i} = \sum_{j=1}^{3}\frac{J_i^j}{J_{x\lambda}}\frac{\partial\varphi}{\partial\lambda_j}.$$

Let us apply this formula to the function $\frac{\partial x^i}{\partial\lambda_1} = T^i$:

$$\sum_{i=1}^{3}\frac{\partial}{\partial x^i}T^i = \sum_{i,j=1}^{3}\frac{J_i^j}{J_{x\lambda}}\frac{\partial}{\partial\lambda_j}T^i = \sum_{i,j=1}^{3}\frac{J_i^j}{J_{x\lambda}}\frac{\partial}{\partial\lambda_1}\left(\frac{\partial x^i}{\partial\lambda_j}\right),$$

J_i^j being the minor relative to the element $\frac{\partial x^i}{\partial\lambda_j}$ of the determinant $J_{x\lambda}$ we have

$$\sum_{i=1}^{3} \frac{\partial}{\partial x^i} T^i = \frac{1}{J_{x\lambda}} \frac{\partial J_{x\lambda}}{\partial \lambda_1}.$$

Thus, the function σ verifies the relation

$$\sigma \frac{\partial J_{x\lambda}}{\partial \lambda_1} + 2 J_{x\lambda} \frac{\partial \sigma}{\partial \lambda_1} = 0$$

which is integrated in immediate way. The general solution is

$$\sigma = \frac{f(\lambda_2, \lambda_3)}{|J_{x\lambda}|^{\frac{1}{2}}},$$

where f denotes an arbitrary function.

For $\lambda_1 = 0$ the determinant $J_{x\lambda}$ vanishes, because the y^i_j are vanishing; the function σ is therefore infinite.

The coefficients $A^{\lambda\mu}$ and their first and second partial derivatives with respect to the x^α being continuous and bounded within the domain V of Σ_0, as well as the functions x^i, y^i_j, z^i_j, we have

$$\lim_{\lambda_1 \to 0} \frac{y^i_j}{\lambda_1} = [A^{ij}]_{\lambda_1=0} = -\delta^j_i. \tag{19.10.2}$$

By dividing the second and third line of $J_{x\lambda}$ by λ_1 we obtain a determinant equal to $\frac{J_{x\lambda}}{(\lambda_1)^2}$; we deduce from the formulas (19.10.1) and (19.10.2)

$$\lim_{\lambda_1 \to 0} \frac{J_{x\lambda}}{(\lambda_1)^2} = \det \begin{pmatrix} -\sin\lambda_2 \cos\lambda_3 & -\sin\lambda_2 \sin\lambda_3 & -\cos\lambda_2 \\ -\cos\lambda_2 \cos\lambda_3 & -\cos\lambda_2 \sin\lambda_3 & \sin\lambda_2 \\ +\sin\lambda_2 \sin\lambda_3 & -\sin\lambda_2 \cos\lambda_3 & 0 \end{pmatrix} = -\sin\lambda_2.$$

As a matter of fact:

$$\lim_{\lambda_1 \to 0} T^i = -\sum_{j=1}^{3} \delta^j_i p^0_j = -p^0_i$$

$$\lim_{\lambda_1 \to 0} \frac{1}{\lambda_1} \frac{\partial x^i}{\partial \lambda_u} = \lim_{\lambda_1 \to 0} \sum_{j=1}^{3} \frac{y^i_j}{\lambda_1} \frac{\partial p^0_j}{\partial \lambda_u} = -\sum_{j=1}^{3} \delta^i_j \frac{\partial p^0_j}{\partial \lambda_u}.$$

We will take for auxiliary function σ the function

$$\sigma = \left| \frac{\sin \lambda_2}{J_{x\lambda}} \right|^{\frac{1}{2}}.$$

We will then have $\lim_{\lambda_1 \to 0} \sigma \lambda_1 = 1$.

19.11   Derivatives of the Functions σ_s^r

Equation (19.7.1) contain, on the one hand the values on Σ_0 of the unknown functions u_r, of their partial derivatives as well as the functions p_i, y and z, on the other hand the functions σ_s^r and their first and second partial derivatives.

Let us study therefore the partial derivatives of the first two orders of the functions σ and ω_s^r.

Derivatives of σ:

$$\sigma = \left| \frac{\sin \lambda_2}{J_{x\lambda}} \right|^{\frac{1}{2}}$$

is a function of the trigonometric lines of λ_u ($u = 2, 3$), of the functions x^α (through the intermediate effect of the $A^{\lambda\mu}$) and of the functions p_i, y_j^i. The first and second partial derivatives of σ with respect to the x^i will be therefore expressed with the help of the functions listed and of their first and second partial derivatives.

(1°) First derivatives: We have seen that the partial derivatives with respect to the x^i of a function whatsoever φ, defined on Σ_0, satisfy the identity

$$\frac{\partial \varphi}{\partial x^i} = \sum_{j=1}^{3} \frac{J_i^j}{J_{x\lambda}} \frac{\partial \varphi}{\partial \lambda_j}, \tag{19.11.1}$$

where $\frac{J_i^j}{J_{x\lambda}}$ is a given function of $\cos \lambda_u$, $\sin \lambda_u$, x^α, p_i, y_i^j; the partial derivatives with respect to λ_1 of the functions x^i, p_i, y_i^j, are the quantities T^i, R_i, T_j^i which are expressed through these functions themselves and through z_i^j; the partial derivatives with respect to λ_u of these functions x^i, p_i, y_i^j being expressible by means of their derivatives with respect to the overabundant parameters p_h^0, denoted here by y_h^i, z_h^i, y_{ih}^j, and by means of $\cos \lambda_u$, $\sin \lambda_u$.

The function σ admits therefore within V, under the assumptions made, of first partial derivatives with respect to the x^i which are expressible by means of the functions x^α (with the intermediate help of the $[A^{\lambda\mu}]$ and of the $\left[\frac{\partial A^{\lambda\mu}}{\partial x^\alpha} \right]$ and of the functions

$$p_i, y_j^i, z_j^i, y_{jh}^i$$

and of $\cos \lambda_u$, $\sin \lambda_u$).

(2°) Second derivatives: Another application of the formula (19.11.1) shows, in analogous fashion, that σ admits within V of second partial derivatives, which are expressible by means of the functions x^α (with the intermediate action of the $A^{\lambda\mu}$ and their first and second partial derivatives) and of the functions

$$p_i, y_j^i, z_j^i, y_{ih}^j, z_{ih}^j, y_{ihk}^j$$

and of $\cos \lambda_u, \sin \lambda_u$.

Derivatives of the w_s^r: The identity (19.11.1) makes it possible moreover to show that the functions w_s^r, solutions of the Eq. (19.8.1), admit within V of first and second partial derivatives with respect to the variables x^i if these functions admit, within V, of first and second partial derivatives with respect to the variables λ_u; it suffices for that purpose that they admit of first and second partial derivatives with respect to the overabundant variables p_i^0.

We shall set

$$\frac{\partial w_s^r}{\partial p_i^0} = w_{si}^r, \quad \frac{\partial^2 w_s^r}{\partial p_i^0 \partial p_j^0} = w_{sij}^r.$$

If these functions are continuous and bounded within V they satisfy, under the assumptions made, the integral equations obtained by derivation under the summation symbol of the Eq. (19.9.1) with respect to the p_i^0. Let respectively

$$1^o) \quad w_{si}^r = \int_0^{\lambda_1} \left(\sum_{t=1}^n Q_t^r w_{si}^t + Q w_{si}^r + \Omega_{si}^r \right) d\lambda,$$

where

$$\Omega_{si}^r = \sum_{t=1}^n \frac{\partial Q_t^r}{\partial p_i^0} w_s^t + \frac{\partial Q}{\partial p_i^0} w_s^r$$

is a polynomial of the functions w_s^r, p_i, y_j^i, z_j^i as well as of the values on Σ_0 of the coefficients $A^{\lambda\mu}, B_s^{r\lambda}$ of the equations $[E]$ and of their partial derivatives with respect to the x^α up to the orders two and one, respectively (quantities that are themselves functions of the functions $x^\alpha(\lambda_j)$).

$$2^o) \quad w_{sij}^r = \int_0^{\lambda_1} \left(\sum_{t=1}^n Q_t^r w_{sij}^t + Q w_{sij}^r + \Omega_{sij}^r \right) d\lambda,$$

where

$$\Omega_{sij}^r = \sum_{t=1}^n \frac{\partial Q_t^r}{\partial p_j^0} w_{si}^t + \frac{\partial Q}{\partial p_j^0} w_{si}^r + \frac{\partial \Omega_{si}^r}{\partial p_j^0}.$$

is a polynomial of the functions ω_s^r, ω_{si}^r, p_i, y_i^j, z_i^j, y_{ih}^j, z_{ih}^j as well as of the values on Σ_0 of the coefficients $A^{\lambda\mu}$, $B_s^{r\lambda}$ and of their partial derivatives with respect to the x^α up to the orders three and two, respectively.

The first and second partial derivatives of the ω_s^r with respect to the variables x^i are expressed by means of the functions x^α (with the help of the coefficients $A^{\lambda\mu}$ and of their first partial derivatives), p_i, y_j^i, z_j^i, y_{jh}^i, z_{jh}^i, ω_s^r, ω_{si}^r and ω_{sij}^r.

Summary We have shown that the auxiliary functions σ_s^r exist and admit within V of first and second partial derivatives with respect to the variables x^i under the following assumptions:

(1°) The coefficients $A^{\lambda\mu}$ and $B_s^{r\lambda}$ have partial derivatives continuous and bounded up to the orders four and two, respectively, within the domain D containing V.

(2°) The integral equations for the unknown functions x^α, p_i and ω_s^r have a unique, continuous, bounded solution and admitting within V partial derivatives with respect to the p_i^0, continuous and bounded up to the second order. This result can be proved by assuming that the partial derivatives of order four and two, respectively, of the functions $A^{\lambda\mu}$ and $B_s^{r\lambda}$ verify some Lipschitz conditions.

The functions σ_s^r and their first and second partial derivatives with respect to the x^i are then expressed only through some functions X and Ω, X denoting any whatsoever of the functions x^α, p_i, y_i^j, z_i^j, y_{ih}^j, z_{ih}^j, y_{ihk}^j, z_{ihk}^j, and Ω any whatsoever among the functions ω_s^r, ω_{si}^r, ω_{sij}^r.

The functions X and Ω satisfy integral equations of the form

$$X = \int_0^{\lambda_1} E(X)\mathrm{d}\lambda + X_0,$$

$$\Omega = \int_0^{\lambda_1} F(X, \Omega)\mathrm{d}\lambda + \Omega_0,$$

where X_0 and Ω_0 denote the given values of the functions X and Ω for $\lambda_1 = 0$.

$E(X)$ is a polynomial of the functions X and of the values on Σ_0 of the coefficients $A^{\lambda\mu}$ and of their partial derivatives up to the fourth order (functions of the functions x^α).

$F(X, \Omega)$ is a polynomial of the functions X and Ω, and of the values on Σ_0 of the coefficients $A^{\lambda\mu}$, $B_s^{r\lambda}$ and of their partial derivatives up to the orders three and two, respectively.

19.12   Behaviour in the Neighbourhood of the Vertex

We are going to study the quantities occurring in the integrals of the fundamental relations (19.7.1), and for this purpose we will look in a more precise way for the expression of the partial derivatives of the functions σ and ω_s^r with respect to the variables x^i by means of the functions X and Ω. The behaviour of these functions in the neighbourhood of $\lambda_1 = 0$ (vertex of the characteristic conoid Σ_0) will make it possible for us to look for the limit of Eq. (19.7.1) for $\eta = 0$: the function $x^4(\lambda_1, \lambda_2, \lambda_3)$ being, within the domain Λ, a continuous function of the three variables λ_i, $\eta = x^4 - x_0^4$ approaches actually zero with λ_1. We will provide the details of the calculations, that we will need in the following, when we will try to solve the system of integral equations obtained.

We will use essentially in the studies of the behaviour in the neighbourhood of $\lambda_1 = 0$, the following fact which results from the assumptions made and from the equations verified by the functions y_j^i, y_{jh}^i, y_{jhk}^i, ω_{si}^r and ω_{sij}^r.

The functions $\frac{y_j^i}{\lambda_1}$, $\frac{y_{jh}^i}{\lambda_1}$, $\frac{y_{jhk}^i}{\lambda_1}$, and $\frac{\omega_{si}^r}{\lambda_1}$, $\frac{\omega_{sij}^r}{\lambda_1}$ are continuous and bounded functions of $\lambda_1, \lambda_2, \lambda_3$ within the domain V. We will denote any whatsoever of these functions by $\tilde{X}$ and $\tilde{\Omega}$.

19.13 Behaviour in the Neighbourhood of $\lambda_1 = 0$

(1°) We have already shown (Sect. 19.10) that the quantity $\frac{J_{x\lambda}}{(\lambda_1)^2}$ is a polynomial of the functions X (here p_i only), $\tilde{X}$ (here $\frac{y_j^i}{\lambda_1}$ only), of the coefficients $A^{\lambda\mu}$ and of the $\sin \lambda_u$, $\cos \lambda_u$ ($u = 2, 3$). It is therefore a continuous bounded function of $\lambda_1, \lambda_2, \lambda_3$ within V. We have seen that the value of this function for $\lambda_1 = 0$ is

$$\lim_{\lambda_1 \to 0} \frac{J_{x\lambda}}{(\lambda_1)^2} = -\sin \lambda_2.$$

In the neighbourhood of $\lambda_1 = 0$ the function $\frac{J_{x\lambda}}{(\lambda_1)^2}$, which will occur in the denominator of the quantities studied in the following, is $\neq 0$, but for $\lambda_2 = 0$ or $\lambda_2 = \pi$. To remove this difficulty we will show that the polynomial $J_{x\lambda}$ is divisible by $\sin \lambda_2$ and we will make sure that the function $D = \frac{J_{x\lambda}}{(\lambda_1)^2 \sin \lambda_2}$ appears in the denominators we consider.

Let us therefore consider on the conoid Σ_0 the following change of variables:

$$\mu_i \equiv \lambda_1 p_i^0. \tag{19.13.1}$$

We set

$$d \equiv \frac{D(\mu_1, \mu_2, \mu_3)}{D(\lambda_1, \lambda_2, \lambda_3)} = \det \begin{pmatrix} p_1^0 & p_2^0 & p_3^0 \\ \lambda_1 \frac{\partial p_1^0}{\partial \lambda_2} & \lambda_1 \frac{\partial p_2^0}{\partial \lambda_2} & \lambda_1 \frac{\partial p_3^0}{\partial \lambda_2} \\ \lambda_1 \frac{\partial p_1^0}{\partial \lambda_3} & \lambda_1 \frac{\partial p_2^0}{\partial \lambda_3} & \lambda_1 \frac{\partial p_3^0}{\partial \lambda_3} \end{pmatrix} = (\lambda_1)^2 \sin \lambda_2$$

and

$$J_{x\lambda} \equiv \frac{D(x^1, x^2, x^3)}{D(\lambda_1, \lambda_2, \lambda_3)}.$$

Since

$$\frac{D(x^1, x^2, x^3)}{D(\lambda_1, \lambda_2, \lambda_3)} = \frac{D(x^1, x^2, x^3)}{D(\mu_1, \mu_2, \mu_3)} \frac{D(\mu_1, \mu_2, \mu_3)}{D(\lambda_1, \lambda_2, \lambda_3)}$$

we have

$$J_{x\lambda} = D(\lambda_1)^2 \sin \lambda_2, \qquad (19.13.2)$$

where the determinant D has elements

$$\frac{\partial x^i}{\partial \mu_j} = \frac{\partial x^i}{\partial \lambda_1} \frac{\partial \lambda_1}{\partial \mu_j} + \sum_{h=1}^{3} \sum_{u=2}^{3} \frac{\partial x^i}{\partial p_h^0} \frac{\partial p_h^0}{\partial \lambda_u} \frac{\partial \lambda_u}{\partial \mu_j}.$$

It results directly from the equalities (19.13.1) and from the identity

$$\sum_{i=1}^{3} (\mu_i)^2 = (\lambda_1)^2$$

that

$$\frac{\partial \lambda_1}{\partial \mu_j} = p_j^0 \text{ and } \frac{\partial p_h^0}{\partial \lambda_u} = \frac{1}{\lambda_1} \frac{\partial \mu_h}{\partial \lambda_u}.$$

On the other hand, we have

$$\frac{\partial \lambda_1}{\partial \mu_j} \frac{\partial \mu_h}{\partial \lambda_1} + \sum_{u=2}^{3} \frac{\partial \lambda_u}{\partial \mu_j} \frac{\partial \mu_h}{\partial \lambda_u} = \delta_j^h.$$

The elements of D are therefore

$$\frac{\partial x^i}{\partial \mu_j} = T^i p_j^0 + \sum_{h=1}^{3} \frac{y_h^i}{\lambda_1} (\delta_j^h - p_j^0 p_h^0).$$

The polynomial $\frac{J_{x\lambda}}{(\lambda_1)^2}$ is therefore divisible by $\sin \lambda_2$, the quotient D being a polynomial of the same functions $X, \tilde{X}$ as $\frac{J_{x\lambda}}{(\lambda_1)^2}$ is of $\sin \lambda_u$, $\cos \lambda_u$ (or, more precisely, of the three p_i^0).

D is a continuous bounded function of $\lambda_1, \lambda_2, \lambda_3$ within V whose value for $\lambda_1 = 0$ is $\lim_{\lambda_1 \to 0} D = -1$. As a matter of fact:

$$\lim_{\lambda_1 \to 0} \frac{\partial x^i}{\partial \mu_j} = -p_i^0 p_j^0 - \delta_j^i + p_i^0 p_j^0 = -\delta_j^i.$$

Remark $\frac{J_{x\lambda}}{(\lambda_1)^2}$ being a homogeneous polynomial of the second degree of the functions $\frac{y_i^j}{\lambda_1}$, the same is true of the polynomial D, and the quantity $(\lambda_1)^2 D$ is a polynomial of the functions X (p_i and y_i^j), of the coefficients $A^{\lambda\mu}$ and of the three p_i^0, homogeneous of the second degree with respect to the y_i^j. One can easily verify these results by evaluating the product $D^+ d^+$ where

$$d^+ = \det \begin{pmatrix} p_1^0 & p_2^0 & p_3^0 \\ \frac{\partial p_1^0}{\partial \lambda_2} & \frac{\partial p_2^0}{\partial \lambda_2} & \frac{\partial p_3^0}{\partial \lambda_2} \\ \frac{\partial p_1^0}{\partial \lambda_3} & \frac{\partial p_2^0}{\partial \lambda_3} & \frac{\partial p_3^0}{\partial \lambda_3} \end{pmatrix} = \frac{d}{(\lambda_1)^2}$$

and where D^+ is the determinant whose elements are

$$T^i p_j^0 - \sum_{h=1}^{3} y_h^i (\bar{\delta}_j^h - p_j^0 p_h^0).$$

One finds

$$D^+ d^+ = J_{x\lambda},$$

and the quantity $(\lambda_1)^2 D = D^+$ possesses therefore the stated properties.

The polynomial D is, in absolute value, bigger than a number assigned in a domain W: D is actually a continuous and bounded function of λ_1 in the domain Λ (where λ_2 and λ_3 vary over a compact) which takes the value -1 for $\lambda_1 = 0$. There exists therefore a number ε_2 such that, in the domain Λ_2, neighbourhood of $\lambda_1 = 0$ of the domain Λ, defined by

$$|\lambda_1| \leq \varepsilon_2, \ 0 \leq \lambda_2 \leq \pi, \ 0 \leq \lambda_3 \leq 2\pi,$$

one has for example

$$|D + 1| \leq \frac{1}{2} \text{ therefore } |D| \geq \frac{1}{2}.$$

We will denote by W the domain of Σ_0 corresponding to the domain Λ_2.

(2°) **Behaviour of the minors of $J_{x\lambda}$.**

(a) Minors relative to the elements of the first line of $J_{x\lambda}$: J_i^1 is, as $J_{x\lambda}$ itself, a homogeneous polynomial of second degree with respect to the functions y_i^j, and $\frac{J_i^1}{(\lambda_1)^2}$ is a polynomial of the functions $X(p_i)$, $\tilde{X}\left(\frac{y_i^j}{\lambda_1}\right)$, of the coefficients $[A^{\lambda\mu}]$ and of $\sin \lambda_u$, $\cos \lambda_u$; it is therefore a continuous and bounded function of $\lambda_1, \lambda_2, \lambda_3$ in V.

In order to study the quantity $\frac{J_i^1}{J_{x\lambda}} = \frac{\partial \lambda_1}{\partial x^i}$, which will occur in the following, we shall put it in the form of a rational fraction with denominator D ($\neq 0$ in W).

We have

$$\frac{J_i^1}{J_{x\lambda}} = \frac{\partial \lambda_1}{\partial x^i} = \sum_{j=1}^{3} \frac{\partial \lambda_1}{\partial \mu_j} \frac{\partial \mu_j}{\partial x^i} = \sum_{j=1}^{3} p_j^0 \frac{D_i^j}{D} \tag{19.13.3}$$

(one has denoted by D_i^j the minor relative to the element $\frac{\partial x^i}{\partial \mu_j}$ of the determinant D).

The quantity $\frac{J_i^1}{J_{x\lambda}}$ is therefore a continuous and bounded function of the three variables $\lambda_1, \lambda_2, \lambda_3$ in W. Let us compute the value of this function for $\lambda_1=0$, one finds

$$\lim_{\lambda_1 \to 0} \frac{J_i^1}{J_{x\lambda}} = -p_i^0,$$

a result that one might have expected. Indeed:

$$\lim_{\lambda_1 \to 0} \frac{\partial \lambda_1}{\partial x^i} = \lim_{\lambda_1 \to 0} \frac{\partial x^4}{\partial x^i},$$

or one has constantly, over Σ_0, $\frac{\partial x^4}{\partial x^i} = -p_i$.

Remark One deduces from the formulas (19.13.2) and (19.13.3) that

$$J_i^1 = \sum_{j=1}^{3} (\lambda_1)^2 \sin \lambda_2 p_j^0 D_i^j.$$

One then sees that the quantity $(\lambda_1)^2 \sum_{j=1}^{3} p_j^0 D_i^j$ is a polynomial of the functions p_i, y_i^j, of the coefficients $[A^{\lambda \mu}]$ and of the three p_h^0, homogeneous of second degree with respect to the y_i^j.

(b) Minors relative to the second and third line of $J_{x\lambda}$: J_i^u is a polynomial of the functions $X(p_i, y_i^j)$, $[A^{\lambda \mu}]$ and of $\sin \lambda_u$, $\cos \lambda_u$, homogeneous of first degree with respect to the functions y_i^j.

$\frac{J_i^u}{\lambda_1}$ is a continuous and bounded function of $\lambda_1, \lambda_2, \lambda_3$ in V.

Let us study the quantity $\sum_u \frac{\partial p_h^0}{\partial \lambda_u} \frac{J_i^u}{J_{x\lambda}}$. One has

$$\sum_u \frac{\partial p_h^0}{\partial \lambda_u} \frac{J_i^u}{J_{x\lambda}} = \sum_u \frac{\partial p_h^0}{\partial \lambda_u} \frac{\partial \lambda_u}{\partial x^i} = \sum_u \sum_{j=1}^{3} \frac{1}{\lambda_1} \frac{\partial \mu_h}{\partial \lambda_u} \frac{\partial \lambda_u}{\partial \mu_j} \frac{\partial \mu_j}{\partial x^i} = \sum_{j=1}^{3} \frac{1}{\lambda_1} (\delta_j^h - p_j^0 p_h^0) \frac{D_i^j}{D}.$$

We see that the quantity $\lambda_1 \sum_u \frac{\partial p_h^0}{\partial \lambda_u} \frac{J_i^u}{J_{x\lambda}}$ is a rational fraction with nonvanishing denominator (in the domain W) of the functions $X(p_i)$, $\tilde{X}\left(\frac{y_i^j}{\lambda_1}\right)$, $[A^{\lambda \mu}]$ and of the three p_i^0. It is therefore a continuous and bounded function of $\lambda_1, \lambda_2, \lambda_3$ in the domain W; the value of this function for $\lambda_1 = 0$ is computed as follows. One has on one hand

$$\frac{\partial x^h}{\partial \lambda_u} = \sum_{j=1}^{3} \frac{\partial x^h}{\partial p_j^0} \frac{\partial p_j^0}{\partial \lambda_u} = \sum_{j=1}^{3} y_j^h \frac{\partial p_j^0}{\partial \lambda_u},$$

from which $\lim_{\lambda_1 \to 0} \frac{1}{\lambda_1} \frac{\partial x^h}{\partial \lambda_u} = -\sum_{j=1}^{3} \delta_j^h \frac{\partial p_j^0}{\partial \lambda_u} = -\frac{\partial p_h^0}{\partial \lambda_u}.$

One knows on the other hand that

$$\frac{J_i^u}{J_{x\lambda}} = \frac{\partial \lambda_u}{\partial x^i},$$

from which $\lim_{\lambda_1 \to 0} \lambda_1 \sum_u \dfrac{\partial p_h^0}{\partial \lambda_u} \dfrac{J_i^u}{J_{x\lambda}} = -\lim_{\lambda_1 \to 0} \sum_u \dfrac{\partial x^h}{\partial \lambda_u} \dfrac{\partial \lambda_u}{\partial x^i} = -\delta_i^h + \lim_{\lambda_1 \to 0} \dfrac{\partial x^h}{\partial \lambda_1} \dfrac{\partial \lambda_1}{\partial x^i},$

from which eventually

$$\lim_{\lambda_1 \to 0} \lambda_1 \sum_u \frac{\partial p_h^0}{\partial \lambda_u} \frac{J_i^u}{J_{x\lambda}} = -\delta_i^h + p_i^0 p_h^0.$$

Remark By a reasoning analogous to the one of previous remarks, one sees that the quantity $\lambda_1 \sum_{j=1}^3 (\delta_j^h - p_j^0 p_h^0) D_i^j$ is a polynomial homogeneous of first degree with respect to the y_i^j, of the functions $X(p_i, y_i^j)$, $[A^{\lambda\mu}]$, p_i^0.

19.14 First Derivatives

The first partial derivatives of an arbitrary function φ satisfy, in light of the identity (19.11.1) and of the results of the previous section, the relation

$$\frac{\partial \varphi}{\partial x^i} = \frac{\partial \varphi}{\partial \lambda_1} \sum_{j=1}^3 \frac{p_j^0 D_i^j}{D} + \frac{1}{\lambda_1} \sum_{h,j=1}^3 \frac{\partial \varphi}{\partial p_h^0} (\delta_j^h - p_j^0 p_h^0) \frac{D_i^j}{D}.$$

Let us apply this formula to the functions p_h^0 and X:

$$\frac{\partial p_h^0}{\partial x^i} = \frac{1}{\lambda_1} \sum_{j=1}^3 (\delta_j^h - p_j^0 p_h^0) \frac{D_i^j}{D},$$

$$\frac{\partial p_h}{\partial x^i} = R_h \sum_{j=1}^3 \frac{p_j^0 D_i^j}{D} + \sum_{k,j=1}^3 \delta_k^h \frac{1}{\lambda_1} (\delta_j^k - p_j^0 p_k^0) \frac{D_i^j}{D}, \tag{19.14.1}$$

$$\frac{\partial y_h^k}{\partial x^i} = \sum_{j=1}^3 T_h^k \frac{p_j^0 D_i^j}{E} + \frac{1}{\lambda_1} \sum_{l=1}^3 y_{hl}^k (\delta_j^l - p_j^0 p_l^0) \frac{D_i^j}{D},$$

$$\frac{\partial z_h^k}{\partial x^i} = \sum_{j=1}^3 R_h^k p_j^0 \frac{D_i^j}{D} + \frac{1}{\lambda_1} \sum_{j,l=1}^3 z_{hl}^k (\delta_j^l - p_j^0 p_l^0) \frac{D_i^j}{D}.$$

These equations and the analogous equations verified by

$$\frac{\partial y^k_{hl}}{\partial x^i}, \frac{\partial z^k_{hl}}{\partial x^i}, \frac{\partial \omega^r_s}{\partial x^i}, \frac{\partial \omega^r_{si}}{\partial x^i}$$

show that the quantities

$$\lambda_1 \frac{\partial p^0_h}{\partial x^i}, \ \lambda_1 \frac{\partial p_h}{\partial x^i}, \ \lambda_1 \frac{\partial z^k_h}{\partial x^i}, \ \lambda_1 \frac{\partial z^k_{hl}}{\partial x^i}, \ \text{and} \ \frac{\partial y^k_h}{\partial x^i}, \frac{\partial y^k_{hl}}{\partial x^i}, \frac{\partial \omega^r_s}{\partial x^i}, \frac{\partial \omega^r_{si}}{\partial x^i}$$

are rational fractions with denominator D of the functions

$$X, \tilde{X}, \Omega, \tilde{\Omega}, [A^{\lambda\mu}], \left[\frac{\partial A^{\lambda\mu}}{\partial x^\alpha}\right], \left[\frac{\partial^2 A^{\lambda\mu}}{\partial x^\alpha \partial x^\beta}\right], p^0_i.$$

These are bounded and continuous functions, within W, of the three variables $\lambda_1, \lambda_2, \lambda_3$.

19.15   Reverting to the Functions σ^r_s

We will use in the study of partial derivatives with respect to x^i of the functions σ^r_s, of the partial derivatives of polynomials considered in the remarks of Sect. 19.13: $(\lambda_1)^2 D$, $(\lambda_1)^2 \sum_{j=1}^3 p^0_j D^j_i$ and $\lambda_1 \sum_{j=1}^3 (\delta^h_j - p^0_j p^0_h) D^j_i$ are polynomials of the functions $X(p_i, y^j_i)$, $[A^{\lambda\mu}]$, p^0_i, *homogeneous of degree 2, 2 and 1, respectively, with respect to the* y^j_i. The previous results and the identity (19.11.1) show then that the quantities

$$\frac{1}{\lambda_1} \frac{\partial}{\partial x^i}(\lambda^2_1 D), \ \frac{1}{\lambda_1} \frac{\partial}{\partial x^i} \sum_{j=1}^3 ((\lambda_1)^2 p^0_j D^j_i),$$

$$\sum_{i,j=1}^3 \frac{\partial}{\partial x^i}(\lambda_1(\delta^h_j - p^0_j p^0_h)D^j_i)$$

are rational fractions with denominator D of the functions

$$X(p_i, y^j_i, z^j_i), \tilde{X}\left(\frac{y^j_i}{\lambda_1}, \frac{y^j_{ih}}{\lambda_1}\right), [A^{\lambda\mu}], \left[\frac{\partial A^{\lambda\mu}}{\partial x^\alpha}\right], p^0_i.$$

They are therefore continuous and bounded functions of $\lambda_1, \lambda_2, \lambda_3$ in W.

In the study of second partial derivatives of the function σ with respect to the x^i we will use the second partial derivatives $\frac{\partial^2((\lambda_1)^2 D)}{\partial x^i \partial x^j}$. Let us first remark that the first-order partial derivatives of $(\lambda_1)^2 D$ can be written

$$\frac{\partial((\lambda_1)^2 D)}{\partial x^i} = \frac{P_1}{(\lambda_1)^2 D},$$

where P_1 is a polynomial of the functions

$$X(p_i, y_i^j, z_i^j, y_{ih}^j), [A^{\lambda\mu}], \left[\frac{\partial A^{\lambda\mu}}{\partial x^\alpha}\right], p_i^0$$

whose terms are of the third degree at least with respect to the set of functions y_i^j, y_{ih}^j. As a matter of fact, the partial derivatives $\frac{\partial p_h}{\partial x^i}$ and $\frac{\partial p_h^0}{\partial x^i}$ can be put (by multiplying denominator and numerator of the second members of the equations by $(\lambda_1)^2$) in form of rational fractions with denominator $(\lambda_1)^2 D$ and whose numerators are polynomials of the functions

$$X(p_i, y_i^j, z_i^j), [A^{\lambda\mu}], \left[\frac{\partial A^{\lambda\mu}}{\partial x^\alpha}\right], p_i^0$$

whose terms are of first degree at least with respect to the y_i^j, and the partial derivatives $\frac{\partial y_h^k}{\partial x^i}$ can be put in form of rational fractions with denominator $(\lambda_1)^2 D$ and whose numerators are polynomials of the functions

$$X(p_i, y_i^j, z_i^j, y_{hk}^j), [A^{\lambda\mu}], \left[\frac{\partial A^{\lambda\mu}}{\partial x^\alpha}\right], p_i^0$$

homogeneous of second degree with respect to the set of functions y_i^0, y_{hk}^i. The polynomial $(\lambda_1)^2 D$ being homogeneous of first degree with respect to the y_i^j, its first partial derivatives have for sure the desired form.

Let us then consider the second partial derivatives:

$$\frac{\partial^2(\lambda_1^2 D)}{\partial x^i \partial x^j} = \frac{1}{(\lambda_1)^2 D}\frac{\partial p_1}{\partial x^i} = \frac{P_1}{((\lambda_1)^2 D)^2}\frac{\partial((\lambda_1)^2 D)}{\partial x^i}.$$

It turns out from the form of the polynomial P_1 and from the results of Sect. 19.13 that:

(1) $\frac{P_1}{(\lambda_1)^3}$ is a polynomial of the functions

$$X(p_i, y_i^j, z_i^j, y_{hk}^j), X\left(\frac{y_i^j}{\lambda_1}, \frac{y_{ih}^j}{\lambda_1}\right), [A^{\lambda\mu}], \left[\frac{\partial A^{\lambda\mu}}{\partial x^\alpha}\right], p_i^0.$$

(2) $\frac{1}{(\lambda_1)^2}\frac{\partial P_1}{\partial x^i}$ is a rational fraction with denominator D of the functions

$$X(p_i, y_i^j, z_i^j, y_{hk}^j, z_{ih}^j, y_{ihk}^j), \tilde{X}\left(\frac{y_i^j}{\lambda_1}, \frac{y_{ih}^j}{\lambda_1}, \frac{y_{ihk}^j}{\lambda_1}\right), [A^{\lambda\mu}], \dots, \left[\frac{\partial^2 A^{\lambda\mu}}{\partial x^\alpha \partial x^\beta}\right], p_i^0.$$

The derivatives $\frac{\partial^2((\lambda_1)^2 D)}{\partial x^i \partial x^j}$ are therefore rational fractions with denominator D^3 of the functions we have just listed.

19.16   Study of σ and Its Derivatives

(1°) The auxiliary function σ has been defined by $\sigma = \left| \frac{\sin \lambda_2}{J_{x\lambda}} \right|^{\frac{1}{2}}$. We have therefore, by virtue of the equality (19.13.2),

$$\sigma = \frac{1}{|(\lambda_1)^2 D|^{\frac{1}{2}}}.$$

One deduces that, in the domain W, the function $\sigma \lambda_1 = \frac{1}{|D|^{\frac{1}{2}}}$ is the square root of a rational fraction, bounded and non-vanishing, of the function

$$X, \tilde{X}, [A^{\lambda \mu}], p_i^0;$$

it is a continuous and bounded function of the three variables λ_i, whose value for $\lambda_1 = D$ is

$$\lim_{\lambda_1 \to 0} \sigma \lambda_1 = 1. \tag{19.16.1}$$

(2°) The first partial derivatives of σ with respect to the x^i are

$$\frac{\partial \sigma}{\partial x^i} = \frac{\sigma}{2} \frac{1}{(\lambda_1)^2 D} \frac{\partial((\lambda_1)^2 D)}{\partial x^i}.$$

One concludes that, in the domain W, the function

$$(\lambda_1)^2 \frac{\partial \sigma}{\partial x^i} = -\frac{\sigma \lambda_1}{2} \frac{1}{D} \frac{1}{\lambda_1} \frac{\partial((\lambda_1)^2 D)}{\partial x^i}$$

is the product of the square root of a non-vanishing bounded rational fraction with a bounded rational fraction of the functions $X, \tilde{X}, [A^{\lambda \mu}], \left[\frac{\partial A^{\lambda \mu}}{\partial x^\alpha} \right], p_i^0$. It is a continuous and bounded function of $\lambda_1, \lambda_2, \lambda_3$ of which we are going to compute the value for $\lambda_1 = 0$.

The identities $\frac{\partial \sigma}{\partial \lambda_1} = \sum_{i=1}^{3} T^i \frac{\partial \sigma}{\partial x^i}$ and $\frac{\partial \sigma}{\partial p_h^0} = \sum_{i=1}^{3} \frac{\partial \sigma}{\partial x^i} y_h^i$ show that the functions $\lambda_1^2 \frac{\partial \sigma}{\partial \lambda_1}$ and $\lambda_1 \frac{\partial \sigma}{\partial p_h^0}$ are continuous and bounded in W. We can therefore, on the one hand differentiate the equality (19.16.1) with respect to p_h^0, and we find

$$\lim_{\lambda_1 \to 0} \lambda_1 \frac{\partial \sigma}{\partial p_h^0} = 0.$$

On the other hand, we can write

$$\frac{\partial(\sigma(\lambda_1)^2)}{\partial \lambda_1} = 2\lambda_1 \sigma + (\lambda_1)^2 \frac{\partial \sigma}{\partial \lambda_1}$$

and

$$\lim_{\lambda_1 \to 0} \frac{\partial(\sigma(\lambda_1)^2)}{\partial \lambda_1} = \lim_{\lambda_1 \to 0} \lambda_1 \sigma,$$

from which

$$\lim_{\lambda_1 \to 0} (\lambda_1)^2 \frac{\partial \sigma}{\partial \lambda_1} = -\lim_{\lambda_1 \to 0} \lambda_1 \sigma = -1.$$

In order to compute the value for $\lambda_1 = 0$ of the function $(\lambda_1)^2 \frac{\partial \sigma}{\partial x^i}$ we shall use the identity

$$(\lambda_1)^2 \frac{\partial \sigma}{\partial x^i} = (\lambda_1)^2 \frac{\partial \sigma}{\partial \lambda_1} \frac{J_1^i}{J_{x\lambda}} + \lambda_1 \sum_{h,u} \frac{\partial \sigma}{\partial p_h^0} \lambda_1 \frac{\partial p_h^0}{\partial \lambda_u} \frac{J_u^i}{J_{x\lambda}},$$

from which, in light of the previous results (Sect. 19.13),

$$\lim_{\lambda_1 \to 0} (\lambda_1)^2 \frac{\partial \sigma}{\partial x^i} = p_i^0. \tag{19.16.2}$$

(3°) The second partial derivatives of σ with respect to the x^i are

$$\frac{\partial^2 \sigma}{\partial x^i \partial x^j} = -\frac{\sigma}{2} \frac{1}{(\lambda_1)^2 D} \frac{\partial^2((\lambda_1)^2 D)}{\partial x^i \partial x^j} - \frac{1}{2(\lambda_1)^2 D} \frac{\partial \sigma}{\partial x^j} \frac{\partial((\lambda_1)^2 D)}{\partial x^i}$$

$$+ \frac{\sigma}{2((\lambda_1)^2 D)^2} \frac{\partial(\lambda_1^2 D)}{\partial x^i} \frac{\partial((\lambda_1)^2 D)}{\partial x^j}.$$

(a) It is easily seen that in the domain W the function $(\lambda_1)^3 \frac{\partial^2 \sigma}{\partial x^i \partial x^j}$ is the product of the square root of a non-vanishing bounded rational fraction with a bounded rational fraction (having denominator D^4) of the functions

$$X, \tilde{X}, [A^{\lambda\mu}], \left[\frac{\partial A^{\lambda\mu}}{\partial x^\alpha}\right], \left[\frac{\partial^2 A^{\lambda\mu}}{\partial x^\alpha \partial x^\beta}\right], p_i^0.$$

It is a continuous and bounded function of the three variables λ_i. We are going to compute the value for $\lambda_1 = 0$ of the function $(\lambda_1)^3 \sum_{i=0}^{3} \frac{\partial^2 \sigma}{\partial x^{i2}}$ which, only, we will need: the second derivatives of σ do not occur actually in the fundamental equations except for the quantity $\sum_{i,j=1}^{3} [A^{ij}] \frac{\partial^2 \sigma}{\partial x^i \partial x^j}$, and one has

$$\lim_{\lambda_1 \to 0} \sum_{i,j=1}^{3} [A^{ij}](\lambda_1)^3 \frac{\partial^2 \sigma}{\partial x^i \partial x^j} = \lim_{\lambda_1 \to 0} (\lambda_1)^3 \sum_{i=1}^{3} \frac{\partial^2 \sigma}{\partial (x^i)^2}.$$

We will evaluate this limit as the limit (19.16.2). We shall find on the one hand, by differentiating the equality (19.16.2),

$$\lim_{\lambda_1 \to 0} (\lambda_1)^2 \frac{\partial}{\partial p_h^0} \left(\frac{\partial \sigma}{\partial x^i} \right) = \delta_h^i,$$

on the other hand

$$\frac{\partial}{\partial \lambda_1} \left[(\lambda_1)^3 \frac{\partial \sigma}{\partial x^i} \right] = 3(\lambda_1)^2 \frac{\partial \sigma}{\partial x^i} + (\lambda_1)^3 \frac{\partial}{\partial \lambda_1} \left(\frac{\partial \sigma}{\partial x^i} \right),$$

from which

$$\lim_{\lambda_1 \to 0} (\lambda_1)^3 \frac{\partial}{\partial \lambda_1} \left(\frac{\partial \sigma}{\partial x^i} \right) = \lim_{\lambda_1 \to 0} \left(-2(\lambda_1)^2 \frac{\partial \sigma}{\partial x^i} \right) = -p_i^0.$$

We find therefore, by using the identity

$$\sum_{i=1}^{3} (\lambda_1)^3 \frac{\partial^2 \sigma}{\partial (x^i)^2} = (\lambda_1)^3 \sum_{i=1}^{3} \frac{\partial}{\partial \lambda_1} \left(\frac{\partial \sigma}{\partial x^i} \right) \frac{J_i^1}{J_{x\lambda}} + (\lambda_1)^3 \sum_{h=1}^{3} \sum_{u} \frac{\partial}{\partial p_h^0} \left(\frac{\partial \sigma}{\partial x^i} \right) \frac{\partial p_h^0}{\partial \lambda_u} \frac{J_i^u}{J_{x\lambda}}$$

and the results of previous paragraphs, that

$$\lim_{\lambda_1 \to 0} \sum_{i=1}^{3} (\lambda_1)^3 \frac{\partial^2 \sigma}{\partial x^{i2}} = 0.$$

Let us show that the function

$$(\lambda_1)^2 \sum_{i,j=1}^{3} [A^{ij}] \frac{\partial^2 \sigma}{\partial x^i \partial x^j}$$

is a continuous and bounded function of the three variables λ_i, in the neighbourhood of $\lambda_1 = 0$ (which will make it possible for us to prove that the quantity under the sign $\int \int \int$ in (19.7.1) is bounded in W).

We have seen that $(\lambda_1)^3 \sum_{i,j=1}^{3} \frac{\partial^2 \sigma}{\partial x^i \partial x^j} [A^{ij}]$ is the product of a square root of a non-vanishing bounded rational fraction $\left(\frac{1}{D} \right)$ with a rational fraction having denominator D^4, whose numerator, polynomial of the functions

$$X, \tilde{X}, [A^{\lambda\mu}], \left[\frac{\partial A^{\lambda\mu}}{\partial x^\alpha} \right], \left[\frac{\partial^2 A^{\lambda\mu}}{\partial x^\alpha \partial x^\beta} \right], p_i^0,$$

vanishes for the values of these functions corresponding to $\lambda_1 = 0$. We have

$$\sum_{i,j=1}^{3} (\lambda_1)^3 [A^{ij}] \frac{\partial^2 \sigma}{\partial x^i \partial x^j} = \frac{P\left(X, \tilde{X}, [A^{\lambda\mu}], \left[\frac{\partial A^{\lambda\mu}}{\partial x^\alpha}\right], \left[\frac{\partial^2 A^{\lambda\mu}}{\partial x^\alpha \partial x^\beta}\right], p_i^0\right)}{D^4} \frac{1}{|D|^{\frac{1}{2}}}$$

with

$$P_0 = P\left(X_0, \tilde{X}_0, \pm\delta_\lambda^\mu, \left[\frac{\partial A^{\lambda\mu}}{\partial x^\alpha}\right]_0, \left[\frac{\partial^2 A^{\lambda\mu}}{\partial x^\alpha \partial x^\beta}\right]_0, p_i^0\right) = 0.$$

We then write:

$$(\lambda_1)^3 \sum_{i,j=1}^{3} [A^{ij}] \frac{\partial^2 \sigma}{\partial x^i \partial x^j} = \frac{(P - P_0)}{D^4} \frac{1}{|D|^{\frac{1}{2}}}. \tag{19.16.3}$$

By applying the Taylor formula (for P) one sees that the quantity (19.16.3) is a polynomial of the functions $X - X_0, \tilde{X} - \tilde{X}_0, A^{\lambda\mu} \pm \delta_\lambda^\mu \ldots$, whose terms are of first degree at least with respect to the set of these functions.

To show that $(\lambda_1)^2 \sum_{i,j=1}^{3} [A^{ij}] \frac{\partial^2 \sigma}{\partial x^i \partial x^j}$ is a continuous and bounded function of $\lambda_1, \lambda_2, \lambda_3$ in the domain W, it is enough to show that the same holds for the functions

$$\frac{(X - X_0)}{\lambda_1}, \frac{(\tilde{X} - \tilde{X}_0)}{\lambda_1}, \frac{[A^{\lambda\mu}] - \delta_\mu^\lambda}{\lambda_1}, \ldots, \frac{\left[\frac{\partial^2 A^{\lambda\mu}}{\partial x^\alpha \partial x^\beta}\right] - \left[\frac{\partial^2 A^{\lambda\mu}}{\partial x^\alpha \partial x^\beta}\right]_0}{\lambda_1}.$$

The functions X verify

$$X = \int_0^{\lambda_1} E(X)d\lambda + X_0,$$

$\frac{(X-X_0)}{\lambda_1}$ is therefore a continuous and bounded function of the λ_i in V:

$$|X - X_0| \le \lambda_1 M. \tag{19.16.4}$$

The coefficients $A^{\lambda\mu}$ possessing in (D) partial derivatives continuous and bounded up to the fourth order with respect to the x^α, the x^α fulfilling the inequalities (19.16.4), we see that

$$[A^{\lambda\mu}] \pm \delta_\lambda^\mu \le \lambda_1 A, \ldots, \left[\frac{\partial^3 A^{\lambda\mu}}{\partial x^\alpha \partial x^\beta \partial x^\gamma}\right] - \left[\frac{\partial^3 A^{\lambda\mu}}{\partial x^\alpha \partial x^\beta \partial x^\gamma}\right]_0 \le \lambda_1 A. \tag{19.16.5}$$

Let us consider $\frac{(X-X_0)}{\lambda_1}$. The corresponding X functions are $y_i^j, y_{ih}^j, y_{ihk}^j$ which verify the equation

$$X = \int_0^{\lambda_1} E(X)d\lambda,$$

$E(X)$ being a polynomial of the functions X, of the $A^{\lambda\mu}$ and of their partial derivatives up to the third order

$$\left[\frac{\partial A^{\lambda\mu}}{\partial x^\alpha}\right], \ldots, \frac{\partial^3 A^{\lambda\mu}}{\partial x^\alpha \partial x^\beta \partial x^\gamma}.$$

We have

$$\tilde{X} - \tilde{X}_0 = \frac{\int_0^{\lambda_1}(E(X) - E(X)_0)d\lambda}{(\lambda_1)^2}.$$

The Taylor formula applied to the polynomial E shows that $E(X) - E(X)_0$ is a polynomial of the functions

$$X_0, \delta_\lambda^\mu, \ldots, \left[\frac{\partial^3 A^{\lambda\mu}}{\partial x^\alpha \partial x^\beta \partial x^\gamma}\right]_0$$

and of the functions

$$X - X_0, [A^{\lambda\mu}] - \delta_\lambda^\mu, \ldots, \left(\left[\frac{\partial^3 A^{\lambda\mu}}{\partial x^\alpha \partial x^\beta \partial x^\gamma}\right] - \left[\frac{\partial^3 A^{\lambda\mu}}{\partial x^\alpha \partial x^\beta \partial x^\gamma}\right]_0\right)$$

whose terms are of first degree at least with respect to this last set of terms.

All these functions being bounded in V and satisfying (19.16.4) and (19.16.5) we see easily that $\frac{(\tilde{X}-\tilde{X}_0)}{\lambda_1}$ is continuous and bounded in V.

The function $(\lambda_1)^2 \sum_{i,j=1}^3 [A^{ij}]\frac{\partial^2\sigma}{\partial x^i \partial x^j}$ is therefore continuous and bounded in W.

19.17   Derivatives of the ω_s^r

We are going to prove that the first and second partial derivatives of the ω_s^r with respect to the x^i are, as σ and its partial derivatives, simple algebraic functions of the functions X and Ω, $\tilde{X}$ and $\tilde{\Omega}$, and of the values on the conoid Σ_0 of the coefficients of the given equations and of their partial derivatives.

(1°) The first partial derivatives of the ω_s^r with respect to the x^i are expressed as functions of their partial derivatives with respect to the λ_i

$$\frac{\partial\omega_s^r}{\partial x^i} = \sum_{j=1}^3 \frac{\partial\omega_s^r}{\partial\lambda_j}\frac{J_i^j}{J_{x\lambda}},$$

therefore

$$\frac{\partial\omega_s^r}{\partial x^i} = \left(\sum_{t=1}^n Q_t^r\omega_s^t + Q\omega_s^r\right)\sum_{j=1}^3 \frac{P_j^0 D_i^j}{D} + \sum_{h,j=1}^3 \frac{\omega_{sh}^r}{\lambda_1}\frac{(\delta_j^h - P_j^0 P_h^0)D_i^j}{D}. \tag{19.17.1}$$

The first partial derivatives of the ω_s^r with respect to the x^i are therefore rational fractions with denominator D of the functions

$$X(P_i, y_i^j), \ \Omega(\omega_s^r), \ \tilde{X}\left(\frac{y_i^j}{\lambda_1}\right), \ \tilde{\Omega}\left(\frac{\omega_{sh}^r}{\lambda_1}\right), \ [A^{\lambda\mu}], \ \left[\frac{\partial A^{\lambda\mu}}{\partial x^\alpha}\right], \ [B^{s\lambda}] \ \text{and} \ P_i^0.$$

These are continuous and bounded functions in W.

(2°) We will compute the second partial derivatives of the ω_s^r with respect to the x^i by writing $\frac{\partial \omega_s^r}{\partial x^i}$ in the form $\frac{\partial \omega_s^r}{\partial x^i} = \frac{P_2}{(\lambda_1)^2 D}$.

The equality (19.17.1) and the remarks of Sect. 19.13 show that P_2 is a homogeneous polynomial of second degree with respect to the set of functions y_i^j, ω_s^r. We have, by differentiating the previous equality,

$$\frac{\partial^2 \omega_s^r}{\partial x^i \partial x^j} = \frac{1}{(\lambda_1)^2 D} \frac{\partial P_2}{\partial x^j} - \frac{P_2}{((\lambda_1)^2 D)^2} \frac{\partial((\lambda_1)^2 D)}{\partial x^i}.$$

These functions $\lambda_1 \frac{\partial^2 \omega_s^r}{\partial x^i \partial x^j}$ are rational fractions with denominator D^3 of the functions

$$X, \Omega, \tilde{X}, \tilde{\Omega}, [A^{\lambda\mu}], \left[\frac{\partial A^{\lambda\mu}}{\partial x^\alpha}\right], \left[\frac{\partial^2 A^{\lambda\mu}}{\partial x^\alpha \partial x^\beta}\right], [B_r^{s\lambda}], \left[\frac{\partial B_r^{s\lambda}}{\partial x^\alpha}\right].$$

(The results of Sect. 19.13 make it possible actually to prove that $\frac{P_2}{(\lambda_1)^2}$ and $\frac{1}{\lambda_1} \frac{\partial P_2}{\partial x^j}$ are a polynomial and a rational fraction, respectively, with denominator D, of these functions.) These are therefore continuous and bounded functions in W.

19.18 Kirchhoff Formulae

We can now study in more precise way the fundamental equations (19.7.1) and look for their limit as η approaches zero.

These equations read as:

$$\int\int_V \int \sum_{r=1}^n ([u_r]L_s^r + \sigma_s^r[f_r])dx^1 \, dx^2 \, dx^3$$

$$+ \int\int_{S_0} \sum_{i=1}^3 E_s^i \cos(n, x^i)dS = \int\int_{S_\eta} \sum_{i=1}^3 E_s^i \cos(n, x^i)dS. \quad (19.18.1)$$

Integral relations involving the parameter λ_i. We have seen that the functional determinant $D = \frac{D(x^i)}{D(\lambda_j)}$ is equal to -1 for $\lambda_1 = 0$. The correspondence between the parameters x^i and λ_j is therefore surjective in a neighbourhood of the vertex M_0 of Σ_0. One derives from this that the correspondence between the parameters x^i and λ_j is one-to-one in a domain $(\Lambda)_\eta$ defined by

$$\eta \le \lambda_1 \le \varepsilon_3, \ 0 \le \lambda_2 \le \pi, \ 0 \le \lambda_3 \le 2\pi,$$

where ε_3 is a given number and where η is arbitrarily small.

To the domain $(\Lambda)_\eta$ of variations of the λ_i parameters there corresponds, in a one-to-one way, a domain W_η of Σ_0, because the correspondence between $(x^4, \lambda_2, \lambda_3)$ and $(\lambda_1, \lambda_2, \lambda_3)$ is one-to-one. We shall then assume that the coordinate x_0^4 of the vertex M_0 of Σ_0 is sufficiently small to ensure that the domain $V_\eta \subset V$, previously considered, is interior to the domains W and W_η. We can, under these conditions, compute the integrals by means of the parameters λ_i, the integrals that we are going to obtain being convergent.

19.19   Evaluation of the Area and Volume Elements

First, we have

$$dV = dx^1 \ dx^2 \ dx^3 = d\lambda_1 \ d\lambda_2 \ d\lambda_3.$$

Let us compute now dS and $\cos(n, x^i)$.

The surfaces S_0 and S_η are $x^4 = c^{te}$ surfaces drawn on the characteristic conoid Σ_0. They therefore satisfy the differential relation

$$\sum_{i=1}^{3} p_i \ dx^i = 0,$$

from which one deduces

$$\cos(n, x^i) = \frac{p_i}{\left(\sum_{j=1}^{3} (p_j)^2 \right)^{\frac{1}{2}}}.$$

In order to evaluate dS we shall write a second expression of the volume element dV in which the surfaces $S(x^4 = c^{te})$ and the bicharacteristics (where only λ_1 is varying) come into play

$$dV = \cos \nu \ |T|^{\frac{1}{2}} d\lambda_1 \ dS,$$

where $|T|^{\frac{1}{2}} d\lambda_1$ denotes the length element of the bicharacteristic, and ν is the angle formed by the bicharacteristic with the normal to the surface S at the point considered.

A system of directional parameters of the tangent to the bicharacteristic being

$$T^h = \sum_{j=1}^{3} [A^{hj}] p_j + [A^{h4}],$$

we have

$$\cos \nu \, |T|^{\frac{1}{2}} = \sum_{h=1}^{3} \left\{ \sum_{j=1}^{3} [A^{hj}]p_j + [A^{h4}] \right\} \cos(n, x^h),$$

from which, by comparing the two expressions of dV,

$$\cos(n, x^i)dS = \frac{J_{x\lambda}p_i d\lambda_2 \, d\lambda_3}{\sum_{h,j=1}^{3} [A^{hj}]p_j p_h + \sum_{h=1}^{3} [A^{h4}]p_h} = \frac{-J_{x\lambda}p_i}{[A^{44}] + \sum_{j=1}^{3} [A^{j4}]p_j} d\lambda_2 \, d\lambda_3.$$

19.20 Limit as $\eta \to 0$ of the Integral Relations

The integral relations (19.18.1) read, in terms of the λ_i parameters, as

$$\int \int_{V_\eta} \int \sum_{r=1}^{n} ([u_r]L_s^r + \sigma_s^r[f_r])d\lambda_1 \, d\lambda_2 \, d\lambda_3$$

$$- \int_0^{2\pi} \int_0^{\pi} \frac{\sum_{i=1}^{3} E_s^i J_{x\lambda}p_i}{[A^{44}] + \sum_{i=1}^{3} [A^{i4}]p_i} d\lambda_{2|x^4=0} \, d\lambda_3$$

$$= - \int_0^{2\pi} \int_0^{\pi} \left\{ \frac{\sum_{i=1}^{3} E_s^i J_{x\lambda}p_i}{[A^{44}] + \sum_{i=1}^{3} [A^{i4}]p_i} \right\}_{x^4=x_0^4-\eta} d\lambda_2 \, d\lambda_3. \qquad (19.20.1)$$

The previous results prove that the quantities to be integrated are continuous and bounded functions of the variables λ_i. They read actually as:

$$(\lambda_1)^2 \sum_{r=1}^{n} \{[u_r]L_s^r + \sigma_s^r[f_r]\} \frac{J_{x\lambda}}{\lambda_1^2} \text{ and } (\lambda_1)^2 \sum_{i=1}^{n} E_s^i \frac{J_{x\lambda}}{(\lambda_1)^2} \frac{p_i}{T^4}.$$

E_s^i and L_s^r being given by the equalities (19.6.2), the quantities considered are continuous and bounded in W if the functions u_r and $\frac{\partial u_r}{\partial x^\alpha}$ are continuous and bounded in D.

The two members of Eq. (19.20.1) tend therefore towards a finite limit when η approaches zero. The triple integral tends to a finite limit, equal to the value of this integral taken over the portion V_0 of hypersurface of the conoid Σ_0 in between the vertex M_0 and the initial surface $x^4 = 0$ (because this integral is convergent). Let us evaluate the limit of the double integral of the second member. From previous sections, all terms of the quantity $(\lambda_1)^2 E_s^i$ approach uniformly zero with λ_1, exception being made for the term

$$-(\lambda_1)^2 \sum_{r=1}^{n} \sum_{j=1}^{3} [u_r][A^{ij}]\omega_s^r \frac{\partial \sigma}{\partial x_j},$$

whose limit for $\lambda_1 = 0$ is

$$\sum_{r=1}^{n} \sum_{j=1}^{3} [u_r(x_0^\alpha)]\delta_i^j \delta_s^r p_j^0 = u_s(x_0^\alpha)p_i^0.$$

Hence one obtains

$$\lim_{\lambda_1 \to 0} \frac{\sum_{i=1}^{3} E_s^i J_{x\lambda} p_i}{[A^{44}] + \sum_{i=1}^{3} [A^{i4}]p_i} = -u_s(x_0^\alpha) \sin \lambda_2.$$

The second member of Eq. (19.20.1) tends therefore, when η approaches zero, to the limit

$$\int_0^{2\pi} \int_0^\pi u_s(x_0^\alpha) \sin \lambda_2 \, d\lambda_2 \, d\lambda_3 = 4\pi u_s(x_0^\alpha).$$

19.21   Reverting to the Kirchhoff Formulae

We arrive in such a way to the following formulae:

$$4\pi u_s(x_0^\alpha) = \int \int_{V_\eta} \int \sum_{r=1}^{n} ([u_r]L_s^r + \sigma_s^r[f_r])J_{x\lambda} \, d\lambda_1 \, d\lambda_2 \, d\lambda_3$$

$$+ \int_0^{2\pi} \int_0^\pi \left\{ \frac{\sum_{i=1}^{3} E_s^i J_{x\lambda} p_i}{T^4} \right\}_{x^4=0} d\lambda_2 \, d\lambda_3. \qquad (19.21.1)$$

In order to compute the second member of these Kirchhoff formulae it will be convenient to take for parameters, on the hypersurface of the conoid Σ_0, the three independent variables $x^4, \lambda_2, \lambda_3$.

Equation (19.21.1) then read, the limits of integration intervals being evident:

$$4\pi u_s(x_j) = \int_{x_0^4}^0 \int_0^{2\pi} \int_0^\pi \sum_{r=1}^{n} ([u_r]L_s^r + \sigma_s^r[f_r])\frac{J_{x\lambda}}{T^4} dx^4 d\lambda_2 d\lambda_3$$

$$+ \int_0^{2\pi} \int_0^\pi \left\{ \frac{\sum_{i=1}^{3} E_s^i J_{x\lambda} p_i}{T^4} \right\}_{x^4=0} d\lambda_2 d\lambda_3. \qquad (19.21.2)$$

The quantity under sign of triple integral is expressed by means of the functions $[u]$ and of the functions $X(\lambda_1, \lambda_2, \lambda_3)$ and $\Omega(\lambda_1, \lambda_2, \lambda_3)$, solutions of the integral equations (19.2.2) and (19.9.1).

We shall obtain the expression of the X and Ω as functions of the new variables $x^4, \lambda_2, \lambda_3$ by replacing λ_1 with its value defined by the Eq. (19.3.2), function of the $x^4, \lambda_2, \lambda_3$.

Let us point out that these functions satisfy the integral equations

$$X(x^4, \lambda_2, \lambda_3) = \int_{x_0^4}^{x^4} \frac{E(X)}{T^4} dw^4 + X_0(x_0^4, \lambda_2, \lambda_3)$$

$$\Omega(x^4, \lambda_2, \lambda_3) = \int_{x_0^4}^{x^4} \frac{F(X, \Omega)}{T^4} dw^4 + \Omega_0(x_0^4, \lambda_2, \lambda_3).$$

The quantity under sign of double integral is expressed by means of the values for $x^4 = 0$ of the functions $[u]$ and $\left[\frac{\partial u}{\partial x^\alpha}\right]$ (Cauchy data) and of the values for $x^4 = 0$ of the functions X and Ω.

19.22 Summary of the Results

We consider a system of *linear*, second-order partial differential equations in four variables, of the type

$$\sum_{\lambda,\mu=1}^{4} A^{\lambda\mu} \frac{\partial^2 u_r}{\partial x^\lambda \partial x^\mu} + \sum_{s=1}^{n} \sum_{\lambda=1}^{4} B_r^{s\lambda} \frac{\partial u_s}{\partial x^\lambda} + f_r = 0. \qquad [E]$$

Assumptions

(1°) At the point M_0 of coordinates x_0^α the coefficients $A^{\lambda\mu}$ take the following values:

$$A_0^{44} = 1, \ A_0^{i4} = 0, \ A_0^{ij} = -\delta^{ij}.$$

(2°) The coefficients $A^{\lambda\mu}$ and $B_s^{r\lambda}$ have partial derivatives with respect to the x^α, of orders four and two, respectively, continuous and bounded in a domain $D : |x^i - \tilde{x}^i| \leq d, |x^4| \leq \varepsilon$. The coefficients f_r are continuous and bounded.

(3°) The partial derivatives of the $A^{\lambda\mu}$ and $B_s^{r\lambda}$ of orders four and two, respectively, satisfy in D some Lipschitz conditions.

Conclusion *Every solution of the equations [E] continuous, bounded and with first partial derivatives continuous and bounded in D verifies the integral relations (19.21.2) if the coordinates x_0^α of M_0 satisfy inequalities of the form*

$$|x_0^4| \leq \varepsilon_0, \ |x_0^i - \tilde{x}^i| \leq d,$$

defining a domain $D_0 \subset D$.

19.23   Transformation of Variables

We are going to establish formulae analogous to (19.21.2), verified by the solutions
of the given equations $[E]$ at every point of a domain D_0 of space-time, where the
values of coefficients will be restricted uniquely by the requirement of having to
verify some conditions of normal hyperbolicity and differentiability.

Let us therefore consider the system $[E]$ of equations

$$\sum_{\lambda,\mu=1}^{4} A^{\lambda\mu} \frac{\partial^2 u_s}{\partial x^\lambda \partial x^\mu} + \sum_{r=1}^{n} \sum_{\lambda=1}^{4} B_s^{r\lambda} \frac{\partial u_r}{\partial x^\lambda} + f_s = 0.$$

We assume that in the space-time domain D, defined by

$$|x^4| \leq \varepsilon, \ |x^i - \tilde{x}^i| \leq d,$$

where the three $\tilde{x}^i$ are given numbers, the equations $[E]$ are of the normal hyperbolic
type, i.e.

$$A^{44} > 0, \ \text{the quadratic form} \ \sum_{i,j=1}^{3} A^{ij} X_i X_j \ \text{is negative} - \text{definite.}$$

At every point $M_0(x_j)$ of the domain D one can associate to the values $A_0^{\lambda\mu} = A^{\lambda\mu}(x_0^\alpha)$ of the coefficients A a system of real numbers $a_0^{\alpha\beta}$, algebraic functions,
defined and indefinitely differentiable of the $A_0^{\lambda\mu}$, satisfying the identity

$$\sum_{\lambda,\mu=1}^{4} A_0^{\lambda\mu} X_\lambda X_\mu = \left(\sum_{\alpha=1}^{4} a_0^{4\alpha} X_\alpha \right)^2 - \left(\sum_{\alpha=1}^{4} a_0^{i\alpha} X_\alpha \right)^2.$$

We shall denote by $a_{\alpha\beta}^0$ the quotient by the determinant a_0 of elements $a_0^{\alpha\beta}$ of the
minor relative to the element $a_0^{\alpha\beta}$ of this determinant. The quantities $a_{\alpha\beta}^0$ are, like
$a_0^{\alpha\beta}$, algebraic functions defined and indefinitely differentiable of the $A_0^{\lambda\mu}$ in D. (The
square of the determinant a_0, being equal to the absolute value A of the determinant
having elements $A^{\lambda\mu}$, a_0, is different from zero in D.)

Let us perform the linear change of variables

$$y_\alpha \equiv \sum_{\beta=1}^{4} a_{\alpha\beta}^0 x^\beta.$$

The partial derivatives of the unknown functions u_s are covariant under such a change of variables, hence the equations $[E]$ read as

$$\sum_{\alpha,\beta=1}^{4} A^{*\alpha\beta} \frac{\partial^2 u_s}{\partial y^\alpha \partial y^\beta} + \sum_{r=1}^{n}\sum_{\alpha=1}^{4} B_s^{*r\alpha} \frac{\partial u_r}{\partial y^\alpha} + f_s = 0, \tag{19.23.1}$$

with

$$A^{*\alpha\beta} = \sum_{\lambda,\mu=1}^{4} A^{\lambda\mu} a_{\alpha\lambda}^0 a_{\beta\mu}^0, \tag{19.23.2a}$$

$$B_s^{*r\alpha} = \sum_{\lambda=1}^{4} B_s^{r\lambda} a_{\alpha\lambda}^0. \tag{19.23.2b}$$

The coefficients of Eq. (19.23.1) take at the point M_0 the values (19.2.4). As a matter of fact:

$$A_0^{*\alpha\beta} = \sum_{\lambda,\mu=1}^{4} A_0^{\lambda\mu} a_{\alpha\lambda}^0 a_{\beta\mu}^0 = -\sum_{\gamma,\lambda,\mu=1}^{4} a_0^{\gamma\lambda} a_0^{\gamma\mu} a_{\alpha\lambda}^0 a_{\beta\mu}^0 + 2\sum_{\lambda,\mu=1}^{4} a_0^{4\lambda} a_0^{4\mu} a_{\alpha\lambda}^0 a_{\beta\mu}^0$$

$$= -\delta_\alpha^\beta + 2\delta_\alpha^4 \delta_\beta^4,$$

hence one has

$$A^{*44} = 1, \ A^{*i4} = 0, \ A^{*ij} = -\delta^{ij}.$$

We can apply to the equations $[E]$, written in the form (19.23.1), in the variables y^α and for the corresponding point M_0, the results of Sect. 19.22. Let us first point out that the integration parameters so introduced will be y^4, λ_2, λ_3 but that, the surface carrying the Cauchy data being always $x^4 \equiv a_0^{\alpha 4} y^4 = 0$, the integration domains will be determined from M_0 and the intersection of this surface with the characteristic conoid with vertex M_0. We see that it will be convenient, in order to evaluate these integrals, to choose the variables y^α relative to a point M_0 whatsoever in such a way that the initial space section, $x^4 = 0$, is a hypersurface $y^4 = 0$. It will be enough for that purpose to choose the coefficients $a_0^{\alpha\beta}$ (which is legitimate) in such a way that one has $a_0^{i4} = 0$. We shall then have

$$a_{4i}^0 = 0, \ a_{44}^0 = \frac{1}{a_0^{44}} = (A_0^{44})^{-\frac{1}{2}} \text{ and } y_4 = a_{44}^0 x^4,$$

where a_{44}^0 is a bounded positive number.

19.24   Application of the Results

The application of the results obtained so far proves then the existence of a domain $D_0 \subset D$, defined by $|x_0^4| \leq \varepsilon$ (which implies at every point $M_0 \in D_0$, $|y_0^4| \leq \eta$) such that one can write at every point M_0 of D_0 a Kirchhoff formula whose first member is the value at M_0 of the unknown u_s, in terms of the quantities $y_0^\alpha = \sum_{\beta=1}^4 a_{\alpha\beta}^0 x_0^\beta$, and whose second member consists of a triple integral and of a double integral. The quantities to be integrated are expressed by means of the functions $X(y^4, \lambda_2, \lambda_3, y_0^\alpha)$ representing $(y^\alpha, p_i, y_i^j, z_i^j, \ldots, z_{kij}^h)$ and $\Omega(y^4, \lambda_2, \lambda_3)$ $(\omega_s^r, \ldots, \omega_{sij}^r)$, solutions of an equation of the kind

$$ X = \int_{y_0^4}^{y^4} E^*(X)\mathrm{d}Y^4 + X_0, \quad \Omega = \int_{y_0^4}^{y^4} F^*(X, \Omega)\mathrm{d}Y^4 + \Omega_0, \qquad (19.24.1) $$

where the functions E^* and F^* are the functions E and F considered before, but evaluated starting from the coefficients (19.23.2) and from their partial derivatives with respect to the y^α, and where Ω_0, X_0 denote the values for $y^4 = y_0^4$ of the corresponding functions Ω, X.

In order to obtain, under a simpler form, some integral equations holding in the whole domain D_0, we will take on the one hand as integration parameter, in place of y^4, x^4 (which is possible, a_{44}^0 being at every point M_0 of D_0 a given positive number). We shall on the other hand replace those of the auxiliary unknown functions X which are the values (in terms of the three parameters) of the coordinates y^α of a point of the conoid Σ_0 of vertex M_0, with the values of the original coordinates x^α of a point of this conoid.

We shall replace for that purpose those of the integral equations which have in the first member y^α with their linear combinations of coefficients $a_0^{\alpha\beta}$ (bounded numbers), i.e. with the equations of the same kind

$$ \sum_{\beta=1}^4 a_0^{\alpha\beta} y^\beta = x^\alpha = \int_{x_0^4}^{x^4} \sum_{\beta=1}^4 a_0^{\alpha\beta} \frac{T^{*\alpha\beta}}{T^{*4}} a_{44}^0 \mathrm{d}w^4 + x_0^\alpha, $$

and we will replace the quantities under integration signs of all our equations in terms of the x^α in place of the y^β by replacing in these equations the y^β with the linear combinations $\sum_{\alpha=1}^4 a_{\alpha\beta}^0 x^\alpha$ (the $a_{\alpha\beta}^0$ are bounded numbers).

The system of integral equations obtained in such a way has, for every point M_0 of the domain D, solutions as for the previous system, solutions which are of the form

$$ X(x_0^\alpha, x^4, \lambda_2, \lambda_3). $$

19.25 Linear Systems of Second Order

We consider a system of linear, second-order partial differential equations of the type

$$\sum_{\lambda,\mu=1}^{4} A^{\lambda\mu} \frac{\partial^2 u_s}{\partial x^\lambda \partial x^\mu} + \sum_{r=1}^{n}\sum_{\lambda=1}^{4} B_s^{r\lambda} \frac{\partial u_r}{\partial x^\lambda} + f_s = 0. \qquad [E]$$

Assumptions

(1°) In the domain D, defined by

$$|x^4| \le \varepsilon, \ |x^i - \tilde{x}^i| \le d,$$

the quadratic form $\sum_{\lambda,\mu=1}^{4} A^{\lambda\mu} X_\lambda X_\mu$ is of normal hyperbolic type:

$$A^{44} > 0, \ \text{the quadratic form} \sum_{i,j=1}^{3} A^{ij} X_i X_j \text{ is negative} - \text{definite.}$$

(2°) The coefficients $A^{\lambda\mu}$ and $B_s^{r\lambda}$ have partial derivatives with respect to the x^α continuous and bounded, up to the orders four and two, respectively, in the domain D.
(3°) The partial derivatives of the $A^{\lambda\mu}$ and $B_s^{r\lambda}$ of orders four and two, respectively, satisfy, within D, Lipschitz conditions.

Conclusion Every solution of Eqs. $[E]$, possessing in D first partial derivatives with respect to the x^α continuous and bounded, verifies, if x_0^α are the coordinates of a point M_0 of a domain D_0 defined by

$$|x_0^4| \le \varepsilon_0 \le \varepsilon, \ |x_0^i - \tilde{x}^i| \le d_0 \le d,$$

some Kirchhoff formulae whose first members are the values at the point M_0 of the unknown functions u_s and whose second members consist of a triple integral (integration parameters x^4, λ_2, λ_3) and of a double integral (integration parameters λ_2, λ_3). The quantities to be integrated are expressed by means of functions $X(x_0^\alpha, x^4, \lambda_2, \lambda_3)$ and $\Omega(x_0^\alpha, x^4, \lambda_2, \lambda_3)$, themselves solutions of given integral equations (19.24.1), and of the unknown functions $[u_s]$; the quantity under the sign of double integral, which is taken for the zero value of the x^4 parameter, contains, besides the previous functions, the first partial derivatives of the unknown functions $\left[\frac{\partial u_s}{\partial x^\alpha}\right]$ (value over Σ_0 of the Cauchy data). We obtain in such a way a system of integral equations verified in D_0 from the solutions of Eq. $[E]$. We write this system in the following reduced form [64]:

$$X = \int_{x_0^4}^{x^4} E \, dw^4 + X_0,$$

$$4\pi U = \int_{x_0^4}^{0} \int_{0}^{2\pi} \int_{0}^{\pi} H \, dx^4 \, d\lambda_2 \, d\lambda_3 + \int_{0}^{2\pi} \int_{0}^{\pi} I \, d\lambda_2 \, d\lambda_3.$$

Chapter 20
Linear System from a Non-linear Hyperbolic System

20.1 Non-linear Equations

We now consider a system (F) of n second-order partial differential equations, with n unknown functions and four variables, *non-linear* of the following type:

$$\sum_{\lambda,\mu=1}^{4} A^{\lambda\mu}\frac{\partial^2 W_s}{\partial x^\lambda \partial x^\mu} + f_s = 0, \quad s = 1, 2...n. \tag{F}$$

The coefficients $A^{\lambda\mu}$ and f_s are given functions of the four variables x^α, the unknown functions W_s, and of their first derivatives $\frac{\partial W_s}{\partial x^\alpha}$.

The coefficients $A^{\lambda\mu}$ are the same for the n equations.

We point out that the calculations, made in the previous chapter for the linear equations $[E]$, are valid for the non-linear equations (F): it suffices to consider in these calculations the functions W_s as functions of the four variables x^α; the coefficients $A^{\lambda\mu}$ and f_s are then functions of these four variables and the previous calculations are valid, subject of course to considering, in all formulae where there is occurrence of partial derivatives of the coefficients with respect to x^α, these derivations as having been performed. One will have for example

$$\frac{\partial A^{\lambda\mu}}{\partial x^\alpha} = \sum_{s=1}^{n} \frac{\partial A^{\lambda\mu}}{\partial W_s}\frac{\partial W_s}{\partial x^\alpha} + \sum_{s=1}^{n}\sum_{\beta=1}^{4} \frac{\partial A^{\lambda\mu}}{\partial(\partial W_s/\partial x^\beta)}\frac{\partial}{\partial x^\alpha}\left(\frac{\partial W_s}{\partial x^\beta}\right).$$

By applying the previous results one would prove that, under certain assumptions, the solutions of equations (F) satisfy a system of integral equations whose second members contain, besides the auxiliary functions, the integration parameters and the unknown functions, the partial derivatives with respect to the x^α of these unknown functions.

© Springer International Publishing AG 2017
G. Esposito, *From Ordinary to Partial Differential Equations*, UNITEXT 106,
DOI 10.1007/978-3-319-57544-5_20

Thus, we do not apply directly to the equations (F) the results of previous chapters; but we are going to show that, by deriving suitably five times with respect to the variables x^α the given equations (F), and by applying to the obtained equations the results of Chap. 19, one obtains a system of integral equations whose first members are the unknown functions W_s, their partial derivatives with respect to the x^α up to the fifth order and some auxiliary functions X, Ω, and whose second members contain only these functions and the integration parameters.

20.2   Differentiation of the Equations (F)

We assume that in a space-time domain D, centred at the point $\overline{M}$ with coordinates x^i, 0 and defined by

$$|x^i - \overline{x}^i| \le d, \ |x^4| \le \varepsilon$$

and for values of the unknown functions W_s and their first partial derivatives satisfying

$$|W_s - \overline{W}_s| \le l, \quad \left|\frac{\partial W_s}{\partial x^\alpha} - \frac{\overline{\partial W_s}}{\partial x^\alpha}\right| \le l \tag{20.2.1}$$

(where $\overline{W}_s$ and $\overline{\frac{\partial W_s}{\partial x^\alpha}}$ are the values of the functions W_s and $\frac{\partial W_s}{\partial x^\alpha}$ at the point $\overline{M}$) the coefficients $A^{\lambda\mu}$ and f_s admit of partial derivatives with respect to all their aguments up to the fifth order.

We shall then obtain, by differentiating five times the equations (F) with respect to the variables x^α, a system of N equations (N is the product of n times the number of derivatives of order five of a function of four variables) verified, in the domain D, by the solutions of equations (F) which satisfy the inequalities (20.2.1) and possess derivatives with respect to the x^α up to the seventh order.

Let us write this system of N equations. We set

$$\frac{\partial W_s}{\partial x^\alpha} = W_{s\alpha}, \quad \frac{\partial^2 W_s}{\partial x^\alpha \partial x^\beta} = W_{s\alpha\beta}$$

and we denote by U_S the partial derivatives of order five of W_s

$$\frac{\partial^5 W_s}{\partial x^\alpha \partial x^\beta \partial x^\gamma \partial x^\delta \partial x^\varepsilon} \equiv W_{s\alpha\beta\gamma\delta\varepsilon} \equiv U_S, \quad s = 1, 2, \ldots N.$$

Upon differentiating the given equations (F) with respect to any whatsoever of the variables x^α, we obtain n equations of the form

$$\sum_{\lambda,\mu=1}^{4} A^{\lambda\mu} \frac{\partial^2 W_{s\alpha}}{\partial x^\lambda \partial x^\mu} + \sum_{r=1}^{n} \sum_{\lambda,\mu=1}^{4} \frac{\partial A^{\lambda\mu}}{\partial W_r} W_{r\alpha} \frac{\partial W_{s\mu}}{\partial x^\lambda} + \sum_{r=1}^{n} \sum_{\lambda,\mu,\nu=1}^{4} \frac{\partial A^{\lambda\mu}}{\partial W_{r\nu}} \frac{\partial W_{r\nu}}{\partial x^\alpha} \frac{\partial W_{s\mu}}{\partial x^\lambda}$$

$$+ \sum_{\lambda,\mu=1}^{4} \frac{\partial A^{\lambda\mu}}{\partial x^\alpha} \frac{\partial W_{s\mu}}{\partial x^\lambda} + \sum_{r=1}^{n} \frac{\partial f_s}{\partial W_r} W_{r\alpha} + \sum_{r=1}^{n} \sum_{\nu=1}^{4} \frac{\partial f_s}{\partial W_{r\nu}} \frac{\partial}{\partial x^\alpha} W_{r\nu} + \frac{\partial f_s}{\partial x^\alpha} = 0.$$

Let us start again four times this procedure, we obtain the following system of N equations (here we omit for simplicity of notation, as an exception to the rule used anywhere else in the book, the summations over Greek and Latin indices)

$$A^{\lambda\mu} \frac{\partial^2 W_{s\alpha\beta\gamma\delta\varepsilon}}{\partial x^\lambda \partial x^\mu} + \left\{ \frac{\partial A^{\lambda\mu}}{\partial W_r} W_{r\alpha} + \frac{\partial A^{\lambda\mu}}{\partial W_{r\nu}} W_{r\nu\alpha} + \frac{\partial A^{\lambda\mu}}{\partial x^\alpha} \right\} \frac{\partial}{\partial x^\lambda} W_{s\beta\gamma\delta\varepsilon\mu}$$

$$+ \left\{ \frac{\partial A^{\lambda\mu}}{\partial W_r} W_{r\beta} + \frac{\partial A^{\lambda\mu}}{\partial W_{r\nu}} W_{r\nu\beta} + \frac{\partial A^{\lambda\mu}}{\partial x^\beta} \right\} \frac{\partial}{\partial x^\lambda} W_{s\alpha\gamma\delta\varepsilon\mu} \cdots$$

$$+ \left\{ \frac{\partial A^{\lambda\mu}}{\partial W_r} W_{r\varepsilon} + \frac{\partial A^{\lambda\mu}}{\partial W_{r\nu}} W_{r\nu\varepsilon} + \frac{\partial A^{\lambda\mu}}{\partial x^\varepsilon} \right\} \frac{\partial}{\partial x^\lambda} W_{s\alpha\beta\gamma\delta\mu} + \frac{\partial A^{\lambda\mu}}{\partial W_{r\nu}} \frac{\partial W_{r\nu\alpha\beta\gamma\delta}}{\partial x^\varepsilon}$$

$$+ \frac{\partial f_s}{\partial W_{r\nu}} \frac{\partial W_{r\nu\alpha\beta\gamma\delta}}{\partial x^\varepsilon} + F_S = 0, \tag{20.2.2}$$

where F_S is a function of the variables x^α, of the unknown functions W_s and of their partial derivatives up to the fifth order included, but not of the derivatives of higher order.

The fifth derivatives U_S of the functions W_s satisfy therefore, in the domain D and under the conditions specified, a system of N equations of the following type:

$$\sum_{\lambda,\mu=1}^{4} A^{\lambda\mu} \frac{\partial^2 U_S}{\partial x^\lambda \partial x^\mu} + \sum_{T} \sum_{\lambda=1}^{4} B_S^{T\lambda} \frac{\partial U_T}{\partial x^\lambda} + F_S = 0. \tag{20.2.3}$$

The coefficients $A^{\lambda\mu}$, $B_S^{T\lambda}$ and F_S of these equations are polynomials of the coefficients $A^{\lambda\mu}$ and f_s of the given equations (F) and of their partial derivatives with respect to all arguments up to the fifth order, as well as of the unknown functions W_s and of their partial derivatives with respect to the x^α up to the fifth order. The coefficients $A^{\lambda\mu}$ depend only on the variables x^α, the unknown functions W_s and their first partial derivatives $W_{s\alpha}$, the coefficients $B_S^{T\lambda}$ depend only on the variables x^α, the unknown functions W_s and their first and second partial derivatives $W_{s\alpha}$ and $W_{s\alpha\beta}$.

20.3   Application of the Results of Chap. 19

We consider the equations (F) as a system of N linear equations of second order, with unknown functions U_S, and we apply to these equations the results of the previous chapter. We shall obtain a system of integral equations whose first members will be some auxiliary functions Ω, X and the unknown functions U_S; the quantities occurring under the integrals of the second members will be expressed by means of the auxiliary functions X, of the unknown functions U_S and of the value for $x^4 = 0$ of their first partial derivatives $\frac{\partial U_S}{\partial x^\alpha}$, of the integration parameters, as well as of the coefficients $A^{\lambda\mu}$, $B_S^{T\lambda}$ and F_S (viewed as functions of the x^α) and of their partial derivatives up to the orders four, three and zero. Since $A^{\lambda\mu}$, $B_S^{T\lambda}$ and F_S do not involve the partial derivatives of the functions W_s except for the orders up to one, two and five, respectively, the second members of the integral equations considered will not contain, besides the auxiliary functions X, Ω, the functions U_S and the value for $x^4 = 0$ of their first derivatives, and the integration parameters, nothing but the unknown functions W_s and their partial derivatives up to the fifth order included.

Integral equations verified by the functions W_s and their derivatives

If the functions W_s and their partial derivatives up to the fifth order

$$W_{s\alpha}, W_{s\alpha\beta}, ..., W_{s\alpha\beta\gamma\delta\varepsilon} = U_S$$

are continuous and bounded in a space-time domain D $(|x^i - \overline{x}^i| \le d, |x^4| \le \varepsilon)$ they verify in this domain the integral relations

$$W_s(x^\alpha) = \int_0^{x^4} W_{s4}(x^i, t)dt + W_s(x^i, 0), \qquad (20.3.1a)$$

$$\cdots\cdots\cdots$$

$$W_{s\alpha\beta\gamma\delta}(x^\alpha) = \int_0^{x^4} W_{s\alpha\beta\gamma\delta4}(x^i, t)dt + W_{s\alpha\beta\gamma\delta}(x^i, 0). \qquad (20.3.1b)$$

By adjoining to the system of integral equations, previously considered, the system (20.3.1), we shall then be able to obtain a system of integral equations, verified, under certain assumptions, by the solutions of the given equations (F), whose second members will only contain the functions occurring in the first members.

20.4   Cauchy Data

We shall write this system of integral equations for the purpose of solving, for the given equations (F), the Cauchy problem: the search for solutions W_s of the equations (F) which take, as well as their first partial derivatives, some values given in a domain (d) of the initial hypersurface $x^4 = 0$:

$$W_s(x^i, 0) = \varphi_s(x^i),$$

$$\frac{\partial W_s}{\partial x^4}(x^i, 0) = \psi_s(x^i),$$

where φ_s and ψ_s are given functions of the three variables x^i in the domain (d). We will prove that, under the assumptions stated below, the data φ_s and ψ_s determine the values in (d) of the partial derivatives up to the sixth order of the solution W_s of the equations (F).

Assumptions

(1°) In the domain (d), defined by

$$|x^i - \overline{x}^i| \leq d,$$

the functions φ_s and ψ_s admit of partial derivatives continuous and bounded with respect to the three variables x^i and satisfy the inequalities

$$|\varphi_s - \overline{\varphi}_s| \leq l_0 \leq l, \; |\psi_s - \overline{\psi}_s| \leq l_0 \leq l, \; \left|\frac{\partial \varphi_s}{\partial x^i} - \frac{\partial \overline{\varphi}_s}{\partial x^i}\right| \leq l_0 \leq l. \quad (20.4.1)$$

(2°) In the domain (d) and for values of the functions

$$W_s = \varphi_s, \quad \frac{\partial W_s}{\partial x^4} = \psi_s \text{ and } \frac{\partial W_s}{\partial x^i} = \frac{\partial \varphi_s}{\partial x^i},$$

satisfying the inequalities (20.4.1), the coefficients $A^{\lambda\mu}$ and f_s have partial derivatives continuous and bounded with respect to all their arguments, up to the fifth order.

(3°) In the domain (d) and for the functions φ_s and ψ_s considered the coefficient A^{44} is different from zero.

It turns out actually from the first assumption that the values in (d) of partial derivatives up to the sixth order, corresponding to a differentiation at most with respect to x^4, of the solutions W_s of the assigned Cauchy problem are equal to the corresponding partial derivatives of the functions φ_s and ψ_s, and are continuous and bounded in (d).

The values in (d) of partial derivatives up to the sixth order of the functions W_s, corresponding to more than one derivative with respect to x^4, are expressed in terms of the previous ones, of the coefficients $A^{\lambda\mu}$ and f_s of the equations (F) and of their partial derivatives up to the fourth order. The third assumption shows actually that the equations (F) make it possible to evaluate, being given within (d) the values of the functions W_s, $W_{s\alpha}$, $W_{s\alpha i}$, the value in (d) of W_{s44}, from which one will deduce by differentiation the value in (d) of the partial derivatives corresponding to two differentiations with respect to x^4. The equations that are derivatives of the equations (F) with respect to the variables x^α (up to the fourth order) make it possible, in analogous manner, to evaluate in (d) the values of partial derivatives up to the sixth order of the functions W_s. It turns out from the three previous assumptions that all functions obtained are continuous and bounded in (d).

We shall set

$$W_{sj}(x^i, 0) = \varphi_{sj}(x^i),$$

$$U_S(x^i, 0) = \Phi_S(x^i),$$

$$\frac{\partial U_S}{\partial x^4}(x^i, 0) = \Psi_S(x^i).$$

20.5   Summary of Results

We consider a system of n partial differential equations of second order, non-linear, of the following kind:

$$\sum_{\lambda,\mu=1}^{4} A^{\lambda\mu} \frac{\partial^2 W_s}{\partial x^\lambda \partial x^\mu} + f_s = 0,$$

where $A^{\lambda\mu}$ and f_s are functions of the W_r and of their first partial derivatives, and of the four variables x^α.

We have seen that, under the assumptions of Sect. 20.2, the seven-times differentiable solutions of the equations (F) satisfy the inequalities (20.2.1), verify the system of N equations

$$\sum_{\lambda,\mu=1}^{4} A^{\lambda\mu} \frac{\partial^2 U_S}{\partial x^\lambda \partial x^\mu} + \sum_{T}\sum_{\alpha=1}^{4} B_S^{T\alpha} \frac{\partial U_T}{\partial x^\alpha} + F_S = 0, \qquad (F')$$

where U_S denotes any whatsoever of the fifth-order partial derivatives of W_s and where $A^{\lambda\mu}$, $B_S^{T\lambda}$ and F_S are functions of the variables x^α, of the functions W_s and of their partial derivatives up to the orders one, two and five, respectively.

We have shown that, under the assumptions of Sect. 20.4, every solution seven times differentiable of the Cauchy problem (with Cauchy data φ_s, ψ_s) takes, as well as its partial derivatives up to the sixth order, some given values continuous and bounded in the considered domain of the initial surface.

We apply to the equations (20.2.3) the results of Chap. 19 and we add to the integral equations obtained the integral equations (20.3.1).

Let us sum up the assumptions made and the results obtained.

Assumptions

(A) In the domain D defined by $|x^i - \bar{x}^i| \le d$, $|x^4| \le \varepsilon$ and for values of the unknown functions satisfying

$$|W_s - \bar{\varphi}_s| \le l, \quad \left|\frac{\partial W_s}{\partial x^4} - \bar{\psi}_s\right| \le l, \quad \left|\frac{\partial W_s}{\partial x^i} - \frac{\overline{\partial \varphi_s}}{\partial x^i}\right| \le l:$$

(1°) The coefficients $A^{\lambda\mu}$ and f_s have partial derivatives with respect to all their arguments up to the fifth order continuous and bounded, the derivatives of order five satisfying some Lipschitz conditions;

(2°) The quadratic form $\sum_{\lambda,\mu=1}^{4} A^{\lambda\mu} X_\lambda X_\mu$ is of normal hyperbolic type: $A^{44} > 0$ and the form $\sum_{i,j=1}^{3} A^{ij} X_i X_j$ is negative-definite.

(B) In the domain of the initial surface $x^4 = 0$, defined by $|x^i - \bar{x}^i| \le d$, the Cauchy data φ_s and ψ_s admit of partial derivatives continuous and bounded up to the orders six and five.

Conclusion If we consider a solution W_s seven times differentiable of the assigned Cauchy problem, possessing partial derivatives with respect to the x^α up to the sixth order, continuous and bounded and satisfying the inequalities (20.2.1) in D, it satisfies in this domain the equations F'. The equations F', viewed as linear equations in the unknown functions U_S, satisfy the assumptions of Chap. 19, and therefore [64]:

There exists a domain $D_0 \subset D$ in which the functions W_s verify the following system of integral equations.

System of integral equations (I)

This system consists of

(1°) equations having in the first member a function X of the three parameters

$$x^4, \lambda_2, \lambda_3$$

(representatives of a point of the characteristic conoid of vertex $M_0(x_0)$) and of the four coordinates x_0^α of a point $M_0 \in D_0$. These functions X are the functions

$$x^i, p^i, y_i^j, z_i^j, y_{ih}^j, z_{ih}^j, y_{ihk}^j, z_{ihk}^j$$

of Chap. 19. These equations are of the form

$$X = \int_{x_0^4}^{x^4} E(X)dw^4 + X_0,$$

where X_0, value of X for $x^4 = x_0^4$, is a given function of $x_0^\alpha, \lambda_2, \lambda_3$;

(2°) equations having in the first member a function

$$\Omega(x_0^\alpha, x_0^4, \lambda_2, \lambda_3)$$

(functions ω_s^r, ω_{si}^r, ω_{sij}^r of Chap. 19), of the form

$$\Omega = \int_{x_0^4}^{x^4} F(X, \Omega)dw^4 + \Omega_0,$$

where Ω_0, value of Ω for $x^4 = x_0^4$, is a given function of x_0^α, λ_2, λ_3;

(3°) equations having in the first member a function W of the four coordinates x^α of a point $M \in D$. The functions W are the functions

$$W_s, W_{s\alpha}, W_{s\alpha\beta}, W_{s\alpha\beta\gamma}, W_{s\alpha\beta\gamma\delta}.$$

The equations are of the form

$$W = \int_0^{x^4} G(W, U)dw^4 + W_0,$$

where W_0, value of W for $x^4 = 0$, is a given function of the three variables x^i.

(4°) equations having in the first member a function U of the four coordinates x_0^α of a point $M_0 \in D_0$. The functions U are the functions U_s, fifth derivatives of W_s. These equations (Kirchhoff formulae) are of the form

$$U = \int_{x_0^4}^0 \int_0^{2\pi} \int_0^\pi H \, dx^4 \, d\lambda_2 \, d\lambda_3 + \int_0^{2\pi} \int_0^\pi I \, d\lambda_2 \, d\lambda_3.$$

The quantities E, F, G, H, I are formally identical to the corresponding quantities evaluated in Chap. 19 for the equations $[E]$ (upon considering the differentiations with respect to the x^α as having been performed). The quantity G is a function W or U. All these quantities are therefore expressed by means of the functions X, Ω, W and U, occurring in the first members of the integral equations considered, and involve the partial derivatives of the $A^{\lambda\mu}$ and f_s with respect to all their arguments, up to the fifth order, and the partial derivatives of the Cauchy data φ_s and ψ_s up to the orders six and five (in the quantity I and by means of W_0).

In order to solve the Cauchy problem for the non-linear equations F we might try to solve, independently of these equations, the system of integral equations verified by the solutions (and to prove afterwards that this solution is indeed a solution of the assigned Cauchy problem). Unfortunately, some difficulties arise for this solution: we have seen in Chap. 19 that the quantities occurring under the integral sign (in particular H) are continuous and bounded, upon assuming differentiability of the coefficients $A^{\lambda\mu}$, viewed as given functions of the variables x^α; these conditions not being realized when the functions $W_s, W_{s\alpha}, \ldots, U_s$ are independent, the quantity $[A^{ij}]\frac{\partial^2 \sigma}{\partial x^i \partial x^j} J_{x\lambda}$ will then fail to be bounded and continuous.

In order to solve the Cauchy problem we shall then pass through the intermediate stage of approximate equations F_1, where the coefficients $A^{\lambda\mu}$ will be some given functions of the x^α, obtained by replacing W_s with a given function $W_s^{(1)}$. The quantities occurring under the integration signs of the integral equations verified by the solutions will then be continuous and bounded if the same holds for the functions $W_s...U_S$ considered as independent. We will then be in a position to solve the integral equations and show that their solution $W_s...U_S$ is solution of the equations F_1, and that $W_{s\alpha}...U_S$ are the partial derivatives of W_s; but we need for that purpose, in the general case, to take as function $W_s^{(1)}$ a function six times differentiable (because the integral equations involve fifth derivatives of the $A^{\lambda\mu}$); the obtained solution W_s being merely five times differentiable, it will be impossible for us to iterate the procedure. The method described will be therefore applicable only if the $A^{\lambda\mu}$ depend uniquely on the W_s and not on the $W_{s\alpha}$: it will then be enough to assume that the approximation function is five times differentiable.

In the general case, where $A^{\lambda\mu}$ is function of W_s and $W_{s\alpha}$, one can solve the Cauchy problem by passing through the intermediate step of approximate equations, not of the equations (F) themselves, but of equations previously differentiated with respect to the x^α and viewed as integro-differential equations in the unknown functions $W_{s\alpha}$.

20.6 Solution of the Cauchy Problem for Non-linear Equations

At this stage, we refer the reader to chapter III of the work in Fourès-Bruhat [64], and we simply describe what she proved therein. She considered a system of non-linear, second-order, hyperbolic partial differential equations with n unknown functions W_s and four variables x^α, of the form

$$E_s = \sum_{\lambda,\mu=1}^{4} A^{\lambda\mu} \frac{\partial^2 W_s}{\partial x^\lambda \partial x^\mu} + f_s = 0, \ s = 1, 2..., n. \tag{E}$$

The f_s are given functions of the unknown W_s, of their first partial derivatives $W_{s\alpha}$ and of the variables x^α. The $A^{\lambda\mu}$ are given functions of the W_s and of the x^α. The Cauchy data are, on the initial surface $x^4 = 0$,

$$W_s(x^i, 0) = \varphi_s(x^i), \ W_{s4}(x^i, 0) = \psi_s(x^i).$$

On the system (E) and the Cauchy data she made the following assumptions:

(1°) In the domain (d), defined by $|x^i - \overline{x}^i| \le d$, φ_s and ψ_s possess partial derivatives up to the orders five and four, continuous, bounded and satisfying Lipschitz conditions.

(2°) For the values of the W_s satisfying

$$|W_s - \varphi_s| \le l, \ |W_{si} - \varphi_{si}| \le l, \ |W_{s4} - \psi_s| \le l$$

and in the domain D defined by

$$|x^i - \overline{x}^i| \leq d, \ |x^4| \leq \varepsilon :$$

(a) $A^{\lambda\mu}$ and f_s possess partial derivatives up to the fourth order, continuous, bounded and satisfing Lipschitz conditions.

(b) The quadratic form $\sum_{\lambda,\mu=1}^{4} A^{\lambda\mu} X_\lambda X_\mu$ is of the normal hyperbolic type: $A^{44} > 0$, $\sum_{i,j=1}^{3} A^{ij}\xi_i\xi_j$ is negative-definite.

She then proved that the Cauchy problem (φ_s, ψ_s) admits a unique solution, possessing partial derivatives continuous and bounded up to the fourth order, in relations with equations (E) in a domain (truncated cone with base d):

$$|x^i - \overline{x}^i| \leq d, \ |x^4| \leq \eta(x^i).$$

For a modern review of the Cauchy problem in general relativity, we refer the reader to the work in Ringström [123].

Chapter 21
Cauchy Problem for General Relativity

21.1 The Equations of Einstein's Gravity

The metric components $g_{\alpha\beta}$ (or potentials) of an Einstein universe satisfy, in the domains without matter and in absence of electromagnetic field, the ten partial differential equations of second order of the exterior case (see (4.4.13))

$$R_{\alpha\beta} \equiv \sum_{\lambda=1}^{4}\left[\partial_\lambda \Gamma\{\lambda,[\alpha,\beta]\} - \partial_\alpha\Gamma\{\lambda,[\lambda,\beta]\}\right] + \sum_{\lambda,\mu=1}^{4}\left[\Gamma\{\lambda,[\lambda,\mu]\}\,\Gamma\{\mu,[\alpha,\beta]\}\right.$$

$$\left. -\Gamma\{\mu,[\lambda,\alpha]\}\,\Gamma\{\lambda,[\mu,\beta]\}\right] = 0,$$

where one has set ∂_λ for $\frac{\partial}{\partial x^\lambda}$ and where the x^λ are a system of four space-time coordinates whatsoever.

The ten equations are not independent because the $R_{\alpha\beta}$ satisfy the four conservation conditions (Bianchi identities)

$$\sum_{\lambda=1}^{4}\nabla_\lambda S^{\lambda\mu} \equiv 0 \text{ where } S^{\lambda\mu} \equiv R^{\lambda\mu} - \frac{1}{2}(g^{-1})^{\lambda\mu}R.$$

The problem of determinism, in the theory of relativistic gravitation, is formulated, for an exterior space-time in the form of the Cauchy problem relative to the system of partial differential equations $R_{\alpha\beta} = 0$ and with initial data (potentials and first derivatives) carried by any hypersurface S.

The study of the values on S of the consecutive partial derivatives of the potentials has shown that, if S is nowhere tangent to a characteristic manifold, and if the Cauchy data satisfy four given conditions, the Cauchy problem admits, with respect to the system of equations $R_{\alpha\beta} = 0$, in the analytic case, a solution. This solution is

© Springer International Publishing AG 2017
G. Esposito, *From Ordinary to Partial Differential Equations*, UNITEXT 106,
DOI 10.1007/978-3-319-57544-5_21

unique, i.e., if there exist two solutions, they coincide up to a change of coordinates (conserving S pointwise and the values on S of the Cauchy data).

If S is defined by the equation $x^4 = 0$, the four conditions that the initial data must verify are the four equations $S^4_\lambda = 0$ which are expressed in terms of the data only.

21.2   Vacuum Einstein Equations and Isothermal Coordinates

The coordinate x^λ is said to be isothermal if the potentials satisfy the following first-order partial differential equation:

$$F^\lambda \equiv \frac{1}{\sqrt{-g}} \sum_{\mu=1}^{4} \frac{\partial(\sqrt{-g}\,(g^{-1})^{\lambda\mu})}{\partial x^\mu} = 0.$$

The vacuum Einstein equations read as, in whatever coordinates,

$$R_{\alpha\beta} \equiv -G_{\alpha\beta} - L_{\alpha\beta} = 0$$

with

$$G_{\alpha\beta} \equiv \frac{1}{2} \sum_{\lambda,\mu=1}^{4} (g^{-1})^{\lambda\mu} \frac{\partial^2 g_{\alpha\beta}}{\partial x^\lambda \partial x^\mu} + H_{\alpha\beta},$$

where $H_{\alpha\beta}$ is a polynomial of the $g_{\lambda\mu}$, $g^{\lambda\mu}$ and of their first derivatives and

$$L_{\alpha\beta} \equiv \frac{1}{2} \sum_{\mu=1}^{4} \left[g_{\beta\mu} \partial_\alpha F^\mu + g_{\alpha\mu} \partial_\beta F^\mu \right]. \qquad (21.2.1)$$

We see that, if the four coordinates are isothermal, every equation $R_{\alpha\beta} = 0$ does not contain second derivatives besides those of $g_{\alpha\beta}$. The system of Einstein equations takes then the form of the systems studied in Chap. 20.

We can, without restricting the generality of the hypersurface S, assume that the initial data satisfy, besides the four conditions $S^4_\lambda = 0$, the conditions of isothermy:

$$F^\mu \equiv \frac{1}{\sqrt{-g}} \sum_{\lambda=1}^{4} \frac{\partial(\sqrt{-g}\,(g^{-1})^{\lambda\mu})}{\partial x^\lambda} = 0 \text{ for } x^4 = 0. \qquad (21.2.2)$$

Once a space-time and a hypersurface $S(x^4 = 0)$ are given, there always exists a coordinate change $\check{x}^\lambda = f(x^\mu)$, with $\check{x}^4 = 0$ for $x^4 = 0$, such that the potentials

$\check{g}_{\alpha\beta}$ verify the conditions (21.2.2). We shall solve this Cauchy problem for the equations $G_{\alpha\beta} = 0$, verified by the potentials in isothermal coordinates, and we shall prove afterwards that the potentials obtained define indeed a space-time, related to isothermal coordinates, and verify the equations of gravitation $R_{\alpha\beta} = 0$.

21.3   Solution of the Cauchy Problem for the Equations $G_{\alpha\beta} = 0$

We shall apply to the system

$$G_{\alpha\beta} \equiv \sum_{\lambda,\mu=1}^{4} (g^{-1})^{\lambda\mu} \frac{\partial^2 g_{\alpha\beta}}{\partial x^\lambda \partial x^\mu} + H_{\alpha\beta} = 0$$

the results of Chap. 20 by setting $(g^{-1})^{\lambda\mu} = A^{\lambda\mu}$, $H_{\alpha\beta} = f_s$, $g_{\alpha\beta} = W_s$. Let us make on the Cauchy data the following assumptions:

Assumptions

In a domain (d) of the initial surface S, $x^4 = 0$, defined by

$$|x^i - \overline{x}^i| \leq d :$$

(1°) The Cauchy data φ_s and ψ_s possess partial derivatives continuous and bounded up to the orders five and four, respectively.

(2°) In the domain (d) and for these Cauchy data the quadratic form

$$\sum_{\lambda,\mu=1}^{4} (g^{-1})^{\lambda\mu} X_\lambda X_\mu$$

is of normal hyperbolic type: $(g^{-1})^{44} > 0$, $\sum_{i,j=1}^{3} (g^{-1})^{ij} X_i X_j$ is negative-definite (let us remark in particular that the determinant g of the $g_{\lambda\mu}$ is $\neq 0$).

We deduce from these assumptions the existence of a number l such that for $|g_{\alpha\beta} - \overline{\varphi}_s| \leq l$ one has $g \neq 0$ and we see that, for some unknown functions $g_{\alpha\beta} = W_s$, the inequalities

$$|W_s - \overline{\varphi}_s| \leq l, \quad \left| \frac{\partial W_s}{\partial x^i} - \frac{\partial \overline{\varphi}_s}{\partial x^i} \right| \leq l, \quad \left| \frac{\partial W_s}{\partial x^4} - \overline{\psi}_s \right| \leq l \tag{21.3.1}$$

are satisfied. The coefficients of the equations $G_{\alpha\beta} = 0$ (which are here independent of the variables x^α) satisfy, as the Cauchy data, the assumptions of Chap. 20, i.e.:

(1°) The coefficients $A^{\lambda\mu} = (g^{-1})^{\lambda\mu}$ and $f_s = H_{\alpha\beta}$ admit partial derivatives with respect to all their arguments up to the fourth order continuous and bounded and satisfying Lipschitz conditions ($(g^{-1})^{\lambda\mu}$ and $H_{\alpha\beta}$ are rational fractions with denominator g of the $g_{\lambda\mu} = W_s$, and of the $g_{\lambda\mu} = W_s$ and $\frac{\partial W_s}{\partial x^\alpha}$, respectively).

(2°) The quadratic form $\sum_{\lambda,\mu=1}^{4} A^{\lambda\mu} X_\lambda X_\mu$ is of normal hyperbolic type: $A^{44} > 0$, $\sum_{i,j=1}^{3} A^{ij} X_i X_j$ is negative-definite.

We can thus apply to the system $G_{\alpha\beta} = 0$, for the Cauchy problem here considered, the conclusion of Chap. 20, which is stated as follows.

There exists a number $\varepsilon(x^i) \neq 0$ such that, in the domain

$$|x^i - \overline{x}^i| < d, \ |x^4| \leq \varepsilon(x^i)$$

the Cauchy problem relative to the equations $G_{\alpha\beta} = 0$ admits a solution which has partial derivatives continuous and bounded up to the fourth order and which verifies the inequalities (21.3.1).

21.4   The Solution of $G_{\alpha\beta} = 0$ Verifies the Conditions of Isothermy

(1°) *The solution found of the system $G_{\alpha\beta} = 0$ verifies the four equations*

$$\partial_4 F^\mu = 0 \text{ for } x^4 = 0.$$

We have assumed indeed that the initial data satisfy the conditions

$$S_\lambda^4 = 0 \tag{21.4.1}$$

and

$$F^\mu = 0 \tag{21.4.2}$$

for $x^4 = 0$. Hence we have

$$S_\lambda^4 \equiv -\sum_{\mu=1}^{4} (g^{-1})^{4\mu} \left\{ G_{\lambda\mu} - \frac{1}{2} g_{\lambda\mu} \sum_{\alpha,\beta=1}^{4} (g^{-1})^{\alpha\beta} G_{\alpha\beta} + L_{\lambda\mu} - \frac{1}{2} g_{\lambda\mu} \sum_{\alpha,\beta=1}^{4} (g^{-1})^{\alpha\beta} L_{\alpha\beta} \right\}.$$

The solution of the system $G_{\alpha\beta} = 0$ verifies therefore, taking into account the expression (21.2.1) of $L_{\alpha\beta}$, the equations

$$-\frac{1}{2} \sum_{\alpha,\mu=1}^{4} (g^{-1})^{4\mu} g_{\lambda\alpha} \partial_\mu F^\alpha - \frac{1}{2} \partial_\lambda F^4 + \frac{1}{2} \sum_{\alpha=1}^{4} \delta_\lambda^4 \partial_\alpha F^\alpha = 0 \text{ for } x^4 = 0,$$

from which, by virtue of (21.4.2) ($F^\mu = 0$ and $\partial_\lambda F^\mu = 0$),

$$-\frac{1}{2}(g^{-1})^{44} \sum_{\alpha=1}^{4} g_{\lambda\alpha}\partial_4 F^\alpha = 0 \text{ for } x^4 = 0.$$

We see eventually that the solution found of the system $G_{\alpha\beta} = 0$ verifies the four equations

$$\partial_4 F^\mu = 0 \text{ for } x^4 = 0.$$

(2°) *The solution found of $G^{\alpha\beta} = 0$ verifies $F^\mu = 0$.*
This property is going to result from the conservation conditions. The metric components $g_{\alpha\beta}$ satisfy indeed the four Bianchi identities

$$\sum_{\lambda=1}^{4} \nabla_\lambda \left(R^{\lambda\mu} - \frac{1}{2}(g^{-1})^{\lambda\mu} R \right) = 0,$$

where $R^{\lambda\mu}$ is the Ricci tensor corresponding to this metric. A solution of the system $G_{\alpha\beta} = 0$ verifies therefore the four equations

$$\sum_{\lambda=1}^{4} \nabla_\lambda \left(L^{\lambda\mu} - \frac{1}{2}(g^{-1})^{\lambda\mu} L \right) = 0,$$

where $L^{\lambda\mu} = \sum_{\alpha,\beta=1}^{4}(g^{-1})^{\alpha\lambda}(g^{-1})^{\beta\mu} L_{\alpha\beta}$ and $L = \sum_{\alpha,\beta=1}^{4}(g^{-1})^{\alpha\beta} L_{\alpha\beta}$.
It turns out from the expression (21.2.1) of $L_{\alpha\beta}$ that these equations read as

$$\frac{1}{2} \sum_{\alpha,\lambda=1}^{4} (g^{-1})^{\alpha\lambda} \nabla_\lambda(\partial_\alpha F^\mu) + \frac{1}{2} \sum_{\beta,\lambda=1}^{4} (g^{-1})^{\beta\mu} \nabla_\lambda(\partial_\beta F^\lambda)$$

$$-\frac{1}{2} \sum_{\alpha,\lambda=1}^{4} (g^{-1})^{\lambda\mu} \nabla_\lambda(\partial_\alpha F^\alpha) = 0,$$

from which, by developing and simplifying,

$$\frac{1}{2} \sum_{\alpha,\lambda=1}^{4} (g^{-1})^{\alpha\lambda} \frac{\partial^2 F^\mu}{\partial x^\alpha \partial x^\lambda} + P^\mu(\partial_\alpha F^\lambda) = 0,$$

where P is a linear combination of the $\partial_\alpha F^\lambda$ whose coefficients are polynomials of the $(g^{-1})^{\alpha\beta}$, $g_{\alpha\beta}$ and of their first derivatives.

We notice therefore that the four quantities F^μ (formed with the $g_{\alpha\beta}$ solutions of $G_{\alpha\beta} = 0$) verify four partial differential equations of the type previously studied. The coefficients $A^{\lambda\mu} = (g^{-1})^{\lambda\mu}$ and $f_s = P_\mu$ verify, in D, the assumptions of Chap. 20. The quantities F^μ are by hypothesis vanishing on the domain (d) of $x^4 = 0$, and we have proved that the same was true of their first derivatives $\partial_\alpha F^\mu$. We deduce then from the uniqueness theorem that, in D, we have

$$F^\mu = 0 \text{ and } \partial_\alpha F^\mu = 0.$$

The metric components, solutions of the Cauchy problem formulated with respect to the system $G_{\alpha\beta} = 0$, verify therefore effectively in (D) the conditions of isothermy and represent the potentials of an Einstein space-time, solutions of the vacuum Einstein equations $R_{\alpha\beta} = 0$.

21.5   Uniqueness of the Solution

In order to prove that there exists only one exterior space-time corresponding to the initial conditions given on S, one has to prove that every solution of the Cauchy problem formulated in such a way with respect to the equations $R_{\alpha\beta} = 0$ can be deduced by a change of coordinates from the solution of this Cauchy problem relative to the equations $G_{\alpha\beta} = 0$. We know that this last solution is unique.

Let us therefore consider a solution $g_{\alpha\beta}$ of the Cauchy problem relative to the equations $R_{\alpha\beta} = 0$ and look for a transformation of coordinates

$$\check{x}^\alpha = f^\alpha(x^\beta).$$

By conserving S pointwise and in such a way that the potentials in the new system of coordinates, let them be $\check{g}_{\alpha\beta}$, verify the four equations

$$\check{F}^\lambda = 0,$$

we know that the four quantities $\check{F}^\lambda$ are invariants which verify the identities

$$\check{F}^\lambda \equiv \check{\triangle}_2 \check{x}^\lambda = \triangle_2 f^\lambda.$$

In order for the equations $\check{F}^\lambda = 0$ to be verified it is therefore necessary and sufficient that the functions f^α satisfy the equations

$$\triangle_2 f^\alpha \equiv \sum_{\lambda,\mu=1}^{4} (g^{-1})^{\lambda\mu} \left(\frac{\partial^2 f^\alpha}{\partial x^\lambda \partial x^\mu} - \sum_{\rho=1}^{4} \Gamma\left\{\rho, [\lambda, \mu]\right\} \frac{\partial f^\alpha}{\partial x^\rho} \right) = 0 \qquad (21.5.1)$$

which are partial differential equations of second order, linear, normal hyperbolic in the domain (D).

If we take for values of the functions f^α and of their first derivatives, upon S, the following values (for them the change of coordinates conserves S pointwise)

$$
\begin{aligned}
f^4 &= 0, \ \partial_\alpha f^4 = \delta^4_\alpha, \\
f^i &= x^i, \ \partial_\alpha f^i = \delta^i_\alpha,
\end{aligned}
\tag{21.5.2}
$$

for $x^4 = 0$, we see that the Cauchy problems formulated in such a way admit in (D) solutions possessing their partial derivatives up to the fourth order continuous and bounded.

We have thus defined a change of coordinates $\check{x}^\lambda = f^\lambda(x^\alpha)$ such that, in the new system of coordinates, the potentials $\check{g}_{\alpha\beta}$ verify the conditions of isothermy $\check{F}^\lambda = 0$. It remains to prove that *this change of coordinates determines in a unique way the Cauchy data $\check{g}_{\alpha\beta}(x^4 = 0)$ and $\check{\partial}_4 \check{g}_{\alpha\beta}(x^4 = 0)$, in terms of the original data $g_{\alpha\beta}(x^4 = 0)$ and $\partial_4 g_{\alpha\beta}(x^4 = 0)$.*

We know that, $g_{\alpha\beta}$ being the components of a covariant rank-two tensor

$$
g_{\alpha\beta} = \sum_{\lambda,\mu=1}^{4} \check{g}_{\lambda\mu}\left(\partial_\alpha f^\lambda\right)\left(\partial_\beta f^\mu\right),
\tag{21.5.3}
$$

from which, in light of (21.5.2),

$$
g_{\alpha\beta} = \check{g}_{\alpha\beta} \qquad \partial_i g_{\alpha\beta} = \check{\partial}_i \check{g}_{\alpha\beta} \text{ for } x^4 = \check{x}^4 = 0.
$$

It remains to evaluate the derivatives of the potentials with respect to x^4 and $\check{x}^4$ for $x^4 = \check{x}^4 = 0$. Since φ is an arbitrary function of a space-time point we have

$$
\partial_4 \varphi = \sum_{\lambda=1}^{4} \left(\check{\partial}_\lambda \varphi\right)\left(\partial_4 f^\lambda\right),
$$

from which

$$
\partial_4 \varphi = \check{\partial}_4 \varphi \text{ for } x^4 = \check{x}^4 = 0.
\tag{21.5.4}
$$

We find, on the other hand, by deriving the equality (21.5.3) with respect to x^4

$$
\partial_4 g_{\alpha\beta} = \sum_{\lambda,\mu=1}^{4} \left[(\partial_4 \check{g}_{\lambda\mu})(\partial_\alpha f^\lambda)(\partial_\beta f^\mu) + \check{g}_{\lambda\mu}\left((\partial^2_{\alpha4} f^\lambda)(\partial_\beta f^\mu) + (\partial^2_{\beta4} f^\mu)(\partial_\alpha f^\lambda)\right)\right],
$$

from which

$$\partial_4 g_{\alpha\beta} = \partial_4 \breve{g}_{\alpha\beta} + \sum_{\lambda=1}^{4} (\breve{g}_{\lambda\beta}) \left(\partial_{\alpha 4}^2 f^\lambda \right) + \sum_{\mu=1}^{4} (\breve{g}_{\mu\alpha}) \left(\partial_{\beta 4}^2 f^\mu \right) \text{ for } x^4 = 0. \quad (21.5.5)$$

We deduce also from the initial values (21.5.2) of the f^λ:

$$\partial_{\alpha i}^2 f^\lambda = 0 \text{ for } x^4 = 0.$$

The f^λ verify on the other hand the conditions of isothermy (21.5.1), from which

$$(g^{-1})^{44} \partial_{44}^2 f^\lambda = \sum_{\alpha,\beta=1}^{4} (g^{-1})^{\alpha\beta} \Gamma \left\{ \lambda, [\alpha, \beta] \right\} \text{ for } x^4 = 0.$$

$\partial_{44}^2 f^\lambda$ is hence determined in a unique way by the original Cauchy data; this is also equally true of $\partial_4 \breve{g}_{\alpha\beta}$ for $x^4 = 0$.

We have thus led the reader to the following result of [64]:

Theorem 21.1 *Once a solution $g_{\alpha\beta}$ of the Cauchy problem is given in relation to the equations $R_{\alpha\beta} = 0$ (the initial data satisfying upon S the differentiability assumptions previously stated) there exists a change of coordinates, conserving S pointwise, such that the potentials $\breve{g}_{\alpha\beta}$ in the new system of coordinates verify everywhere the conditions of isothermy and represent the solution, unique, of a Cauchy problem, determined in a unique way, relative to the equations $G_{\alpha\beta} = 0$.*

We conclude therefore, for gravitational physics:

Theorem 21.2 *There exists one and only one exterior space-time corresponding to the initial conditions assigned upon S [64].*

Chapter 22
Causal Structure and Global Hyperbolicity

22.1 Causal Structure of Space-Time

Let (M, g) be a space-time, and let $p \in M$. The *chronological future* of p is defined as

$$I^+(p) \equiv \{q \in M : p << q\}, \tag{22.1.1}$$

i.e. $I^+(p)$ is the set of all points q of M such that there is a future-directed timelike curve from p to q. Similarly, one defines the *chronological past* of p

$$I^-(p) \equiv \{q \in M : q << p\}. \tag{22.1.2}$$

The *causal future* of p is then defined by

$$J^+(p) \equiv \{q \in M : p \leq q\}, \tag{22.1.3}$$

and similarly for the *causal past*

$$J^-(p) \equiv \{q \in M : q \leq p\}, \tag{22.1.4}$$

where $a \leq b$ means that there exists a future-directed non-spacelike curve from a to b. The causal structure of (M, g) is the collection of past and future sets at all points of M together with their properties. Following Hawking and Ellis [76], we shall here provide the following basic definitions:

Definition 22.1 A set Σ is *achronal* if no two points of Σ can be joined by a timelike curve.

Definition 22.2 A point p is an *endpoint* of the curve λ if λ enters and remains in any neighbourhood of p.

© Springer International Publishing AG 2017
G. Esposito, *From Ordinary to Partial Differential Equations*, UNITEXT 106,
DOI 10.1007/978-3-319-57544-5_22

Definition 22.3 Let Σ be a spacelike or null achronal three-surface in M. The *future Cauchy development* (or future domain of dependence) $D^+(\Sigma)$ of Σ is the set of points $p \in M$ such that every past-directed timelike curve from p without past endpoint intersects Σ.

Definition 22.4 The *past Cauchy development* $D^-(\Sigma)$ of Σ is defined by interchanging future and past in definition D9.III. The total Cauchy development of Σ is then given by $D(\Sigma) = D^+(\Sigma) \cup D^-(\Sigma)$.

Definition 22.5 The *future Cauchy horizon* $H^+(\Sigma)$ of Σ is given by

$$H^+(\Sigma) \equiv \left\{ X : X \in D^+(\Sigma), I^+(X) \cap D^+(\Sigma) = \phi \right\}. \tag{22.1.5}$$

Similarly, the *past Cauchy horizon* $H^-(\Sigma)$ is defined as

$$H^-(\Sigma) \equiv \left\{ X : X \in D^-(\Sigma), I^-(X) \cap D^-(\Sigma) = \phi \right\}. \tag{22.1.6}$$

Definition 22.6 The *edge* of an achronal set Σ is given by all points $p \in \overline{\Sigma}$ such that any neighbourhood U of p contains a timelike curve from $I^-(p, U)$ to $I^+(p, U)$ that does not meet Σ.

Our definitions of Cauchy developments differ indeed from the ones of Hawking and Ellis, because they look at past-inextendible curves which are timelike or null, whereas we agree with [69, 108] in not including null curves in the definition. We are now going to discuss three fundamental causality conditions: strong causality, stable causality and global hyperbolicity.

22.2 Strong Causality

The underlying idea for the definition of strong causality is that there should be no point p such that every small neighbourhood of p intersects some timelike curve more than once [75]. Namely, the space-time (M, g) does not *almost contain* closed timelike curves. In rigorous terms, strong causality is defined as follows:

Definition 22.7 *Strong causality* holds at $p \in M$ if arbitrarily small neighbourhoods of p exist which each intersect no timelike curve in a disconnected set.

A very important characterization of strong causality can be given by defining at first the Alexandrov topology.

Definition 22.8 In the *Alexandrov topology*, a set is open if and only if it is the union of one or more sets of the form [76] $I^+(p) \cap I^-(q)$, $p, q \in M$.

Thus any open set in the Alexandrov topology will be open in the manifold topology. Now, the following fundamental result holds [108]:

Theorem 22.1 *The following three requirements on a space-time* (M, g) *are equivalent:*

(1) (M, g) *is strongly causal;*
(2) *the Alexandrov topology agrees with the manifold topology;*
(3) *the Alexandrov topology is Hausdorff.*

22.3 Stable Causality

Strong causality is not enough to ensure that space-time is not just about to violate causality. The situation can be considerably improved if stable causality holds. For us to be able to properly define this concept, we must discuss the problem of putting a topology on the space of all Lorentz metrics on a four-manifold M. Essentially three possible topologies seem to be of major interest [75]: compact-open topology, open topology, fine topology.

(i) Compact-Open Topology

$\forall n = 0, 1, \ldots, r$, let ϵ_n be a set of continuous positive functions on M, U be a compact set $\subset M$ and g the Lorentz metric under study. We then define: $G(U, \epsilon_n, g) =$ set of all Lorentz metrics $\widetilde{g}$ such that

$$|g - \widetilde{g}|_n < \epsilon_n \text{ on } U \ \forall n, \tag{22.3.1}$$

where [68]

$$|g - \widetilde{g}|_n \equiv \sqrt{\sum_{a_i, b_j, r, s, u, v} \left[\nabla_{a_1} \ldots \nabla_{a_n} (g_{rs} - \widetilde{g}_{rs}) \right] \left[\nabla_{b_1} \ldots \nabla_{b_n} (g_{uv} - \widetilde{g}_{uv}) \right] h^{a_1 b_1} \ldots h^{sv}}, \tag{22.3.2}$$

∇_a being the covariant derivative operator on M and

$$\sum_{a,b=1}^{4} h_{ab} \mathrm{d}x^a \otimes \mathrm{d}x^b$$

any positive-definite metric on M.

In the compact-open topology, open sets are obtained from the $G(U, \epsilon_i, g)$ through the operations of arbitrary union and finite intersection.

(ii) Open Topology

We no longer require U to be compact, and we take $U = M$ in section (i).

(iii) Fine Topology

We define $H(U, \epsilon_i, g) =$ set of all Lorentz metrics $\widetilde{g}$ such that

$$|g - \tilde{g}|_i < \epsilon_i, \tag{22.3.3}$$

and $\tilde{g} = g$ out of the compact set U. Moreover, we set $G'(\epsilon_i, g) = \cup H(U, \epsilon_i, g)$. A sub-basis for the fine topology is then given by the neighbourhoods $G'(\epsilon_i, g)$ [75].

Now, the underlying idea for stable causality is that space-time must not contain closed timelike curves, and we still fail to find closed timelike curves if we open out the null cones. In view of the former definitions, this idea can be formulated as follows:

Definition 22.9 A metric g satisfies the *stable causality* condition if, in the C^0 open topology, an open neighbourhood of g exists no metric of which has closed timelike curves.

The Minkowski, Friedmann-Robertson-Walker, Schwarzschild and Reissner-Nordstrom space-times are all stably causal. If stable causality holds, the differentiable and conformal structure can be determined from the causal structure, and space-time cannot be compact (because in a compact space-time there exist closed timelike curves). A very important characterization of stable causality is given by the following theorem [76]:

Theorem 22.2 *A space-time (M, g) is stably causal if and only if a cosmic-time function exists on M, i.e. a function whose gradient is everywhere timelike.*

22.4   Global Hyperbolicity

Global hyperbolicity plays a key role in developing a rigorous theory of geodesics in Lorentzian geometry and in proving singularity theorems. Its ultimate meaning can be seen as requiring the existence of Cauchy surfaces, i.e. spacelike hypersurfaces which each non-spacelike curve intersects exactly once. In fact some authors [138] take this property as the starting point in discussing global hyperbolicity. Indeed, Leray's original idea was that the set of non-spacelike curves from p to q must be compact in a suitable topology [89]. We shall here follow [69, 76], defining and proving in part what follows.

Definition 22.10 A space-time (M, g) is *globally hyperbolic* if

(a) strong causality holds;
(b) $J^+(p) \cap J^-(q)$ is compact $\forall p, q \in M$.

Theorem 22.3 *In a globally hyperbolic space-time, the following properties hold:*

(1) $J^+(p)$ and $J^-(p)$ are closed $\forall p$;
(2) $\forall p, q$, the space $C(p, q)$ of all non-spacelike curves from p to q is compact in a suitable topology;
(3) there exist Cauchy surfaces.

Proof of (1) It is well-known that, if (X, F) is a Hausdorff space and $A \subset X$ is compact, then A is closed. In our case, this implies that $J^+(p) \cap J^-(q)$ is closed. Moreover, it is not difficult to see that $J^+(p)$ itself must be closed. In fact, otherwise we could find a point $r \in \overline{J^+(p)}$ such that $r \notin J^+(p)$. Let us now choose $q \in I^+(r)$. We would then have $r \in \overline{J^+(p) \cap J^-(q)}$ but $r \notin J^+(p) \cap J^-(q)$, which implies that $J^+(p) \cap J^-(q)$ is not closed, contradicting what we found before. Similarly we also prove that $J^-(p)$ is closed.

Remark a stronger result can also be proved, i.e., if (M, g) is globally hyperbolic and $K \subset M$ is compact, then $J^+(K)$ is closed.

Proof of (3) The proof will use the following ideas:

Step 1. We define a function f^+, and we prove that global hyperbolicity implies continuity of f^+ on M [76].

Step 2. We consider the function

$$f : p \in M \to f(p) \equiv \frac{f^-(p)}{f^+(p)}, \tag{22.4.1}$$

and we prove that the $f = constant$ surfaces are Cauchy surfaces.

Step 1. The function f^+ we are looking for is given by $f^+ : p \in M \to$ volume of $J^+(p, M)$. This can only be done with a suitable choice of measure. The measure is chosen in such a way that the total volume of M is equal to 1. For f^+ to be continuous on M, it is sufficient to show that f^+ is continuous on any non-spacelike curve γ. In fact, let $r \in \gamma$, and let $\{x_n\}$ be a sequence of points on γ in the past of r. We now define

$$T \equiv \cap J^+(x_n, M). \tag{22.4.2}$$

If f^+ were not upper semi-continuous on γ in r, there would be a point $q \in T - J^+(r, M)$, with $r \notin J^-(q, M)$. But on the other hand, the fact that $x_n \in J^-(q, M)$ implies that $r \in J^-(q, M)$, which is impossible in view of global hyperbolicity. The absurd proves that f^+ is upper semi-continuous. In the same way (by exchanging the role of past and future) we can prove lower semi-continuity, and thus continuity. It becomes then trivial to prove the continuity of the function $f^+ : p \in M \to$ volume of $I^+(p, M)$. From now on, we shall mean by f^+ the volume function of $I^+(p, M)$.

Step 2. Let Σ be the set of points where $f = 1$, and let $p \in M$ be such that $f(p) > 1$. The idea is to prove that every past-directed timelike curve from p intersects Σ, so that $p \in D^+(\Sigma)$. In a similar way, if $f(p) < 1$, one can then prove that $p \in D^-(\Sigma)$ (which finally implies that Σ is indeed a Cauchy surface). The former result can be proved as follows [69].

Step 2a. We consider any past-directed timelike curve μ without past endpoint from p. In view of the continuity of f proved in step 1, such a curve μ must intersect Σ, provided one can show that there exists $\epsilon \to 0^+$: $f_{on\,\mu} = \epsilon$, where ϵ is arbitrary.

Step 2b. Given $q \in M$, we denote by U a subset of M such that $U \subset I^+(q)$. The subsets U of this form cover M. Moreover, any U cannot be in $I^-(r) \, \forall r \in \mu$. This is forbidden by global hyperbolicity. In fact, suppose for absurd that $q \in \cap_{r \in \mu} I^-(r)$. We then choose a sequence $\{t_i\}$ of points on μ such that

$$t_{i+1} \in I^-(t_i), \quad \exists i : z \in I^-(t_i) \, \forall z \in \mu.$$

For all i, we also consider a timelike curve μ' such that

(1) μ' starts at p;
(2) $\mu' = \mu$ to t_i;
(3) μ' continues to q.

Global hyperbolicity plays a role in ensuring that the sequence $\{t_i\}$ has a limit curve Ω, which by construction contains μ. On the other hand, we know this is impossible. In fact, if μ were contained in a causal curve from p to q, it should have a past endpoint, which is not in agreement with the hypothesis. Thus, having proved that $\exists r \in \mu : U \not\subset I^-(r)$, we find that $f^-(r) \to 0$ when r continues into the past on μ, which in turn implies that μ intersects Σ as we said in step 2a [69].

The proof of (2) is not given here, and can be found in Hawking and Ellis [76]. Global hyperbolicity plays a key role in proving singularity theorems because, if p and q lie in a globally hyperbolic set and $q \in J^+(p)$, there exists a non-spacelike geodesic from p to q whose length is greater than or equal to that of any other non-spacelike curve from p to q. The proof that arbitrary, sufficiently small variations in the metric do not destroy global hyperbolicity can be found in Geroch [69], and has been improved by Benavides Navarro and Minguzzi [8]. Globally hyperbolic spacetimes are also peculiar in that for them the Lorentzian distance function defined in Sect. D2.4 is finite and continuous as the Riemannian distance function. The relation between strong causality, finite distance function and global hyperbolicity is proved on p. 107 of Beem and Ehrlich [7].

Part V
Parabolic Equations

Chapter 23
The Heat Equation

23.1 A Summary on Linear Equations in Two Independent Variables

The following section is devoted to the fundamental solution of the heat equation in one space dimension, but since so far we have not said anything about parabolic equations, we begin by summarizing the standard properties of linear partial differential equations in two independent variables. Our discussion begins therefore by studying the equation

$$\left[a\frac{\partial^2}{\partial x^2} + 2b\frac{\partial^2}{\partial x\partial y} + c\frac{\partial^2}{\partial y^2} + d\frac{\partial}{\partial x} + e\frac{\partial}{\partial y} + f \right] u(x,y) = g(x,y), \qquad (23.1.1)$$

where all coefficients a, b, c, d, e, f are assumed to be functions of x and y only. Our first aim is to reduce Eq. (23.1.1), by means of a change of variables

$$\xi = \xi(x,y), \qquad (23.1.2)$$

$$\eta = \eta(x,y), \qquad (23.1.3)$$

to one of the three canonical forms ([28, 109])

$$\left[\frac{\partial^2}{\partial \xi^2} - \frac{\partial^2}{\partial \eta^2} \right] u(\xi,\eta) + f_1(u, u_\xi, u_\eta) = 0, \qquad (23.1.4)$$

$$\frac{\partial^2}{\partial \xi^2} u(\xi,\eta) + f_2(u, u_\xi, u_\eta) = 0, \qquad (23.1.5)$$

$$\left[\frac{\partial^2}{\partial \xi^2} + \frac{\partial^2}{\partial \eta^2} \right] u(\xi,\eta) + f_3(u, u_\xi, u_\eta) = 0, \qquad (23.1.6)$$

© Springer International Publishing AG 2017
G. Esposito, *From Ordinary to Partial Differential Equations*, UNITEXT 106,
DOI 10.1007/978-3-319-57544-5_23

where, hereafter, subscripts of u denote the independent variable with respect to which the derivative is taken. Equations (23.1.4)–(23.1.6) are viewed as *canonical forms* of the original equation in that they correspond to *particularly simple choices of the coefficients of the second derivatives of u to which one can ascribe an invariant meaning* [28, 67].

The canonical forms (23.1.4)–(23.1.6) can be achieved if and only if

$$b^2 - ac > 0, \tag{23.1.7}$$

$$b^2 - ac = 0, \tag{23.1.8}$$

and

$$b^2 - ac < 0, \tag{23.1.9}$$

respectively. If such conditions are fulfilled, the original Eq. (23.1.1) is said to be of the *hyperbolic type* (e.g. the wave equation of Part IV), *parabolic type* (e.g. the heat equation) and *elliptic type* (e.g. the Laplace equation of Part II), respectively. We are now going to show that the type of Eq. (23.1.1) cannot be altered by a real-valued change of variables. For this purpose, we point out that the rules for evaluating derivatives of composite functions lead to the identity

$$au_{xx} + 2bu_{xy} + cu_{yy} = Au_{\xi\xi} + 2Bu_{\xi\eta} + Cu_{\eta\eta} + f_4(u, u_\xi, u_\eta), \tag{23.1.10}$$

where

$$A \equiv a\xi_x^2 + 2b\xi_x\xi_y + c\xi_y^2, \tag{23.1.11}$$

$$B \equiv a\xi_x\eta_x + b\xi_x\eta_y + b\xi_y\eta_x + c\xi_y\eta_y, \tag{23.1.12}$$

$$C \equiv a\eta_x^2 + 2b\eta_x\eta_y + c\eta_y^2. \tag{23.1.13}$$

One thus finds

$$B^2 - AC = (b^2 - ac)\left(\xi_x\eta_y - \xi_y\eta_x\right)^2, \tag{23.1.14}$$

where $\left(\xi_x\eta_y - \xi_y\eta_x\right)$ is the Jacobian of the transformation expressed by (23.1.2) and (23.1.3). In other words, for all those changes of independent variables whose Jacobian does not vanish, the sign of the discriminant $b^2 - ac$ remains invariant and hence the classification given above is invariant as well.

23.2   Fundamental Solution of the Heat Equation

The simplest parabolic equation whose solution is not trivial is the heat equation obtained by Fourier. In one space dimension, here denoted by x, it reads as

$$\left(\frac{\partial}{\partial t} - \frac{\partial^2}{\partial x^2}\right) u = 0. \tag{23.2.1}$$

The characteristics are defined by $t =$ constant. Since Eq. (23.2.1) is linear, it admits solutions of the form

$$u = e^{ax+bt}, \tag{23.2.2}$$

where a and b are some constants. By insertion into the original equation, one finds immediately the relation $a^2 = b$. In particular, one can take $a = i\alpha$, $\alpha \in \mathbf{R}$, which shows that

$$e^{-\alpha^2 t} \cos(\alpha x), \quad e^{-\alpha^2 t} \sin(\alpha x)$$

are solutions. Hence the integral

$$u_1 = \int_{-\infty}^{\infty} e^{-\alpha^2 t + i\alpha x} \, d\alpha, \quad t > 0 \tag{23.2.3}$$

is also a solution, as one checks immediately by evaluating the derivatives of this function. In the argument of the exponential in the integrand, one can complete the square by exploiting the identity

$$-\alpha^2 t + i\alpha x = -t\left(\alpha - \frac{ix}{2t}\right)^2 - \frac{x^2}{4t}, \tag{23.2.4}$$

which yields

$$u_1 = e^{-\frac{x^2}{4t}} \int_{-\infty}^{\infty} e^{-tv^2} \, d\alpha, \quad v \equiv \alpha - \frac{ix}{2t}. \tag{23.2.5}$$

Fig. 23.1 Integration contour $ABCD$ for the fundamental solution of the heat equation

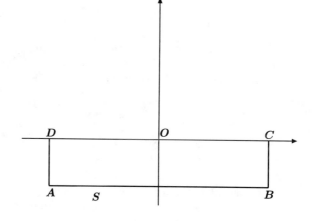

If one considers the last integral in the plane of the complex variable $\alpha = \alpha' + i\alpha''$, it is equal to

$$J = \int e^{-t\alpha^2} \, d\alpha, \qquad (23.2.6)$$

evaluated on the straight line S of equation $\mathrm{Im}(\alpha) = \alpha'' = -\frac{x}{2t}$ from $-\infty$ to $+\infty$. The integral J evaluated on the contour $ABCD$ of Fig. 23.1 is vanishing, because the integrand is a holomorphic function in this rectangle. If the vertex B moves infinitely away on the right, the integral (23.2.6), whose modulus is majorized by

$$\left| \frac{x}{2t} \right| e^{-t\lambda}, \ \lambda \equiv \overline{OC}^2 - \frac{x^2}{4t^2}$$

approaches 0. In the same fashion, the integral taken along AD approaches 0 when D moves infinitely far away towards the left. It follows that the integral (23.2.6) evaluated on S is equal to this integral evaluated on the real axis, i.e.

$$\int_{-\infty}^{\infty} e^{-t\alpha^2} \, d\alpha = \frac{1}{\sqrt{t}} \int_{-\infty}^{\infty} e^{-t^2} \, dt = \sqrt{\frac{\pi}{t}}. \qquad (23.2.7)$$

Thus, the function $\frac{u_1}{\sqrt{\pi}} = \frac{e^{-\frac{x^2}{4t}}}{\sqrt{t}}$ is a solution of the heat equation, and in the same way

$$\frac{e^{-\frac{(x-\xi)^2}{4t}}}{\sqrt{t}}, \ t > 0$$

is a solution for all values of ξ. It follows furthermore that, if f is an arbitrary continuous and bounded function of ξ, the function

$$u(x, t) = \frac{1}{2\sqrt{\pi t}} \int_{-\infty}^{\infty} f(\xi) e^{-\frac{(x-\xi)^2}{4t}} \, d\xi, \ t > 0 \qquad (23.2.8)$$

is also a solution of the heat equation. One can again check it by a direct calculation. One might clearly replace t by $t - t_0$, with $t > t_0$, and make weaker assumptions on $f(\xi)$. In particular, it is sufficient that $f(\xi)$ is integrable and satisfies the majorization [134]

$$|f(\xi)| < k' e^{k\xi^2}, \qquad (23.2.9)$$

k and k' being some given positive numbers, for the function u defined by (23.2.8) to be a solution of (23.2.1), for all x and provided that t is positive and sufficiently small. The integral formula (23.2.8) provides a solution of the following problem in the mathematical theory of heat conduction: *to find an integral of Eq. (23.2.1) which takes some continuous values on a characteristic $t = t_0$.* One can indeed assume that the characteristic is $t = 0$. On the right of $t = 0$ we assign $u = f(x)$, $f(x)$ verifying the condition (23.2.9), and we are going to show that, when the point $M(x, t)$ approaches a point $(\xi', 0)$ of the axis $t = 0$, the function $u(x, t)$ defined by (23.2.8) approaches $f(\xi')$.

Proof If $f(\xi) = 1$, Eq. (23.2.8) provides $u = 1$. One has therefore

$$u(x, t) - f(\xi') = \int_{-\infty}^{\infty} \left[f(\xi) - f(\xi') \right] K(\xi, x, t) d\xi, \tag{23.2.10}$$

where the kernel K is defined by

$$K(\xi, x, t) \equiv \frac{e^{-\frac{(x-\xi)^2}{4t}}}{2\sqrt{\pi t}}, \tag{23.2.11}$$

and satisfies

$$\int_{-\infty}^{\infty} K(\xi, x, t) d\xi = 1. \tag{23.2.12}$$

Given a positive ε and $f(\xi)$ continuous at the point ξ', one can find η such that

$$|f(\xi) - f(\xi')| < \varepsilon$$

if $|\xi - \xi'| < \eta$. One can split the integral on the right-hand side of (23.2.10) into three integrals: an integral J_1 taken from $\xi' - \eta$ to $\xi' + \eta$, another J_2 taken from $-\infty$ to $\xi' - \eta$, and the third J_3 taken from $\xi' + \eta$ to $+\infty$. One has, for all x and $t > 0$,

$$|J_1| < \varepsilon \int_{\xi'-\eta}^{\xi'+\eta} K(\xi, x, t) d\xi < \varepsilon \int_{-\infty}^{+\infty} K(\xi, x, t) d\xi = \varepsilon. \tag{23.2.13}$$

We are going to show that, if x and t are sufficiently close to ξ' and 0, respectively, the integrals J_2 and J_3 are also as small as is desired, which achieves the proof. It is enough to consider, for example, J_2. On bearing in mind (23.2.9) one finds, m being fixed

$$|f(\xi) - f(\xi')| < m e^{k\xi^2}, \tag{23.2.14}$$

hence

$$2\sqrt{\pi t} |J_2| < m \int_{\xi'+\eta}^{\infty} e^{\nu} d\xi, \quad \nu \equiv k\xi^2 - \frac{(\xi - x)^2}{4t}. \tag{23.2.15}$$

One has, in every integration interval,

$$\nu < k\xi^2 - \frac{\left(\xi - \xi' - \frac{\eta}{2}\right)^2}{4t} < -\frac{\left(\xi - \xi' - \frac{\eta}{2}\right)^2}{16t}, \tag{23.2.16}$$

provided that $|x - \xi'| < \frac{\eta}{2}$ and that t is sufficiently small, i.e. $t < t_1$. Upon replacing ν by its upper bound and setting

$$\xi \equiv \xi' + \frac{\eta}{2} + 4\sqrt{t}T, \tag{23.2.17}$$

one obtains

$$2\sqrt{\pi t}|J_3| < m \int_{T'}^{\infty} e^{-T^2} 4\sqrt{t}\, dT, \quad T' \equiv \frac{\eta}{8\sqrt{t}}, \tag{23.2.18}$$

therefore

$$|J_3| < \frac{2m}{\sqrt{\pi}} \int_{T'}^{\infty} e^{-T^2}\, dT, \tag{23.2.19}$$

and the second member is also as small as is desired provided that T' is sufficiently large, i.e. t is sufficiently small. Q.E.D.

Analyticity of the solution. The solution defined by the integral formula (23.2.8) is an analytic function of the real variables x and t, for $t > 0$. This can be seen as follows. If A is positive and finite, the integral

$$J_A(x, t) \equiv \int_{-A}^{A} f(\xi) e^{-\frac{(x-\xi)^2}{4t}}\, d\xi \tag{23.2.20}$$

defines an analytic function of x and t taken to be complex, if the real part of t is positive (but one might also assume $\mathrm{Re}(t) < 0$). In order to prove it, one can consider the exponential which can be expanded in an entire series in $(x - x_0)$ and $(t - t_0)$ about x_0 and t_0 ($\mathrm{Re}(t_0) > 0$). This series is majorized by

$$e^w, \quad w \equiv \frac{(|x - x_0| + |x_0| + |\xi|)^2}{4(t_0 - |t - t_0|)}, \quad |t - t_0| < y_0, \tag{23.2.21}$$

and, since $|\xi| \le A$, uniform convergence is achieved. Since $|f(\xi)|$ is bounded, the integral (23.2.20) leads to a uniformly convergent series, which can be integrated term by term. If one assumes $|t - t_0| < \alpha t_0$, $2\alpha^2 < 1$, and if x belongs to a bounded domain, the integral (23.2.20) converges uniformly when A approaches infinity. By taking for A the sequence of integer values, one obtains a sequence of holomorphic functions of x and t which converge uniformly. The theorem of Weierstrass for sequences of holomorphic functions can be extended to the case of functions of two variables by using the double integral of Poincaré. It follows that the function u defined by (23.2.8) is an analytic function of x and t for $t > 0$, this function being on the other hand an entire function of $(x - x_0)$, for all x_0.

Chapter 24
The Nash Theorem on Parabolic Equations

24.1 The Moment Bound

The state of knowledge on non-linear equations in the late fifties is well summarized by what Nash writes in the Introduction of his seminal paper [105]. He recalls therein that successful treatment of non-linear partial differential equations depends on a priori estimates that control the behaviour of solutions. These estimates are theorems about linear equations with variable coefficients, and they can give a certain compactness to the class of possible solutions. Such a compactness is necessary for iterative or fixed-point techniques, such as the Schauder–Leray methods. Alternatively, the a priori estimates may establish continuity or smoothness of generalized solutions. The strongest estimates provide qualitative information on the continuity of solutions without making quantitative assumptions about the continuity of coefficients.

In Nash [105], the author studied *linear parabolic* equations of the form

$$\sum_{i,j=1}^{n} \frac{\partial}{\partial x^i}\left[C^{ij}(x^1,\ldots,x^n,t)\frac{\partial T}{\partial x^j}\right] = \frac{\partial T}{\partial t}, \qquad (24.1.1)$$

i.e.

$$\mathrm{div}\left(C(x,t)\cdot\mathrm{grad}\ T\right) = T_t, \qquad (24.1.2)$$

where the C^{ij} form a symmetric real matrix $C(x,t)$ for each point x and time t. He assumed the existence of universal bounds $c_2 \geq c_1 > 0$ on the eigenvalues of C, so that any eigenvalue θ_ν satisfies

$$c_1 \leq \theta_\nu \leq c_2. \qquad (24.1.3)$$

This is said to be the *uniform ellipticity* assumption (cf. (7.4.3)). More precisely, Nash made two basic assumptions:

(i) The C^{ij} are uniformly of class C^∞.

© Springer International Publishing AG 2017
G. Esposito, *From Ordinary to Partial Differential Equations*, UNITEXT 106,
DOI 10.1007/978-3-319-57544-5_24

(i) The C^{ij} can be expressed in the form

$$C^{ij}(x, t) = \sqrt{c_1 c_2}\, \delta^{ij} \text{ if } |x| \geq r_0, \qquad (24.1.4)$$

where r_0 is some large constant. He considered only solutions $T(x, t)$ bounded in x for each t for which the solution is defined, i.e.

$$\max_x |T(x, t)| < \infty. \qquad (24.1.5)$$

Upon making these restrictions, any bounded measurable function $T(x, t_0)$ of x given at an initial time t_0 determines a unique continuation $T(x, t)$ defined for all $t \geq t_0$ and of class C^∞ for $t > t_0$. Moreover, $T(x, t)$ approaches $T(x, t_0)$ almost everywhere as t approaches t_0, and $\max_x |T(x, t)|$ is non-increasing in t. The fundamental solutions of parabolic equations, that we discuss below, are well known to exist and have the general properties just stated (see Chap. 23, [42], [44]).

The use of fundamental solutions turns out to be very helpful with equations of the form (24.1.2), and the Nash work was built around a step-by-step control of the properties of fundamental solutions, and most of his results concern them directly. A fundamental solution $T(x, t)$ has a source point x_0 and starting time t_0, and is defined and positive for $t > t_0$. Moreover,

$$\int T(x, t)\mathrm{d}x = 1 \,\,\forall t > t_0, \qquad (24.1.6)$$

where $\mathrm{d}x$ is the volume element in n-dimensional Euclidean space. As t approaches t_0, the fundamental solution gets concentrated around its source point. The limit of $T(x, t)$ is zero unless $x = x_0$, in which case it diverges and equals $+\infty$. A fundamental solution can be viewed as representing the concentration of a diffusant that is spreading from an initial concentration of unit weight at x_0 at time t_0.

Another important concept is the one of *characterizing function* $S(x, t, \overline{x}, \overline{t})$, that makes it possible to unify all fundamental solutions. For fixed values $\overline{x}$ and $\overline{t}$, and as a function of x and t, S is a fundamental solution of Eq. (24.1.2) with source point $\overline{x}$ and starting time $\overline{t}$. Conversely, once x and t are fixed, S is a fundamental solution of the adjoint equation

$$\mathrm{div}_{\overline{x}}\left(C(\overline{x}, t) \cdot \mathrm{grad}_{\overline{x}} S\right) = -\frac{\partial S}{\partial \overline{t}}, \qquad (24.1.7)$$

where the time variable runs backwards. This property makes it possible to use estimates for fundamental solutions in two ways on S.

The dependence of a bounded solution $T(x, t)$ on bounded initial data $T(x, t_0)$ can be expressed through the characterizing function in the form

$$T(x, t) = \int S(x, t, \overline{x}, t_0) T(\overline{x}, t_0)\mathrm{d}\overline{x}. \qquad (24.1.8)$$

In particular, one finds a reproductive property of fundamental solutions, expressed by the equation

$$S(x^2, t^2, x_0, t_0) = \int S(x^2, t^2, x^1, t^1) S(x^1, t^1, x_0, t_0) dx^1. \tag{24.1.9}$$

Nash began his analysis by considering a special fundamental solution $T = T(x, t) = S(x, t, 0, 0)$ with source at the origin and starting time zero. He used such a fundamental solution to build the functional

$$E \equiv \int T^2 \, dx, \tag{24.1.10}$$

which, by virtue of Eq. (24.1.2), obeys the equation

$$\frac{\partial E}{\partial t} = 2 \int T \, T_t \, dx = 2 \int T \operatorname{div}(C \cdot \operatorname{grad} T) dx = -2 \int \operatorname{grad} T \cdot C \cdot \operatorname{grad} T \, dx, \tag{24.1.11}$$

where an integration by parts has been performed in order to write the last equality. Moreover, by virtue of the uniform ellipticity (24.1.3), one can write that, for any vector V,

$$c_1 |V|^2 \leq V \cdot C \cdot V \leq c_2 |V|^2, \tag{24.1.12}$$

and hence

$$-\frac{\partial E}{\partial t} \geq 2c_1 \int |\operatorname{grad} T|^2 \, dx, \tag{24.1.13}$$

upon choosing $V \equiv \operatorname{grad} T$ in the inequality (24.1.12).

With the help of (24.1.13) and a lower bound for the integral $\int |\operatorname{grad} T|^2 \, dx$ in terms of E, Nash was able to bound E from the above, obtaining therefore his first a priori estimate. In order to bound the integral of the squared modulus of grad T, he exploited a general inequality, valid for any function $u : x \to u(x)$ in n-dimensional Euclidean space. He assumed that the function u is smooth and well-behaved at infinity, at which stage E. M. Stein provided the proof outlined below [105].

The Fourier transform of the function u is

$$v(y) \equiv (2\pi)^{-\frac{n}{2}} \int e^{ix \cdot y} u(x) dx, \tag{24.1.14}$$

and it has the well-known property

$$\int |v|^2 dy = \int |u|^2 dx. \tag{24.1.15}$$

The Fourier transform of the partial derivative $\frac{\partial u}{\partial x^k}$ is $iy_k v$; therefore one finds

$$\int \left| \frac{\partial u}{\partial x^k} \right|^2 dx = \int (y_k)^2 |v|^2 dy, \tag{24.1.16}$$

and

$$\int |\text{grad } u|^2 dx = \sum_{j,k=1}^{n} \int \delta^{jk} \frac{\partial u}{\partial x^j} \frac{\partial u}{\partial x^k} dx = \sum_{k=1}^{n} \int \left(\frac{\partial u}{\partial x^k} \right)^2 dx = \int |y|^2 |v|^2 dy. \tag{24.1.17}$$

Moreover, the Fourier transform v can be majorized according to

$$|v| \leq (2\pi)^{-\frac{n}{2}} \int \left| e^{ix \cdot y} \right| |u| \, dx = (2\pi)^{-\frac{n}{2}} \int |u| \, dx. \tag{24.1.18}$$

Thus, for any positive ρ, one has

$$\int_{|y| \leq \rho} |v|^2 dy \leq \left(\frac{\pi^{\frac{n}{2}} \rho^n}{\Gamma\left(\frac{n}{2} + 1\right)} \right) \left\{ (2\pi)^{-\frac{n}{2}} \int |u| \, dx \right\}^2, \tag{24.1.19}$$

upon using the formula for the volume of a n-sphere, that we already encountered in Part II. On the other hand, when $|y| > \rho$, one finds

$$\int_{|y| > \rho} |v|^2 dy \leq \int_{|y| > \rho} \left| \frac{y}{\rho} \right|^2 |v|^2 \, dy = \rho^{-2} \int |\text{grad } u|^2 \, dx, \tag{24.1.20}$$

where, in the last equality, we have exploited the identity (24.1.17). Now the clever idea is *to choose the value of ρ minimizing the sum of two bounds* (24.1.19) *and* (24.1.20). On setting

$$a_n \equiv \frac{\pi^{\frac{n}{2}}}{\Gamma\left(\frac{n}{2} + 1\right)}, \quad I_n \equiv (2\pi)^{-\frac{n}{2}} \int |u| \, dx, \quad J \equiv \int |\text{grad } u|^2 \, dx, \tag{24.1.21}$$

we are therefore dealing with the function $f : \rho \to f(\rho)$ given by

$$f(\rho) \equiv a_n \rho^n (I_n)^2 + \rho^{-2} J, \tag{24.1.22}$$

whose first and second derivatives read as

$$f'(\rho) = n\rho^{n-1} a_n (I_n)^2 - 2\rho^{-3} J, \quad f''(\rho) = n(n-1)\rho^{n-2} a_n (I_n)^2 + 6\rho^{-4} J. \tag{24.1.23}$$

Thus, the first derivative of f vanishes at $\rho = \widetilde{\rho}$ given by

$$\widetilde{\rho}^{n+2} = \frac{2}{n a_n} \frac{J}{(I_n)^2} \implies f''(\widetilde{\rho}) = 2(n+2)\frac{J}{\widetilde{\rho}^4} > 0, \tag{24.1.24}$$

which proves that, at $\widetilde{\rho}$, we are indeed minimizing the sum $f(\rho)$ of two bounds. This result yields a bound on the integral of $|u|^2$ (see (24.1.15)) in terms of the integrals of $|u|$ and $|\text{grad } u|^2$. When solved for the integral of the latter, this is

$$\int |\text{grad } u|^2 \, dx$$

$$\geq \left(\frac{4\pi n}{(n+2)}\right) \left[\frac{\Gamma\left(\frac{n}{2}+1\right)}{\left(1+\frac{n}{2}\right)}\right]^{\frac{2}{n}} \left[\int |u| \, dx\right]^{-\frac{4}{n}} \left[\int |u|^2 \, dx\right]^{1+\frac{2}{n}}. \quad (24.1.25)$$

If we choose $u = T$ in the minorization (24.1.25) and bear in mind that the space integral of the fundamental solution equals 1 from (24.1.6), we obtain from (24.1.13)

$$-\frac{\partial E}{\partial t} \geq k E^{1+\frac{2}{n}}, \quad k = k(n, c_1, c_2). \quad (24.1.26)$$

This inequality implies that

$$\frac{\partial}{\partial t}\left(E^{-\frac{2}{n}}\right) \geq k \implies E^{-\frac{2}{n}} \geq kt \implies E \leq k\, t^{-\frac{n}{2}}, \quad (24.1.27)$$

where we have exploited the relation $\lim_{t \to 0} E = \infty$.

The majorization (24.1.27) is the first bound in the Nash analysis. From it and the reproductive property of fundamental solutions, he obtained

$$T(x, t) = \int S\left(x, t, \overline{x}, \frac{t}{2}\right) S\left(\overline{x}, \frac{t}{2}, 0, 0\right) d\overline{x}, \quad (24.1.28)$$

and hence the square of the fundamental solution can be majorized according to

$$(T(x, t))^2 \leq \int \left[S\left(x, t, \overline{x}, \frac{t}{2}\right)\right]^2 d\overline{x} \cdot \int \left[S\left(\overline{x}, \frac{t}{2}, 0, 0\right)\right]^2 d\overline{x} \leq \left[k\left(\frac{t}{2}\right)^{-\frac{n}{2}}\right]^2. \quad (24.1.29)$$

This implies, in turn, that

$$T \leq kt^{-\frac{n}{2}}, \quad (24.1.30)$$

which is a pointwise bound, stronger that (24.1.23), that involves instead E, the integral of T^2.

The key estimate obtained by Nash controls the *moment* of a fundamental solution, defined by

$$M \equiv \int rT \, dx = \int |x| \, T \, dx. \quad (24.1.31)$$

His first major goal was to prove that $M \leq k\sqrt{t}$.

Nash defined also the *entropy* associated to a fundamental solution, in the form

$$Q \equiv - \int T \, \log(T) \, dx. \tag{24.1.32}$$

By virtue of the pointwise bound (24.1.30) he obtained

$$Q \geq \int \min_x \left[-\log(T) \right] (T \, dx) \geq -\log\left(kt^{-\frac{n}{2}}\right) \int T \, dx, \tag{24.1.33}$$

which implies

$$Q \geq \pm k + \frac{n}{2} \log(t), \tag{24.1.34}$$

by virtue of (24.1.6).

The desired derivation of a bound on M requires a lower bound on M in terms of Q. This inequality depends only on the fact that the fundamental solution T is never negative and has unit space integral over n-dimensional Euclidean space. First of all one can point out that, for any fixed λ,

$$\min_T \left(T \log(T) + \lambda T \right) = -e^{-\lambda-1}. \tag{24.1.35}$$

In particular, one can take $\lambda = ar + b$, where $r = |x|$ and a, b are arbitrary constants. Integration over n-dimensional Euclidean space yields therefore

$$\int \left[T \log(T) + (ar + b)T \right] dx \geq -e^{-b-1} \int e^{-ar} \, dx, \tag{24.1.36}$$

or, in terms of the entropy Q and moment M,

$$-Q + aM + b \geq -e^{-b-1} a^{-n} D_n, \quad D_n \equiv 2^n \pi^{\frac{(n-1)}{2}} \Gamma\left[\frac{(n+1)}{2}\right], \tag{24.1.37}$$

where the constant D_n is obtained from the surface of the $(n-1)$-sphere. Now one can choose a and b in such a way that

$$a \equiv \frac{n}{M}, \quad e^{-b} \equiv \frac{e}{D_n} a^n. \tag{24.1.38}$$

The inequality in (24.1.37) becomes therefore

$$-Q + n + b \geq -1 \Longrightarrow n + 1 \geq Q + \log\left(\frac{e}{D_n}\right) + n \log(a)$$

$$= Q + 1 - \log(D_n) + n \log(n) - n \log(M) \Longrightarrow n \log(Me) \geq Q + n \log(n) - \log(D_n)$$

$$\implies M \geq \left(\frac{n}{eD_n^{\frac{1}{n}}}\right) e^{\frac{Q}{n}} = k e^{\frac{Q}{n}}. \tag{24.1.39}$$

This completes the Carleman proof reported in Nash [105].

The following step in the Nash analysis is a dynamic inequality, that relates the rates of change with time of moment M and entropy Q. Indeed, by differentiation of Eq. (24.1.32) that defines the entropy, one obtains (bearing in mind the Eq. (24.1.2) obeyed by the fundamental solution)

$$\frac{\partial Q}{\partial t} = -\int \left(1 + \log(T)\right)\frac{\partial T}{\partial t}dx = -\int \left(1 + \log(T)\right)\mathrm{div}\left(C \cdot \mathrm{grad}\ T\right)dx$$

$$= \int \mathrm{grad}(\log(T)) \cdot C \cdot \mathrm{grad}\ T\ dx, \tag{24.1.40}$$

where integration by parts has been exploited to write the second line. In this equation, one can use the identity

$$\mathrm{grad}\ T = T\ \mathrm{grad}\ \log(T)$$

to obtain

$$\frac{\partial Q}{\partial t} = \int \mathrm{grad}(\log(T)) \cdot C \cdot \mathrm{grad}(\log(T))(T\ dx). \tag{24.1.41}$$

The integrand on the right-hand side suggests exploiting the minorization involving vectors V

$$V \cdot c_2 C \cdot V \geq V \cdot C^2 \cdot V = |C \cdot V|^2, \tag{24.1.42}$$

which leads to (with $V = \mathrm{grad}(\log(T))$)

$$c_2 \frac{\partial Q}{\partial t} \geq \int |C \cdot \mathrm{grad}(\log(T))|^2\ (T\ dx) \geq \left[\int |C \cdot \mathrm{grad}(\log(T))|\ (T\ dx)\right]^2$$

$$\geq \left[\int |C \cdot \mathrm{grad}\ T|\ dx\right]^2. \tag{24.1.43}$$

The second minorization on the first line holds by virtue of the Cauchy-Schwarz inequality in the form

$$\int_0^1 f^2\ du \geq \left[\int_0^1 f\ du\right]^2, \tag{24.1.44}$$

where the integration measure du corresponds to $T\ dx$ in (24.1.43).

In analogous way, one obtains for the moment M the evolution equation

$$\frac{\partial M}{\partial t} = -\int \mathrm{grad}\ r \cdot C \cdot \mathrm{grad}\ T\ dx, \tag{24.1.45}$$

and the majorization

$$\left| \frac{\partial M}{\partial t} \right| \leq \int |\mathrm{grad}\, r|\ |C \cdot \mathrm{grad}\, T|\ \mathrm{d}x \leq \int |C \cdot \mathrm{grad}\, T|\ \mathrm{d}x. \qquad (24.1.46)$$

At this stage, by virtue of inequalities (24.1.43) and (24.1.46) one obtains

$$c_2 \frac{\partial Q}{\partial t} \geq \left(\frac{\partial M}{\partial t} \right)^2. \qquad (24.1.47)$$

Nash defined the entropy integral (24.1.32) precisely in order to get the inequality (24.1.47). He then pointed out that the three inequalities obtained, written again for the sake of clarity

$$Q \geq \pm k + \frac{n}{2} \log(t), \qquad (24.1.34)$$

$$M \geq k\, e^{\frac{Q}{n}}, \qquad (24.1.39)$$

$$c_2 \frac{\partial Q}{\partial t} \geq \left(\frac{\partial M}{\partial t} \right)^2, \qquad (24.1.47)$$

and the limiting behaviour

$$\lim_{t \to 0} M = 0, \qquad (24.1.48)$$

are sufficient to bound from the above and from below both M and Q, in their time dependence. The original functional equation is no longer needed. Indeed, from $M(0) = 0$ and (24.1.47) one finds

$$M \leq \int_0^t \sqrt{c_2 \frac{\partial Q}{\partial \tau}}\ \mathrm{d}\tau, \qquad (24.1.49)$$

and hence, from (24.1.39),

$$k\, e^{\frac{Q}{n}} \leq M \leq \int_0^t \sqrt{c_2 \frac{\partial Q}{\partial \tau}}\ \mathrm{d}\tau. \qquad (24.1.50)$$

One can now define

$$n R \equiv Q \mp k - \frac{n}{2} \log(t) \qquad (24.1.51)$$

in such a way that the condition $R \geq 0$ corresponds to (24.1.34). From (24.1.51) one obtains

$$\frac{\partial Q}{\partial t} = n\frac{\partial R}{\partial t} + \frac{n}{2t} \implies k\sqrt{t}e^R \leq M \leq \sqrt{n}c_2 \int_0^t \sqrt{\frac{1}{2\tau} + \frac{\partial R}{\partial \tau}} \, d\tau. \quad (24.1.52)$$

When both a and $a + b$ are positive one has

$$a > 0, (a + b) > 0 \implies \sqrt{a + b} \leq \sqrt{a} + \frac{b}{2\sqrt{a}}, \quad (24.1.53)$$

and hence the integral in (24.1.52) can be majorized according to

$$\int_0^t \sqrt{\frac{1}{2\tau} + \frac{\partial R}{\partial \tau}} \, d\tau \leq \int_0^t \frac{1}{\sqrt{2\tau}} d\tau + \int_0^t \sqrt{\frac{\tau}{2}}\frac{\partial R}{\partial \tau} d\tau$$
$$\leq \sqrt{2t} + R\sqrt{\frac{t}{2}} - \int_0^t \frac{R}{\sqrt{8\tau}} d\tau \leq \sqrt{2t} + R\sqrt{\frac{t}{2}}, \quad (24.1.54)$$

where integration by parts and non-negativity of R have been exploited. By virtue of (24.1.54), the inequalities in (24.1.52) read as

$$k\sqrt{t}e^R \leq kM \leq \sqrt{2t} + R\sqrt{\frac{t}{2}} \implies ke^R \leq \frac{kM}{\sqrt{t}} \leq \sqrt{2}\left(1 + \frac{R}{2}\right). \quad (24.1.55)$$

Of course, ke^R increases faster in R than $\sqrt{2}\left(1 + \frac{R}{2}\right)$, so that R must be bounded from the above. This implies in turn that $\frac{M}{\sqrt{t}}$ is bounded both from the above and from below, i.e.

$$k\sqrt{t} \leq M \leq k\sqrt{t}. \quad (24.1.56)$$

Upon choosing best possible constants in (24.1.34) and (24.1.39), and defining

$$b_n \equiv \sqrt{\frac{n}{2t}} \left(\frac{\sqrt{\pi}}{\Gamma\left[\frac{(n+1)}{2}\right]}\right)^{\frac{1}{n}} \geq 2^{-\frac{1}{2n}}, \quad (24.1.57)$$

$$\lambda \equiv \frac{1}{2}\log\left(\frac{c_2}{c_1}\right) - \log(b_n) \leq \frac{1}{2n}\log(2) + \frac{1}{2}\log\left(\frac{c_2}{c_1}\right), \quad (24.1.58)$$

Nash obtained

$$b_n\sqrt{2c_1nt} \leq M \leq \sqrt{2c_2nt}\left[1 + \min\left(\lambda, \sqrt{\frac{\lambda}{2}}\right)\right]. \quad (24.1.59)$$

His λ is relatively small and his b_n approaches 1 as $n \to \infty$, hence the bounds become sharper with increasing n. In the case of the heat equation, where $C^{ij} = c\delta^{ij}$ and $c_1 = c_2 = c$, one finds $M = \sqrt{2nct}$.

24.2　The G Bound

This nomenclature may appear to be unfamiliar, but it expresses the need for a result that limits the extent to which a fundamental solution can be very small over a large volume of space close to its source point. This result will be used in turn to prove that there exists some overlap

$$\int \min(T_1, T_2) dx$$

of two fundamental solutions with nearby source points.

Let the fundamental solution T be given by the characterizing function S with arguments $(x, t, 0, 0)$, and define

$$U(\xi, t) \equiv t^{\frac{n}{2}} T(\xi \sqrt{t}, t). \tag{24.2.1}$$

By virtue of this definition, U has unit integral

$$\int U \, d\xi = 1, \tag{24.2.2}$$

and if μ is the constant such that $M \leq \mu \sqrt{t}$ from the previous section, one finds

$$\int |\xi| U \, d\xi \leq \mu. \tag{24.2.3}$$

The linear parabolic equation (24.1.2) reads now, in terms of U, as

$$2t \frac{\partial U}{\partial t} = nU + \xi \cdot \operatorname{grad} U + 2\operatorname{div}(C \cdot \operatorname{grad} U). \tag{24.2.4}$$

On denoting by δ a small positive constant, we now come to terms with the desired function G considered by Nash, i.e.

$$G \equiv \int e^{-|\xi|^2} \log(U + \delta) d\xi. \tag{24.2.5}$$

This function exploits in a clever way the properties of two basic functions of real analysis, because it is sensitive to areas where $|\xi|$ is not large and U is small. In such cases, $\log(U + \delta)$ approaches $-\infty$, which is not mitigated by the exponential which remains finite. Thus, G tends to become strongly negative in these areas. We are going to see that Nash obtained a lower bound on G of the form

$$G \geq -k\sqrt{-\log(\delta)},$$

for sufficiently small δ. This bound is interesting because it limits the possibility for U to be small in a large portion of the region where $|\xi|$ is not large.

As a first step, differentiation with respect to time of the definition (24.2.5) and use of Eq. (24.2.4) lead to a formula for the first time derivative of G as a sum of three terms, i.e.

$$2t\frac{\partial G}{\partial t} = \sum_{k=1}^{3} H_k, \tag{24.2.6}$$

where

$$H_1 \equiv n \int e^{-|\xi|^2} \frac{U}{(U+\delta)} d\xi \geq 0, \tag{24.2.7}$$

and

$$H_2 \equiv \int e^{-|\xi|^2} \xi \cdot \operatorname{grad} \log(U+\delta) d\xi = -\int \operatorname{div}\left[\xi e^{-|\xi|^2}\right] \log(U+\delta) d\xi. \tag{24.2.8}$$

In the formula for H_2 one can integrate by parts to find

$$H_2 = -\int e^{-|\xi|^2}(\operatorname{div}\xi) \log(U+\delta) d\xi + \int e^{-|\xi|^2}(2|\xi|\operatorname{grad}|\xi|) \cdot \xi \log(U+\delta) d\xi$$

$$= -nG + 2\int e^{-|\xi|^2}|\xi|^2 \left[\log(\delta) + \log\left(1 + \frac{U}{\delta}\right)\right] d\xi, \tag{24.2.9}$$

which implies the lower bound

$$H_2 \geq -nG + 2(\log(\delta)) \int |\xi|^2 e^{-|\xi|^2} d\xi \geq -nG + n\pi^{\frac{n}{2}} \log(\delta). \tag{24.2.10}$$

The third function, H_3, is defined by

$$H_3 \equiv 2\int e^{-|\xi|^2} \frac{\operatorname{div}(C \cdot \operatorname{grad} U)}{(U+\delta)} d\xi = -2\int \operatorname{grad}\left[\frac{e^{-|\xi|^2}}{(U+\delta)}\right] \cdot C \cdot \operatorname{grad} U \, d\xi$$

$$= 4\int e^{-|\xi|^2} \frac{[|\xi|\operatorname{grad}|\xi| \cdot C \cdot \operatorname{grad} U]}{(U+\delta)} d\xi + 2\int e^{-|\xi|^2} \frac{[\operatorname{grad} U \cdot C \cdot \operatorname{grad} U]}{(U+\delta)^2} d\xi$$

$$= H_3' + H_3'', \tag{24.2.11}$$

where the functions H_3' and H_3'' are defined by

$$H_3' \equiv 4\int e^{-|\xi|^2} \xi \cdot C \cdot \operatorname{grad} \log(U+\delta) d\xi, \tag{24.2.12}$$

$$H_3'' \equiv 2\int e^{-|\xi|^2} \operatorname{grad} \log(U+\delta) \cdot C \cdot \operatorname{grad} \log(U+\delta) d\xi. \tag{24.2.13}$$

Now one can use the Schwarz inequality to obtain a majorization of H_3' by means of H_3'', i.e.

$$
\begin{aligned}
(H_3')^2 &\leq \left\{ 4 \int e^{-|\xi|^2} \xi \cdot C \cdot \xi \, d\xi \right\} \\
&\quad \times \left\{ 4 \int e^{-|\xi|^2} \operatorname{grad} \log(U+\delta) \cdot C \cdot \operatorname{grad} \log(U+\delta) d\xi \right\} \\
&\leq \left\{ 4c_2 \int |\xi|^2 e^{-|\xi|^2} d\xi \right\} 2 H_3'' \\
&\leq \left(4c_2 \frac{n}{2} \pi^{\frac{n}{2}} \right) 2 H_3'' = 4nc_2 \pi^{\frac{n}{2}} H_3''.
\end{aligned}
\tag{24.2.14}
$$

Hence one finds for $|H_3'|$ the upper bound

$$
|H_3'| \leq k \sqrt{H_3''},
\tag{24.2.15}
$$

while H_3'' has the lower bound

$$
H_3'' \geq 2c_1 \int e^{-|\xi|^2} |\operatorname{grad} \log(U+\delta)|^2 \, d\xi.
\tag{24.2.16}
$$

In light of the lower bounds for the H_k's obtained so far, one obtains a lower bound for the time derivative of G, i.e.

$$
\begin{aligned}
2t \frac{\partial G}{\partial t} &\geq H_1 + H_2 + H_3'' - |H_3'| \\
&\geq 0 + \left(-nG + n\pi^{\frac{n}{2}} \log(\delta) \right) + H_3'' - k\sqrt{H_3''} \\
&\geq k \log(\delta) - nG - k\sqrt{H_3''} + H_3''.
\end{aligned}
\tag{24.2.17}
$$

When H_3'' is bounded below in terms of G, the relation (24.2.17) leads to a lower bound on G.

One can get rid of the gradient occurring in the lower bound (24.2.16) if one recalls that a function f depending on $\xi = (\xi^1, \ldots, \xi^n)$ can be expanded in products of Hermite polynomials, of the form

$$
\prod H_{n(i)}(\xi^i),
$$

where the polynomials are orthonormalized according to the well-known rule (note the same exponential factor occurring n times in (24.2.16))

$$
\int_{-\infty}^{\infty} e^{-s^2} H_n(s) H_l(s) ds = \delta_{nl}.
$$

The coefficients of these products in the similar expansion of the first derivatives $\frac{\partial f}{\partial \xi^i}$ depend in a simple way on the coefficients in the expansion of f. In particular, if

$$\int e^{-|\xi|^2} f \, d\xi = 0,$$

the coefficient of $\prod H_0(\xi^i)$ vanishes, and one obtains

$$\int e^{-|\xi|^2} |\mathrm{grad} f|^2 \, d\xi = \sum_{i=1}^{n} \int e^{-|\xi|^2} \left(\frac{\partial f}{\partial \xi^i} \right)^2 \, d\xi \geq 2 \int e^{-|\xi|^2} f^2 \, d\xi. \quad (24.2.18)$$

The application of the lower bound (24.2.18), with the choice

$$f \equiv \log(U + \delta) - \pi^{-\frac{n}{2}} G, \quad (24.2.19)$$

to the lower bound (24.2.16), leads to

$$H_3'' \geq 4c_1 \int e^{-|\xi|^2} \left[\log(U + \delta) - \pi^{-\frac{n}{2}} G \right]^2 \, d\xi. \quad (24.2.20)$$

Now the function $h(U) \equiv \frac{1}{U} \left[\log(U + \delta) - \pi^{-\frac{n}{2}} G \right]^2$, related in obvious way to the integrand in (24.2.20), takes large values for very small U, then decreases to zero, rises from zero to a local maximum at U equal to a value U_c, and eventually decreases monotonically as U approaches ∞ for $U \geq U_c$. The equation for the maximum point U_c is

$$\log(U_c + \delta) - \pi^{-\frac{n}{2}} G = 2 \frac{U_c}{(U_c + \delta)} \implies U_c < U_0 = e^{2 + \pi^{-\frac{n}{2}} G}. \quad (24.2.21)$$

Thus, the function $h(U)$ is decreasing for $U \geq U_0$. The bound (24.1.30) corresponds to $U \leq k$. Hence the function $h(U)$ has a lower bound of the form

$$k \left[\log(k + \delta) - kG \right]^2$$

for $U \geq U_0$. By applying this to (24.2.20), one can write that

$$H_3'' \geq 4c_1 \int e^{-|\xi|^2} k \left[\log(k + \delta) - kG \right]^2 U^* \, d\xi, \quad (24.2.22)$$

where

$$U^* = U \text{ if } U > U_0, \quad U^* = 0 \text{ if } U \leq U_0. \quad (24.2.23)$$

This means that the contribution to (24.2.20) of the region where $U \leq U_0$ is here ignored, whereas U is taken as large as possible in the remaining region. If G is sufficiently negative, the function $\left[\log(k + \delta) - kG\right]$ remains positive when δ is omitted, so that one has

$$[\log(k) - kG]^2 < [\log(k + \delta) - kG]^2, \tag{24.2.24}$$

and one can simplify again the lower bound for H_3'' to obtain

$$H_3'' \geq k^2(1 - G)^2 \int e^{-|\xi|^2} U^* \, \mathrm{d}\xi. \tag{24.2.25}$$

On defining

$$\lambda \equiv \int U^* \, \mathrm{d}\xi, \tag{24.2.26}$$

and pointing out that

$$\int |\xi| U^* \, \mathrm{d}\xi \leq \int |\xi| U \, \mathrm{d}\xi \leq \mu, \tag{24.2.27}$$

one finds

$$\mu \geq \int_{|\xi| \geq \frac{2\mu}{\lambda}} |\xi| U^* \, \mathrm{d}\xi \geq \frac{2\mu}{\lambda} \int_{|\xi| \geq \frac{2\mu}{\lambda}} U^* \, \mathrm{d}\xi, \tag{24.2.28}$$

and therefore

$$\int_{|\xi| \geq \frac{2\mu}{\lambda}} U^* \, \mathrm{d}\xi \leq \frac{\lambda}{2} \implies \int_{|\xi| \leq \frac{2\mu}{\lambda}} U^* \, \mathrm{d}\xi \geq \lambda - \frac{\lambda}{2} = \frac{\lambda}{2}. \tag{24.2.29}$$

This inequality can be applied to (24.2.25) and leads to a refined lower bound for H_3'', i.e.

$$H_3'' \geq k^2(1 - G)^2 \frac{\lambda}{2} e^{-\frac{4\mu^2}{\lambda^2}}. \tag{24.2.30}$$

However, we cannot be satisfied with this bound unless we can bound λ from below, or, defining

$$\widehat{U} \equiv U - U^*, \tag{24.2.31}$$

bound from the above the integral

$$\int \widehat{U} \, \mathrm{d}\xi = 1 - \lambda. \tag{24.2.32}$$

Of course, one has

$$U \geq \widehat{U} \implies \mu \geq \int |\xi| \, \widehat{U} \, \mathrm{d}\xi. \tag{24.2.33}$$

In light of the moment constraint and of the constraint $\widehat{U} \leq U_0$, the maximum of the integral of $\widehat{U}$ is achieved by having

$$\widehat{U} = U_0 \text{ if } |\xi| \le \rho, \ \widehat{U} = 0 \text{ if } |\xi| > \rho, \qquad (24.2.34)$$

the parameter ρ being defined by the condition

$$\int |\xi| \, \widehat{U} \, d\xi = \int |\xi| \, U_0 \, d\xi = \frac{n \pi^{\frac{n}{2}}}{(n+1)\Gamma\left(\frac{n}{2}+1\right)} \rho^{n+1} U_0 = \mu. \qquad (24.2.35)$$

Hence one finds that

$$1 - \lambda = \int \widehat{U} \, d\xi = \frac{\pi^{\frac{n}{2}}}{\Gamma\left(\frac{n}{2}+1\right)} \rho^n U_0, \qquad (24.2.36)$$

$$1 - \lambda \le U_0 \left(\frac{k\mu}{U_0}\right)^{\frac{n}{(n+1)}} \text{ or } \le k U_0^{\frac{1}{(n+1)}}. \qquad (24.2.37)$$

If U_0 is small enough, then $1 - \lambda$ is small and λ is bounded from below. More precisely, by virtue of the definition (24.2.21) of U_0, one has, for all sufficiently large $-G$,

$$\lambda \ge \frac{1}{2} \implies H_3'' \ge k^2 (1 - G)^2. \qquad (24.2.38)$$

One can now revert to the inequality (24.2.17) for $\frac{\partial G}{\partial t}$ and exploit the lower bound (24.2.38) to find that, for sufficiently negative values of G, one can write

$$2 \frac{\partial G}{\partial \log(t)} = 2t \frac{\partial G}{\partial t} \ge -nG + k \log(\delta) + k^2 \left[(1 - G)^2 - (1 - G) \right]$$
$$\ge k |G|^2 + k \log(\delta). \qquad (24.2.39)$$

Let now $G(c_1, c_2, n)$ be the number such that, when $G \le G_1$, we know for sure that G is small enough to guarantee that (24.2.39) is fulfilled. Let

$$G_2(c_1, c_2, n, \delta) \equiv -k\sqrt{-\log(\delta)} \qquad (24.2.40)$$

be the largest number such that

$$k |G|^2 + k \log(\delta) > 0, \ \forall G < G_2. \qquad (24.2.41)$$

Then the quantity

$$G_3 \equiv \min(G_1, G_2) \qquad (24.2.42)$$

is the smallest possible value of G. If it were possible to find a t_1 at which $G(t_1) = G_3 - \epsilon$, one would have

$$\frac{\partial G}{\partial \log(t)} \geq \epsilon^* \, \forall \, t \leq t_1 \implies G(t) \leq G(t_1) - \epsilon^* \log\left(\frac{t_1}{t}\right) \implies G \to -\infty \text{ as } t \to 0.$$

$$(24.2.43)$$

On the other hand, by virtue of the lower bound $G \geq \pi^{\frac{n}{2}} \log(\delta)$, the assumption $G(t_1) = G_3 - \epsilon$ is untenable, hence one finds

$$G \geq G_3 \implies G \geq -k\sqrt{-\log(\delta)} \qquad (24.2.44)$$

for all sufficiently small values of δ, because $G_2 \leq G_1$ and

$$G_3 = G_2 = -k\sqrt{-\log(\delta)} \qquad (24.2.45)$$

for sufficiently small values of δ.

24.3 The Overlap Estimate

The object of interest is now the overlap of fundamental solutions. These are obtained as particular values of the characterizing function as

$$T_1 = S(x, t, x^1, 0), \quad T_2 = S(x, t, x^2, 0). \qquad (24.3.1)$$

A change of coordinates is performed, defining

$$U_1 = t^{\frac{n}{2}} T_1(\xi\sqrt{t}, t), \quad U_2 = t^{\frac{n}{2}} T_2(\xi\sqrt{t}, t), \qquad (24.3.2)$$

and it is convenient to consider $\xi^1 = \frac{x^1}{\sqrt{t}}, \xi^2 = \frac{x^2}{\sqrt{t}}$. With this notation, the source of the fundamental solution U_i is ξ^i rather than the origin. The application of (24.2.44) yields now

$$\int e^{-|\xi - \xi^i|^2} \log(U_i + \delta)\mathrm{d}\xi = G_i \geq -k\sqrt{-\log(\delta)}, \quad i = 1, 2 \qquad (24.3.3)$$

where δ must be taken sufficiently small. Upon adding the two inequalities (24.3.3), one finds

$$\int \max_i e^{-|\xi - \xi^i|^2} \max_i [\log(U_i + \delta)]\mathrm{d}\xi$$
$$+ \int \min_i e^{-|\xi - \xi^i|^2} \min_i [\log(U_i + \delta)]\mathrm{d}\xi \geq -2k\sqrt{-\log(\delta)}. \qquad (24.3.4)$$

This inequality can be written in more concise form as

$$\int f^* \log(U_{max} + \delta)d\xi + \int \widehat{f} \log(U_{min} + \delta)d\xi \geq -k\sqrt{-\log(\delta)}. \qquad (24.3.5)$$

On assuming $\delta \leq 1$, the first integral in (24.3.5) can be majorized in the form

$$\int f^* \log(U_{max} + \delta)d\xi \leq \int f^*(U_1 + U_2)d\xi \leq \int (U_1 + U_2)d\xi = 2. \qquad (24.3.6)$$

For the second integral in (24.3.5), one can write other majorizations according to

$$\int \widehat{f} \log(U_{min} + \delta)d\xi \leq \log(\delta) \int \widehat{f} \, d\xi + \max[\widehat{f}] \int \log\left(1 + \frac{U_{min}}{\delta}\right) d\xi$$
$$\leq w \log(\delta) + \frac{1}{\delta} \int U_{min} \, d\xi, \qquad (24.3.7)$$

having defined

$$w \equiv \int \min\left[e^{-|\xi - \xi^1|^2}, e^{-|\xi - \xi^2|^2}\right] d\xi. \qquad (24.3.8)$$

Hence Nash obtained the lower bound

$$2 + w \log(\delta) + \frac{1}{\delta} \int \min(U_1, U_2)d\xi \geq -k\sqrt{-\log(\delta)}, \qquad (24.3.9)$$

which implies that

$$\int \min(T_1, T_2)dx = \int \min(U_1, U_2)d\xi \geq \delta\left[-2 - w \log(\delta) - k\sqrt{-\log(\delta)}\right].$$
$$(24.3.10)$$

This lower bound holds for $\delta \leq \delta_1$. Moreover, a value $\delta_2(w)$ exists such that, if $\delta < \delta_2(w)$, the expression in square brackets is positive. If one chooses

$$\delta = \frac{1}{2}\min(\delta_1, \delta_2), \qquad (24.3.11)$$

the right-hand side of (24.3.10) is positive, and one arrives at the lower bound

$$\int \min(T_1, T_2)dx \geq \phi\left(|\xi^1 - \xi^2|\right) \geq \phi\left(\frac{|x^1 - x^2|}{\sqrt{t}}\right), \qquad (24.3.12)$$

because w is a function of $|\xi^1 - \xi^2|$. The function ϕ is decreasing but remains positive; it is determined only by c_1, c_2, n. The inequality (24.3.12) was the first desired estimate of the overlap of fundamental solutions.

24.4   Time Continuity

In his work, starting with iterative use of the lower bound (24.3.12), Nash obtained
the inequality

$$\left|T(x^1, t) - T(x^2, t)\right| \le B \, A_1 \left(\frac{|x^1 - x^2|}{\sqrt{t - t_0}}\right)^\alpha, \tag{24.4.1}$$

where $A_1 = A_1(n, c_1, c_2)$. He then made further progress by exploiting the moment
bound (24.1.56). In other words, on considering a fundamental solution $T(x, t)$ of
(24.1.2) with $|T| \le B$ for $t \ge t_0$, one has for $t' > t > t_0$

$$T(x, t) - T(x, t') = T(x, t) - \int S(x, t', \overline{x}, t) T(\overline{x}, t) \mathrm{d}\overline{x}$$

$$= \int S(x, t', \overline{x}, t) \Big[T(x, t) - T(\overline{x}, t)\Big] \mathrm{d}\overline{x}, \tag{24.4.2}$$

where the equality $\int S \, \mathrm{d}\overline{x} = 1$ has been exploited. One can therefore obtain the upper
bound

$$|T(x, t) - T(x, t')| \le \int S(x, t', \overline{x}, t) \, |T(x, t) - T(\overline{x}, t)| \ \mathrm{d}\overline{x}$$

$$\le \int S(x, t', x + y, t) \, |T(x, t) - T(x + y, t)| \ \mathrm{d}y. \tag{24.4.3}$$

Nash separated this integral into two parts, by introducing a radius ρ; one part where
$|y| \le \rho$ and one where $|y| > \rho$. Hence he obtained

$$\left|T(x, t) - T(x, t')\right| \le I_1 + I_2, \tag{24.4.4}$$

where, upon exploiting $\int S \, \mathrm{d}y = 1$, one has

$$I_1 \equiv \int_{|y| \le \rho} S(x, t', x + y, t) \, |T(x, t) - T(x + y, t)| \ \mathrm{d}y \le B \, A_1 \left(\frac{\rho}{\sqrt{t - t_0}}\right)^\alpha, \tag{24.4.5}$$

while

$$I_2 \equiv \int_{|y| > \rho} S(x, t', x + y, t) \, |T(x, t) - T(x + y, t)| \ \mathrm{d}y$$

$$\le \frac{2B}{\rho} \int_{|y| > \rho} |y| \, S(x, t', x + y, t) \mathrm{d}y \le 2B\mu \frac{\sqrt{t' - t}}{\rho}. \tag{24.4.6}$$

The addition of inequalities (24.4.5) and (24.4.6) yields therefore

$$\left| T(x,t) - T(x,t') \right| \leq BA_1 \left(\frac{\rho}{\sqrt{t-t_0}} \right)^{\alpha} + 2B\mu \frac{\sqrt{t'-t}}{\rho}. \tag{24.4.7}$$

At this stage, ρ is chosen in such a way that the sum is minimized, i.e.

$$\alpha A_1 \rho^{1+\alpha} = 2\mu\sqrt{t'-t}(t-t_0)^{\frac{\alpha}{2}}. \tag{24.4.8}$$

Thus, on defining

$$A_2 \equiv (1+\alpha)A_1 \left(\frac{2\mu}{\alpha A_1} \right)^{\frac{\alpha}{(1+\alpha)}}, \tag{24.4.9}$$

one obtains

$$\left| T(x,t) - T(x,t') \right| \leq B\, A_2 \left[\frac{(t'-t)}{(t-t_0)} \right]^{\frac{\alpha}{2(1+\alpha)}}. \tag{24.4.10}$$

The joint effect of the inequalities (24.4.1) and (24.4.10) yields the fundamental inequality

$$\left| T(x^1, t^1) - T(x^2, t^2) \right|$$
$$\leq B\, A \left\{ \left[\frac{|x^1 - x^2|}{\sqrt{t_1 - t_0}} \right]^{\alpha} + \left[\frac{(t_2 - t_1)}{(t_1 - t_0)} \right]^{\frac{\alpha}{2(1+\alpha)}} \right\}, \tag{24.4.11}$$

with the understanding that

$$A \equiv \max(A_1, A_2). \tag{24.4.12}$$

Moreover, Nash [105] pointed out that elliptic problems can be viewed as a special type of parabolic problem, i.e. one in which the coefficients of the equation are time independent, and a time-independent solution is desired. The Hölder continuity of solutions of uniformly elliptic equations of the form

$$\text{div}(C \cdot \text{grad } T) = 0 \tag{24.4.13}$$

arises then as a corollary of the Nash result for parabolic equations. At about the same time, De Giorgi [35] obtained an independent proof of the result for elliptic equations. We refer the reader to the outstanding original paper by De Giorgi just cited, and to the enlightening discussion of De Giorgi's analysis in Ambrosio [4]. A simplified proof of De Giorgi's theorem was obtained by Moser [103].

Part I
Fuchsian Functions

Chapter 25
The Poincaré Work on Fuchsian Functions

25.1 Properties of Fuchsian Functions

Poincaré tried to understand for many years whether there exist analytic functions, analogous to the elliptic functions, which make it possible to integrate various linear differential equations with algebraic coefficients. He proved eventually that there exist a large class of functions which satisfy these conditions, that he named Fuchsian functions (in honour of the German mathematician Immanuel Lazarus Fuchs, whose work inspired Poincaré) and nowadays called *automorphic functions*.

If z is a complex variable represented by a point in the plane, we denote by K_1 the operation that changes z into $f_1(z)$, and by K_2 the operation changing z into $f_2(z)$, written as (our maps act from left to right, whereas Poincaré used the opposite convention)

$$K_1 z = f_1(z), \quad K_2 z = f_2(z), \quad K_2 K_1 z = f_2[f_1(z)]. \tag{25.1.1}$$

When z remains within a certain region R, $K_1 z$ remains within a certain region S, and we write

$$K_1 R = S. \tag{25.1.2}$$

Following Poincaré, we call *fundamental circle* the circle of unit radius centred at the origin. The *hyperbolic group* is the group of operations that consist in changing z into $\frac{(az+b)}{(cz+d)}$, a, b, c, d being some constants, and which do not modify the fundamental circle (nowadays these are called fractional linear transformations). A *discontinuous group* is meant to be any group that does not contain the infinitesimal operation, which is the operation changing z into a quantity infinitely close to z. A *Fuchsian group* is any discontinuous group contained in the hyperbolic group. A *Fuchsian function* is any uniform (i.e. single-valued and analytic) function of z which is not modified by the operations of a Fuchsian group. Thus, a Fuchsian function of z is

© Springer International Publishing AG 2017
G. Esposito, *From Ordinary to Partial Differential Equations*, UNITEXT 106,
DOI 10.1007/978-3-319-57544-5_25

a meromorphic function all over the interior of the fundamental circle [117] and satisfying the condition

$$f(z) = f\left(\frac{az + b}{cz + d}\right). \tag{25.1.3}$$

To begin with, Poincaré needed first to form all Fuchsian groups; he succeeded in this task with the help of non-Euclidean geometry. He proved that the surface of the fundamental circle can be decomposed (in infinitely many ways) into a countable infinity of regions $R_0, R_1, R_2, ..., R_i, ...$ satisfying the following conditions [111]:

(i) These regions are curvilinear polygons whose sides are circle arcs belonging to circumferences that cut orthogonally the fundamental circle.
(ii) One has, for all i,

$$R_i = K_i R_0, \tag{25.1.4}$$

K_i being an operation of the hyperbolic group.

It is clear that the various K_i operations form a discontinuous group contained in the hyperbolic group, i.e. a Fuchsian group according to the above definition. Poincaré asked himself whether it is possible to perform this decomposition in such a way that the first of these regions, R_0, is a given curvilinear polygon. For this purpose, it is helpful to consider a particular example, i.e. let us imagine we have two curvilinear triangles ABC, BCD whose sides are arcs belonging to circumferences which cut orthogonally the fundamental circle. Let us suppose that the curvilinear angles of these triangles are equal to

$$B\hat{A}C = B\hat{D}C = \frac{\pi}{\alpha}, \quad C\hat{B}A = C\hat{B}D = \frac{\pi}{\beta}, \quad B\hat{C}A = B\hat{C}D = \frac{\pi}{\gamma}, \tag{25.1.5}$$

respectively, where α, β, γ are positive integers (either finite or infinite) such that

$$\frac{1}{\alpha} + \frac{1}{\beta} + \frac{1}{\gamma} < 1. \tag{25.1.6}$$

One can decompose the surface of the fundamental circle into infinitely many regions satisfying the conditions (i) and (ii) and in such a way that R_0 is the quadrangle $ABCD$. To this decomposition there corresponds a Fuchsian group that is denoted by (α, β, γ). Among the Fuchsian groups, some deserve a particular attention, i.e.

(1) The group $(2, 3, \infty)$, which is isomorphic to the group of operations that map z into $\frac{(az+b)}{(cz+d)}$, a, b, c, d being some integers such that $ad - bc = 1$.
(2) Certain groups which are isomorphic to the groups of linear substitutions with integer coefficients, that reproduce an indefinite ternary quadratic form with integer coefficients.

The existence of these groups makes it manifest that profound links exist which unify number theory with the analytic questions relevant for Fuchsian functions.

25.2 Θ-Fuchsian Functions

A Θ-*Fuchsian function* is any function Θ, uniform in the z-variable, and such that (K_i being any operation whatsoever of a Fuchsian group) one has identically

$$\Theta(K_i z) = \Theta(z) \left(\frac{\mathrm{d}}{\mathrm{d}z} K_i z \right)^{-m}, \qquad (25.2.1)$$

m being a positive integer. In other words, for infinitely many values of a, b, c, d such that

$$ad - bc = 1, \qquad (25.2.2)$$

one has identically

$$\Theta \left(\frac{az + b}{cz + d} \right) = \Theta(z)(cz + d)^{2m}. \qquad (25.2.3)$$

Poincaré was able to prove that there exist infinitely many Θ-Fuchsian functions defined by the convergent series

$$\sum_{i=1}^{\infty} H(K_i z) \left(\frac{\mathrm{d}}{\mathrm{d}z} K_i z \right)^{m}.$$

In this series, m is an integer number bigger than 1; K_i is an operation whatsoever of an arbitrary Fuchsian group G; H is a rational function of z.

The ratio of two Θ-Fuchsian functions corresponding to the same Fuchsian group and to the same value of the integer number m is a function F uniform in z, and such that [112]

$$F(K_i z) = F(z). \qquad (25.2.4)$$

It is therefore a Fuchsian function, by virtue of the definition given in Sect. 25.1. In other words, one has identically, for infinitely many values of the constants a, b, c, d,

$$F \left(\frac{az + b}{cz + d} \right) = F(z). \qquad (25.2.5)$$

Poincaré proved two remarkable theorems, which are as follows [112].

Theorem 25.1 *Among two Fuchsian functions having the same group and having as only essential singularities those that result from their definition, there exists an algebraic relation.*

Theorem 25.2 *Every Fuchsian function F makes it possible to integrate a linear equation with algebraic coefficients in the following way. If one defines*

$$x = F(z), \quad y_1 = \sqrt{\frac{\mathrm{d}F}{\mathrm{d}z}}, \quad y_2 = z\sqrt{\frac{\mathrm{d}F}{\mathrm{d}z}}, \tag{25.2.6}$$

the functions y_1 and y_2 satisfy the differential equation

$$\frac{\mathrm{d}^2 y}{\mathrm{d}x^2} = y\varphi(x), \tag{25.2.7}$$

$\varphi(x)$ *being algebraic in x.*

In particular, let us consider the equation

$$\frac{\mathrm{d}^2 y}{\mathrm{d}x^2} = y\left[\frac{\frac{1}{\alpha^2}-1}{4x^2} + \frac{\frac{1}{\beta^2}-1}{4(x-1)^2} + \frac{1+\frac{1}{\gamma^2}-\frac{1}{\alpha^2}-\frac{1}{\beta^2}}{4x(x-1)}\right], \tag{25.2.8}$$

where α, β, γ are positive integers, either finite or infinite, and fulfill the restriction (25.1.6).

If z is the ratio of the integrals, one has $x = f(z)$, f being a Fuchsian function pertaining to the group (α, β, γ). Remarkably, f exists only within the fundamental circle, and can be viewed as the ratio of the two Fuchsian functions

$$\frac{\left(\frac{\mathrm{d}f}{\mathrm{d}z}\right)^m}{[f(z)]^p[f(z)-1]^q}, \quad \frac{\left(\frac{\mathrm{d}f}{\mathrm{d}z}\right)^m f(z)}{[f(z)]^p[f(z)-1]^q}, \tag{25.2.9}$$

where m, p, q are integer numbers satisfying the inequalities, always compatible among themselves,

$$1 - \frac{p}{m} \geq \frac{1}{\alpha}, \quad 1 - \frac{q}{m} \geq \frac{1}{\beta}, \quad \frac{p+q-1}{m} - 1 \geq \frac{1}{\gamma}. \tag{25.2.10}$$

The two functions in (25.2.9), which only exist within the fundamental circle, are holomorphic within this circle.

If $\alpha = \beta = \gamma = \infty$, Eq. (25.2.8) reduces to the equation that determines the periods of $\sin \mathrm{am}x$ in terms of the square of the modulus.

25.3   System of ζ-Fuchsian Functions

Let F be an arbitrary Fuchsian function of the complex variable z, and set $x = F(z)$. A *system of ζ-Fuchsian functions* is any system of functions $\theta_1(z), \ldots, \theta_n(z)$ uniform in z, and such that the determinant

$$\det \begin{pmatrix} \frac{d^{\alpha_1}\theta_1}{dx^{\alpha_1}} & \cdots & \frac{d^{\alpha_n}\theta_1}{dx^{\alpha_n}} \\ \cdots & \cdots & \\ \frac{d^{\alpha_1}\theta_n}{dx^{\alpha_1}} & \cdots & \frac{d^{\alpha_n}\theta_n}{dx^{\alpha_n}} \end{pmatrix} \tag{25.3.1}$$

is a Fuchsian function of z, for whichever values of the integers $\alpha_1, ..., \alpha_n$. It is clear, by construction, that the functions $\theta_1, ..., \theta_n$ satisfy a linear differential equation whose coefficients are algebraic in x.

Poincaré proved that one can form an infinite number of ζ-Fuchsian functions of which he provided various expressions in terms of series, and which make it possible to integrate infinitely many differential equations, and, among others, *all linear differential equations with rational coefficients having only two singular points at finite distance and a singular point at infinity.*

A particular application can be described as follows. Let K and K' be the periods of an elliptic function, and ω the square of its modulus. Let φ be an algorithm such that

$$\omega = \varphi\left(\frac{K + iK'}{K - iK'}\right). \tag{25.3.2}$$

Let us consider a linear differential equation with rational coefficients having as singular points $x = 0$, $x = 1$, $x = \infty$. Set $x = \varphi(z)$, and let $\theta_1(z), ..., \theta_n(z)$ be the integrals of the proposed equation; $\varphi(z)$ is found to be a Fuchsian function, while $\theta_1, ..., \theta_n$ are ζ-Fuchsian functions. Such functions exist only within the fundamental circle (for many years, Poincaré tried to prove that no Fuchsian function can exist anywhere, until he found the properties stated so far); they are holomorphic within this circle, and therefore can be always represented by series analytic in the whole complex plane, whose coefficients can be easily evaluated.

To sum up, Poincaré proved the existence of a very rich class of functions, of which the elliptic functions are only a particular case. They make it possible to integrate a large number of differential equations.

Chapter 26
The Kernel of (Laplacian Plus Exponential)

26.1 Motivations for the Analysis

Among the ordinary differential equations of the form [115]

$$\frac{d^2 v}{dx^2} = \varphi(x, y)v, \tag{26.1.1}$$

where φ is a rational function of two variables x and y related by a given algebraic equation

$$f(x, y) = 0, \tag{26.1.2}$$

with given singular points of Eq. (26.1.1) and assuming that the difference of roots of every determinant equation is a given integer, there exists always a Fuchsian equation, i.e. generating Fuchsian functions. Klein and Poincaré obtained a first proof of this property, based upon the continuity principle, and later on Picard reduced the problem to the integration of the non-linear partial differential equation

$$\triangle u = e^u. \tag{26.1.3}$$

The method used by Picard consisted in establishing first the integrability in a sufficiently small domain, with subsequent extension to the whole plane.

Poincaré wanted to avoid this diversion, and hence he proceeded as follows. He introduced the *Klein surface*, i.e. a closed surface such that every real point of this surface corresponds to an imaginary point of the curve (26.1.2), and conversely. He then set

$$Du \equiv \triangle u \frac{d\Omega}{d\omega}, \tag{26.1.4}$$

where $d\omega$ is a surface element of the Klein surface, while $d\Omega$ is the corresponding element in the x-plane. Equation (26.1.3) is then re-expressed in the form

© Springer International Publishing AG 2017
G. Esposito, *From Ordinary to Partial Differential Equations*, UNITEXT 106,
DOI 10.1007/978-3-319-57544-5_26

$$DU = \Theta \, e^U - \Phi, \tag{26.1.5}$$

where Θ and Φ are two given functions, the former being always positive. The problem of forming a Fuchsian equation becomes therefore the problem of finding the function U which must be everywhere finite. The analysis relies upon certain simple inequalities which result from one remark only: if U is maximum, DU is negative, while if U is minimum, DU is positive.

Poincaré began by integrating the equation

$$Du = \varphi, \tag{26.1.6}$$

where the function φ is given; this integration is only possible if

$$\int_{\text{Klein}} \varphi \, d\omega = 0, \tag{26.1.7}$$

the integral being taken over the whole Klein surface. Equation (26.1.6) has the same form of the Poisson equation studied in potential theory, i.e.

$$\Delta u = \varphi, \tag{26.1.8}$$

which can be integrated by means of the Green function. Equation (26.1.6) can be integrated in analogous way, and *the function that plays the role of the Green function is the real part of an Abelian integral of third kind* that can be easily formed.

As a next step, Poincaré studied the equation

$$Du = \lambda \, \eta \, u - \varphi - \lambda \, \psi, \tag{26.1.9}$$

where η, φ, ψ are three given functions, the first being always positive, while λ is a positive parameter.

Poincaré proved first that Eq. (26.1.9) can be integrated for small values of λ, and that the integral can be expanded in powers of λ. He proved afterwards that, if the equation is integrable for $\lambda = \lambda_0$, it is again integrable for small values of $\lambda - \lambda_0$, and that the integral can be expanded in powers of $\lambda - \lambda_0$. He proved eventually that Eq. (26.1.9) is integrable for all positive values of λ.

A case exists where this method is not applicable. This occurs when the Fuchsian polygon has some vertices on the fundamental circle; in this case there exist points where the functions η and θ diverge as

$$\frac{1}{x^2 \log^2(x)}.$$

In this case, on the other hand, the Picard method equally fails. Poincaré had to work very hard to overcome this technical difficulty. Remarkably, he proved eventually that the desired integral is always finite and can be expanded in powers of λ, but the terms of the expansion may become infinite.

This is indeed the case for the function $x \rightarrow x^\lambda$ which remains finite for $x = 0$, if $\lambda > 0$; it can be expanded in powers of λ

$$x^\lambda = 1 + \lambda \, \log(x) + \frac{\lambda^2}{2}(\log(x))^2 + \cdots , \tag{26.1.10}$$

but the individual terms of the expansion become infinite at $x = 0$.

After having got rid of this difficulty, Poincaré approached the equation

$$Du = \theta \, e^u - \varphi - \lambda \, \psi, \tag{26.1.11}$$

where the functions θ, φ, ψ are known and positive, while λ is a positive parameter.

He assumed of being able to solve Eq. (26.1.11) for $\lambda = 0$; he was then able to form a series out of the powers of λ, and satisfying Eq. (26.1.11). He obtained every term of the series by integrating an equation of the form (26.1.9). By virtue of the inequalities established at the beginning, he could prove easily that the series converges, if λ is sufficiently small. Moreover, a step by another, he could prove that Eq. (26.1.11) can be integrated for all positive values of λ.

To sum up, Eq. (26.1.5) can be integrated if Φ is always positive, and it is easy to prove that it is again integrable provided that

$$\int \Phi \, d\omega > 0. \tag{26.1.12}$$

This condition is necessary and sufficient. It remained to verify that condition (26.1.12) is fulfilled in the applications to Fuchsian functions, and this turned out to be correct in the cases of interest. One might envisage the possibility of a rigorous proof based upon the calculus of variations, because one can easily form a double integral which must be minimum if Eq. (26.1.5) is satisfied. But this kind of argument is not satisfactory, because, this integral being dependent on an arbitrary function, it is not sure it has a minimum in the proper sense of the word. However, in the problem considered, the unknown function $\varphi(x, y)$ must satisfy certain conditions, hence it depends only on a finite number of unknown constants. The function u will therefore depend, itself, on a finite number of unknown constants. Our double integral, which no longer depends on an arbitrary function, but only depends on a certain number of arbitrary parameters, will certainly have a minimum, as was desired.

26.2 The u Function

The occurence of Eq. (26.1.2) suggests studying the general problem of two complex variables Z and Z' related by an algebraic relation

$$f(Z, Z') = 0. \tag{26.2.1}$$

Since Z and Z' are complex, they read as

$$Z = X + iY, \ Z' = X' + iY', \tag{26.2.2}$$

while their complex conjugates are

$$Z_0 = X - iY, \ Z_0' = X' - iY'. \tag{26.2.3}$$

In light of Eq. (26.2.1), the variables Z_0 and Z_0' are related by an equation

$$f_0(Z_0, Z_0') = 0, \tag{26.2.4}$$

where f_0 is a polynomial whose coefficients are complex conjugates of the coefficients of f.

Let us now suppose that Z and Z' are some Fuchsian functions

$$Z = \varphi(z), \ Z' = \varphi'(z) \tag{26.2.5}$$

of the complex variable $z = x + iy$. This implies that Z_0 and Z_0' are Fuchsian functions

$$Z_0 = \varphi_0(z_0), \ Z_0' = \varphi_0'(z_0) \tag{26.2.6}$$

of the complex conjugate variable $z_0 = x - iy$. We note also that, with our notation, the fundamental circle of Chap. 25 has equation

$$x^2 + y^2 = 1 \implies zz_0 = 1. \tag{26.2.7}$$

Let us assume in addition that the variable Z describes in its plane an infinitesimal arc dS, while the variable z describes in its plane the corresponding infinitesimal arc. Let ds be the length of the arc described by z, a length evaluated from the point of view of non-Euclidean geometry. One has [116]

$$dS = \sqrt{dZ \, dZ_0}, \ ds = \frac{\sqrt{dz \, dz_0}}{(1 - zz_0)}. \tag{26.2.8}$$

Let

$$z \to \left(\frac{\alpha z + \beta}{\gamma z + \delta} \right) \tag{26.2.9}$$

be a substitution of the Fuchsian group. When the variable z describes the arc of length ds, the variable $\frac{\alpha z + \beta}{\gamma z + \delta}$ describes an arc of length again equal to ds, and the variable

$$\varphi \left(\frac{\alpha z + \beta}{\gamma z + \delta} \right),$$

which is again equal to Z, describes an arc of length again equal to dS.

The ratio $\frac{dS}{ds}$ is therefore unaffected by the substitutions of the Fuchsian group; it is hence a uniform function of Z, Z', Z_0, Z_0'. Let us therefore define

$$e^u \equiv \frac{ds^2}{dS^2} = \frac{dz}{dZ}\frac{dz_0}{dZ_0}(1 - zz_0)^{-2}, \tag{26.2.10}$$

from which u can be obtained in the form

$$u = -\log\frac{dZ}{dz} - \log\frac{dZ_0}{dz_0} - 2\log(1 - zz_0). \tag{26.2.11}$$

If one remarks that, in light of (26.2.5) and (26.2.6), Z depends only on z while Z_0 depends only on z_0, one finds

$$\frac{\partial^2 u}{\partial Z \partial Z_0} = -2\frac{dz}{dZ}\frac{dz_0}{dZ_0}\frac{\partial^2}{\partial z \partial z_0}\log(1 - zz_0) = 2\frac{dz}{dZ}\frac{dz_0}{dZ_0}(1 - zz_0)^{-2} = 2e^u. \tag{26.2.12}$$

On passing from the variables Z and Z_0 to the variables X and Y, one finds therefore

$$\triangle u = \left(\frac{\partial^2}{\partial X^2} + \frac{\partial^2}{\partial Y^2}\right)u = 4\frac{\partial^2 u}{\partial Z \partial Z_0} = 8e^u. \tag{26.2.13}$$

For the function u to stop being finite, it is necessary that one of the factors

$$\frac{dz}{dZ}, \quad \frac{dz_0}{dZ_0}, \quad 1 - zz_0$$

either vanishes or becomes infinite.

The factor $1 - zz_0$ can never become infinite; if the Fuchsian function is of the *first family*, the factor $1 - zz_0$ cannot vanish, because the polygon R_0 of Chap. 25 has no points on the circumference of the fundamental circle. If, on the contrary, the polygon R_0 has some cycles with vertices of the third kind, the factor $1 - zz_0$ can only vanish at a vertex (see Appendix) belonging to one of these cycles.

Outside the vertices belonging to cycles of the third kind, the function u remains finite unless one of the factors

$$\frac{dz}{dZ}, \quad \frac{dz_0}{dZ_0}$$

vanishes or becomes infinite. Since these two factors are complex conjugate of each other, they vanish or become infinite at the same time.

The condition for u being no longer finite is therefore

$$\frac{dZ}{dz} = 0 \text{ or } \frac{dZ}{dz} = \infty. \tag{26.2.14}$$

One has $\frac{dZ}{dz} = 0$ in two cases:

(i) At the vertices of the polygon R_0 which belong to a cycle of second or third kind.

(ii) If one has at the same time

$$f(Z, Z') = 0, \quad \frac{\partial f}{\partial Z'} = 0, \tag{26.2.15}$$

i.e. if the tangent to the curve $f(Z, Z') = 0$ is parallel to the Z' axis.

If the corresponding point is not a double point of the curve $f = 0$, one has that $\frac{\partial f}{\partial Z'} = 0$, but the derivative $\frac{\partial f}{\partial Z}$ is nowhere vanishing, and hence also $\frac{dZ'}{dz} \neq 0$. Thus, if one defines

$$dS' \equiv \sqrt{dZ'\, dZ'_0} \tag{26.2.16}$$

and

$$e^{u'} \equiv \frac{ds^2}{dS'^2}, \tag{26.2.17}$$

i.e. if u' is the function which, with respect to Z', plays the same role as u with respect to Z, this new function u' does not become infinite at the corresponding point [116].

Let us now assume that the point considered is a double point of $f = 0$; one has therefore at the same time

$$\frac{dZ}{dz} = 0, \quad \frac{dZ'}{dz} = 0. \tag{26.2.18}$$

But let us suppose that

$$Z'' = R(Z, Z'), \tag{26.2.19}$$

R being a rational function of Z and Z'; we can choose this function R in such a way that Z'' takes different values at the double point on the two branches of the curve that grow out of this point. Under these conditions, the derivative $\frac{dZ'}{dz}$ does not vanish, and therefore, if u'' is the function which plays, with respect to Z'', the same role that u' plays with respect to Z', u'' does not become infinite at the corresponding point.

On the other hand, one can always assume that the curve $f = 0$ does not display singular points other than ordinary double points. If, on the contrary, the point z were a vertex of the polygon R_0, belonging to a cycle of second or third kind, all the derivatives

$$\frac{dZ}{dz}, \frac{dZ'}{dz}, \frac{dZ''}{dz}$$

would vanish at once, and all the functions

$$u, u', u''$$

would become infinite at once.

One has

$$Z = \infty \implies \frac{dZ}{dz} = \infty. \tag{26.2.20}$$

But if one sets

$$Z'' = \frac{1}{Z}, \tag{26.2.21}$$

and if u'' is the function which plays, with respect to Z'', the same role played by u with respect to Z, one has $Z'' = 0$, and hence neither $\frac{dZ''}{dz}$ can vanish nor can u'' become infinite.

To sum up, if the Fuchsian function is of first kind, i.e. if it does not possess cycles of second and third kind, the functions $u, u', u'', \ldots$ cannot become infinite at once.

In the general case, these functions can only become infinite at once at the different vertices of the various cycles of second and third kind. Let us now study what happens at a vertex of a cycle of second kind. If the sum of the angles of the cycle is $\frac{2\pi}{n}$, and if a is a vertex of second kind, and A is the corresponding value of Z, the Fuchsian function

$$Z - A = \varphi(z) - A \tag{26.2.22}$$

can be expanded in powers of $z - a$, and the expansion begins by a term in $(z - a)^n$. The expansion of $\frac{dZ}{dz}$ begins by a term in $(z - a)^{n-1}$; one obtains therefore

$$\log(Z - A) = n \log(z - a) + Q, \tag{26.2.23}$$

$$\log \frac{dZ}{dz} = (n - 1) \log(z - a) + Q', \tag{26.2.24}$$

the functions Q and Q' remaining finite for $z = a$, and therefore

$$\log \frac{dZ}{dz} = \frac{(n - 1)}{n} \log(Z - A) + Q'', \tag{26.2.25}$$

$$\log \frac{dZ_0}{dz_0} = \frac{(n - 1)}{n} \log(Z_0 - A_0) + Q_0'', \tag{26.2.26}$$

and eventually, from (26.2.11),

$$u = -\frac{(2n - 2)}{n} \log |Z - A| + q, \tag{26.2.27}$$

Q'', Q_0'' and q remaining finite for $z = a$. Moreover, q can be expanded in powers of $Z - A$, $Z_0 - A_0$, $((Z - A)(Z_0 - A_0))^{\frac{1}{n}}$.

As far as cycles of third kind are concerned one has, in a neighbourhood of such cycles,

$$\frac{1}{(z-a)} = \psi(Z) + b \log(Z - A), \tag{26.2.28}$$

where b is a constant and ψ is a function that can be expanded in powers of $Z - A$. One derives from this

$$\frac{dz}{dZ} = -(z-a)^2 \left[\psi'(Z) + \frac{b}{(Z-A)} \right], \tag{26.2.29}$$

which implies

$$\log \frac{dZ}{dz} = \log(Z - A) - 2\log(z - a) + Q, \tag{26.2.30}$$

Q denoting a quantity that remains finite for $z = a$.

With analogous procedure, one would have

$$\log \frac{dZ_0}{dz_0} = \log(Z_0 - A_0) - 2\log(z_0 - a_0) + Q_0, \tag{26.2.31}$$

Q_0 remaining finite for $z = a$. Since the vertex a is on the fundamental circle, one has $aa_0 = 1$, and hence

$$u = -2\log|Z - A| + 2\log \frac{(z-a)(z_0 - a_0)}{(aa_0 - zz_0)} + q, \tag{26.2.32}$$

q remaining finite for $z = a$.

Note now that one has identically

$$\frac{zz_0 - aa_0}{(z-a)(z_0 - a_0)} = \frac{a}{(z-a)} + \frac{a_0}{(z_0 - a_0)} + 1 = ab \log(Z - A) + a_0 b_0 \log(Z_0 - A_0) + q', \tag{26.2.33}$$

where q' remains finite for $z = a$.

It is easy to verify that the product ab must be real, in such a way that

$$ab = a_0 b_0, \tag{26.2.34}$$

and (26.2.33) takes the form

$$\frac{zz_0 - aa_0}{(z-a)(z_0 - a_0)} = 2ab \log|Z - A| + q' = -\log|Z - A| \, e^{q''}, \tag{26.2.35}$$

the function q'' remaining finite for $z = a$.

Thus, one finds eventually

$$u = -2\log|Z - A| - 2\log\log|Z - A| - 2q'' + q. \tag{26.2.36}$$

More precisely, one has

$$u = -2 \log |Z - A| - 2 \log \left(\log |Z - A| + \psi \right) + \varphi, \tag{26.2.37}$$

the functions ψ and φ being holomorphic in $(Z - A)$ and $(Z_0 - A_0)$.
The findings so far are as follows.

(i) In the neighbourhood of a vertex of second kind, all the differences

$$u + \frac{(2n - 2)}{n} \log |Z - A|, \ u' + \frac{(2n - 2)}{n} \log |Z' - A'|, \ u'' + \frac{(2n - 2)}{n} \log |Z'' - A''|$$

cannot become infinite at once (i.e., at least one of them remains finite).

(ii) In the neighbourhood of a vertex of third kind, all the differences

$$u + 2 \log |Z - A| + 2 \log \log |Z - A|, \ u' + 2 \log |Z' - A'| + 2 \log \log |Z' - A'|,$$

$$u'' + 2 \log |Z'' - A''| + 2 \log \log |Z'' - A''|, \ ...$$

cannot become infinite at once.

It is important to understand the reason of this difference among vertices of second and third kind. For this purpose, let us set

$$\rho^2 \equiv x^2 + y^2, \tag{26.2.38}$$

and suppose, for simplicity, that u depends on ρ only. Equation (26.2.13) can then be written in the form

$$\left(\frac{d^2}{d\rho^2} + \frac{1}{\rho} \frac{d}{d\rho} \right) u = 8e^u. \tag{26.2.39}$$

Suppose that one wants to integrate it by successive approximations in the following way. One assumes that u_0 solves the homogeneous equation

$$\left(\frac{d^2}{d\rho^2} + \frac{1}{\rho} \frac{d}{d\rho} \right) u_0 = 0, \tag{26.2.40}$$

while the subsequent terms obey inhomogeneous equations according to the algorithm

$$\left(\frac{d^2}{d\rho^2} + \frac{1}{\rho} \frac{d}{d\rho} \right) u_1 = 8e^{u_0}, \tag{26.2.41}$$

$$\left(\frac{d^2}{d\rho^2} + \frac{1}{\rho} \frac{d}{d\rho} \right) u_2 = 8 \left(e^{u_0 + u_1} - e^{u_0} \right), \tag{26.2.42}$$

$$\left(\frac{d^2}{d\rho^2} + \frac{1}{\rho} \frac{d}{d\rho} \right) u_3 = 8 \left(e^{u_0 + u_1 + u_2} - e^{u_0 + u_1} \right), \tag{26.2.43}$$

and infinitely many other equations of the general form

$$\left(\frac{d^2}{d\rho^2} + \frac{1}{\rho}\frac{d}{d\rho}\right)u_n = 8\left[\exp\left(\sum_{k=0}^{n-1}u_k\right) - \exp\left(\sum_{k=0}^{n-2}u_k\right)\right], \qquad (26.2.44)$$

so that eventually the desired solution is given by

$$u = \sum_{k=0}^{\infty}u_k. \qquad (26.2.45)$$

Some insight is already gained by limiting ourselves to the first two approximations. One finds, for example,

$$u_0 = p\log(\rho), \quad e^{u_0} = \rho^p, \qquad (26.2.46)$$

$$u_1 = \frac{8\rho^{p+2}}{(p+2)^2}. \qquad (26.2.47)$$

The expression of u_1 becomes misleading when $p = -2$, and one can infer that Eq. (26.2.39) cannot have an integral of the form

$$u = -2\log(\rho) + v, \qquad (26.2.48)$$

the function v remaining finite when $\rho = 0$. On the contrary, there exists an integral of Eq. (26.2.39) of the form

$$u = p\log(\rho) + v, \qquad (26.2.49)$$

if $p \neq -2$.

Equation (26.2.39) admits however the integral

$$u = -2\log(\rho) - 2\log\log(\rho) - \log(4). \qquad (26.2.50)$$

This remark is sufficient to make it possible to understand the difference between vertices of second kind, which correspond to $p \neq -2$, and vertices of third kind, which correspond instead to $p = -2$.

The variable z is the ratio of integrals of a certain Fuchsian equation

$$\frac{d^2}{dZ^2}v = \varphi(Z, Z')v, \qquad (26.2.51)$$

which corresponds to Eq. (26.1.1).

If the *kind* of this equation is given, one knows:

(i) The algebraic relation

$$f(Z, Z') = 0. \qquad (26.2.52)$$

(ii) The values A of Z which correspond to the different cycles with vertices either of second or of third kind.

(iii) The integer number n relative to each vertex of second kind.

26.3 Klein Surfaces

Klein imagined a way of representing algebraic functions whose application remains of high pedagogical value. Following again Poincaré [116], the basic concepts are as follows.

Let us consider an algebraic curve defined by Eq. (26.2.52). One can build a closed surface in such a way that, to every real or imaginary point of this curve there corresponds one and only one real point of the surface. Meanwhile, to a double point of the curve (26.2.52) there correspond two real points of the surface, belonging to the two branches of the curve that cut each other at the double point, respectively.

We shall assume, on the other hand, that the coordinates of the point of the surface are continuous functions of the coordinates of the corresponding point of the curve, and also that these continuous functions possess derivatives of all orders, and hence are of class C^∞.

The Riemann surfaces are only a particular case of Klein surfaces; a Klein surface reduces to a Riemann surface when it is infinitely flattened and is reduced to a certain number of sheets applied upon a plane. But one has to distinguish between *isotropic* and *anisotropic Klein surfaces*.

A Klein surface is said to be *isotropic* if the correspondence between the points of the curve and those of this surface is such that the representation of the Z-plane on the Klein surface is a conformal representation. To the different points of the Z-plane there correspond different points of the curve (26.2.52), and hence different points of the Klein surface. This correspondence defines a transformation of a region of the Z-plane into a region of the Klein surface; for the Klein surface to be isotropic, this transformation must be angle-preserving.

One might decide to use isotropic surfaces only, because the work of Schwarz and Klein makes it possible to state that the different points of an arbitrary algebraic curve can be always represented on an isotropic Klein surface. But it is better not to impose such a restriction.

On the contrary, one can regard it as being clear that, if an arbitrary curve (26.2.52) of genus p and an arbitrary surface $(2p + 1)$-times connected are given, one can establish a one-to-one correspondence between the points of the curve and those of the surface. In order to specify the nature of the correspondence, one can assume that the curve (26.2.52) does not have singularities other than ordinary double points and that it has a regular behaviour at infinity, i.e. its asymptotic directions are all distinct and not parallel to the axes. Having done so, let M be a point on the surface, corresponding to the point Z, Z' of the curve; let P be a point of the Klein surface, corresponding to the point $Z = A$, $Z' = A'$ of the curve. If the distance $M P$ is very small, and if $\frac{\partial f}{\partial Z} \neq 0$ at $Z = A$, $Z' = A'$, the distance MP is of the same order of magnitude as $|Z - A|$. If, on the contrary,

$$\frac{\partial f}{\partial Z} = 0, \quad \frac{\partial f}{\partial Z'} \neq 0, \tag{26.3.1}$$

the distance MP is not of the same order of magnitude as $|Z - A|$; it is, in general, of the same order of magnitude as the square of $|Z - A|$, and in any case of the same order of magnitude as $|Z' - A'|$.

If one has at the same time

$$\frac{\partial f}{\partial Z} = 0, \quad \frac{\partial f}{\partial Z'} = 0, \tag{26.3.2}$$

i.e. if the point $Z = A$, $Z' = A'$ is a double point of the curve (26.2.52), the distance MP is no longer of the same order of magnitude as $|Z - A|$ or $|Z' - A'|$, but one can set

$$Z'' = R(Z, Z'), \tag{26.3.3}$$

R being a rational function of Z and Z' whose numerator and denominator vanish at the double point.

If A'' is one of the two values that Z'' takes at the double point, and, precisely, the value corresponding to the point P, the distance MP is of the same order of magnitude as $|Z'' - A''|$.

Last, if the point P corresponds to a point at infinity of the curve (26.2.52), the distance MP is of the same order of magnitude as $\left|\frac{1}{Z}\right|$.

Let us now consider a function u of the coordinates of the point M; the point M corresponds to a point of the Z-plane whose coordinates are X and Y, real and imaginary part of Z, respectively. Also the function u depends therefore on X and Y. Let dS be an infinitesimal arc in the Z-plane, and $d\sigma$ the corresponding arc of the Klein surface; let $d\Omega$ be an infinitesimal area in the Z-plane, and $d\omega$ the corresponding area on the Klein surface. We draw, through the edge Z of the arc dS, an infinitesimal arc ZZ_1 orthogonal to dS and of length dN. Let MM_1 be the infinitesimal arc of the Klein surface corresponding to the arc ZZ_1 and ending at the edge M of the arc $d\sigma$.

Suppose first that the Klein surface is isotropic; in such a case, denoting by dn the length of the arc MM_1, the arc MM_1 is orthogonal to $d\sigma$ and one has

$$\frac{d\sigma}{dn} = \frac{dS}{dN}, \quad \frac{d\omega}{d\Omega} = \left(\frac{d\sigma}{dS}\right)^2. \tag{26.3.4}$$

Let $u = u(M) = u(Z)$, and let

$$u + \frac{du}{dn}dn = u + \frac{du}{dN}dN = u(M_1) = u(Z_1). \tag{26.3.5}$$

One has

$$\frac{du}{dn} d\sigma = \frac{du}{dN} dS.$$
(26.3.6)

The Green theorem, applied to the Z-plane, leads to

$$\int \Delta u \, d\Omega = \int \frac{du}{dN} dS,$$
(26.3.7)

or, on considering two functions u and v,

$$\int v \, \Delta u \, d\Omega = \int v \frac{du}{dN} dS - \int \left(\frac{\partial u}{\partial X} \frac{\partial v}{\partial X} + \frac{\partial u}{\partial Y} \frac{\partial v}{\partial Y} \right) d\Omega.$$
(26.3.8)

The integrals are extended over all elements $d\Omega$ of a certain area, and to all elements dS of a certain contour bounding this area. The functions u and v are taken to be of class C^1.

Let us now set (see (26.1.4))

$$DU \equiv \Delta u \frac{d\Omega}{d\omega},$$
(26.3.9)

and

$$E(u, v) \equiv \left(\frac{\partial u}{\partial X} \frac{\partial v}{\partial X} + \frac{\partial u}{\partial Y} \frac{\partial v}{\partial Y} \right) \frac{d\Omega}{d\omega},$$
(26.3.10)

from which one finds that E is symmetric and positive-definite, i.e.

$$E(u, v) = E(v, u), \quad E(u, u) > 0.$$
(26.3.11)

Equations (26.3.7) and (26.3.8) can be therefore re-expressed as

$$\int Du \, d\omega = \int \frac{du}{dn} d\sigma,$$
(26.3.12)

and

$$\int v \, Du \, d\omega = \int v \frac{du}{dn} d\sigma - \int E(u, v) d\omega,$$
(26.3.13)

respectively. These equations represent the Green theorem extended to the Klein surface, but the former admits a remarkable physical interpretation. If u represents the temperature of a point of the surface, then $\frac{du}{dn}$ is the derivative of this temperature evaluated along the normal to the arc $d\sigma$. If the surface is homogeneous and isotropic from the point of view of the heat that it is conducting, the product $\frac{du}{dn} d\sigma$ represents the heat flux passing through the surface element $d\sigma$ in the time unit. The integral

$$\int \frac{du}{dn}d\sigma$$

represents the quantity of heat entering the area $\int d\omega$ through its contour, while $Du\, d\omega$ represents the quantity of heat given to the surface element $d\omega$ by the neighbouring elements of the surface.

If the Klein surface is instead anisotropic, the arc MM_1 is no longer orthogonal to the arc $d\sigma$. If one considers on the Z-plane an infinitesimally small circle centred at the point Z, the corresponding figure on the Klein surface is an infinitesimally small ellipse, centred at the point M. The tangents to the arc MM_1 and to the arc $d\sigma$ are then two conjugate diameters of this ellipse. One then has

$$\frac{MM_1}{dN} \neq \frac{d\sigma}{dS}, \tag{26.3.14}$$

but one can define

$$dn = \frac{d\sigma\, dN}{dS}, \tag{26.3.15}$$

and one finds

$$MM_1 = \frac{a'\, dn}{b'}, \tag{26.3.16}$$

where a' and b' are the lengths of the conjugate diameters of our ellipse that are directed along the tangents to MM_1 and $d\sigma$, respectively.

Having modified in such a way the definition of dn, one finds

$$\frac{du}{dn}d\sigma = \frac{du}{dN}dS. \tag{26.3.17}$$

The definitions (26.3.9) of Du and (26.3.10) of $E(u, v)$ are preserved, and one recovers Eqs. (26.3.12) and (26.3.13).

A physical interpretation remains possible; indeed, if the surface is taken to be heat conducting, but with anisotropic conductivity, the flux of heat passing through the surface element $d\sigma$ is again represented by the product $\frac{du}{dn}d\sigma$, the definition of dn having been modified according to Eq. (26.3.15). Moreover, Eq. (26.3.12) implies that

$$\int Du\, d\omega = 0, \tag{26.3.18}$$

when the integral is extended over the whole Klein surface. The physical interpretation of this equation is that the conductivity of the surface leads to heat exchanges among the various parts, but cannot modify the total quantity of heat possessed by the whole surface.

26.4 The *U* Function

Following Poincaré [116], we consider various rational functions of Z, Z', and the corresponding functions u, u', u'' defined in Sect. 26.2. We set now

$$e^U = e^u \frac{d\Omega}{d\omega}. \tag{26.4.1}$$

If the Klein surface were isotropic, one would find

$$e^U = \frac{ds^2}{d\sigma^2}. \tag{26.4.2}$$

In all cases one has

$$e^U = e^{u'} \frac{d\Omega'}{d\omega}, \tag{26.4.3}$$

the area $d\Omega'$ being the area of the Z' plane which corresponds to the area $d\Omega$ of the Z-plane. Indeed, one has

$$e^u = \frac{ds^2}{dS^2}, \quad e^{u'} = \frac{ds^2}{dS'^2}, \quad \frac{d\Omega}{d\Omega'} = \frac{dS^2}{dS'^2}. \tag{26.4.4}$$

One would have, in analogous way,

$$e^U = e^{u''} \frac{d\Omega''}{d\omega}. \tag{26.4.5}$$

The definition of U is therefore independent of the choice of rational function of Z and Z'. Hence one finds

$$Du = \Delta u \frac{d\Omega}{d\omega} = 8e^u \frac{d\Omega}{d\omega} = 8e^U, \tag{26.4.6}$$

and eventually

$$DU = 8e^U + D \log \frac{d\Omega}{d\omega}. \tag{26.4.7}$$

Since $D \log \frac{d\Omega}{d\omega}$ is a known function that can be set equal to $-\Phi$, one can write

$$DU = 8e^U - \Phi. \tag{26.4.8}$$

The definition of U being independent of the choice for Z, it must be possible to write

$$-\Phi = D \log \frac{d\Omega}{d\omega} = D \log \frac{d\Omega'}{d\omega} = D \log \frac{d\Omega''}{d\omega} = \cdots, \tag{26.4.9}$$

and therefore

$$D \log \frac{d\Omega'}{d\Omega} = 0, \quad \triangle \log \frac{d\Omega'}{d\Omega} = 0. \tag{26.4.10}$$

Note that the function Φ cannot become infinite, and indeed it remains finite unless $\log \frac{d\Omega}{d\omega} = \infty$. This infinite value is attained if

$$Z = \infty \text{ or } \frac{\partial f}{\partial Z'} = 0. \tag{26.4.11}$$

But one would need at the same time

$$\log \frac{d\Omega'}{d\omega} = 0 \implies Z' = \infty \text{ or } \frac{\partial f}{\partial Z} = 0. \tag{26.4.12}$$

Thus, there remain doubts only for the points of the Klein surface corresponding to the points at infinity or to the double points of the curve (26.2.52). But one would need furthermore, for all functions Z'', the equation

$$\log \frac{d\Omega'}{d\omega} = \infty. \tag{26.4.13}$$

This does not occur, for a point at infinity, if one sets $Z'' = \frac{1}{Z}$, nor does it occur for a double point, if one imposes Eq. (26.3.3), the numerator and denominator of R vanishing at the same time at the double point. The function Φ is therefore always finite. Q.E.D.

As far as the function U is concerned, if one sets aside the points of the Klein surface which correspond to a vertex of second or third kind of the polygon R_0, one can find a function Z'' such that u'' and $\log \frac{d\Omega''}{d\omega}$ are finite. The function U is therefore finite.

If the Fuchsian function is of first kind, the function U is therefore finite on every Klein surface, and is defined by the functional equation (26.4.8).

Suppose now that a is a vertex of second kind of R_0 and let A, A', A'' be the corresponding values of Z, Z' and Z'', respectively, and P the corresponding point of the Klein surface. One can assume of having chosen the function Z'' in such a way that $\log \frac{d\Omega''}{d\omega}$ remains finite. In such a case the difference

$$u'' + \frac{(2n-2)}{n} \log |Z'' - A''|$$

and, therefore, the difference

$$U + \frac{(2n-2)}{n} \log(MP),$$

remain finite.

In the meantime, a remark is in order. When Z'' approaches A'', the expression

$$u'' + \frac{(2n-2)}{n} \log |Z'' - A''|$$

approaches a finite and determined limit. But when the point M gets closer to P, the expression

$$U + \frac{(2n-2)}{n} \log(MP)$$

remains finite (unless the Klein surface is not isotropic), but does not approach a definite limit. Its limit depends instead on the direction of the line MP. If ρ is the radius of a very small circle located in the Z''-plane, and ε is the diameter of the ellipse corresponding to the Klein surface whose direction coincides with the MP direction, what approaches a definite limit when MP approaches zero is the expression

$$U + \frac{(2n-2)}{n} \log \left(\frac{\rho \, MP}{\varepsilon} \right),$$

that approaches a definite limit when MP approaches zero.

If one defines the function

$$v'' \equiv u'' + \frac{(2n-2)}{n} \log |Z'' - A''|, \tag{26.4.14}$$

one obtains

$$\triangle v'' = \triangle u'' = 8e^{u''} \frac{d\Omega''}{d\Omega} = 8e^{v''} |Z'' - A''|^{\frac{(2-2n)}{n}} \frac{d\Omega''}{d\Omega}. \tag{26.4.15}$$

One finds that the function v'' satisfies an equation of the form

$$\triangle v'' = 8e^{v''} \frac{d\Omega''}{d\Omega}, \tag{26.4.16}$$

where θ is a given function, always positive, but becoming infinite at $Z'' = A''$.

Let us now set

$$e^V \equiv e^{v''} \frac{d\Omega''}{d\omega}. \tag{26.4.17}$$

If one preserves for θ and Φ their meaning

$$\theta = 8 |Z'' - A''|^{\frac{(2-2n)}{n}}, \quad \Phi = -D \log \frac{d\Omega''}{d\omega}, \tag{26.4.18}$$

one finds

$$DV = \theta e^V - \Phi. \tag{26.4.19}$$

At the point P (corresponding to the point $Z'' = A''$), the function θ becomes infinite, but V and Φ remain finite.

Let us look in the same way at a vertex of third kind. The function u'' can be chosen in such a way that

$$u'' + 2 \log |Z'' - A''| + 2 \log \log |Z'' - A''|$$

remains finite, and we set [116]

$$v'' \equiv u'' + 2 \log \left| Z'' - A'' \right| + 2 \log \log \left| Z'' - A'' \right|, \tag{26.4.20}$$

$$\theta \equiv \frac{8}{|Z'' - A''|^2 \log^2 |Z'' - A''|}. \tag{26.4.21}$$

Hence one obtains

$$\Delta v'' = \Delta u'' + 2 \, \Delta \log \log \left| Z'' - A'' \right| = \frac{d\Omega''}{d\Omega} \left(8 e^{u''} - \frac{2}{\left| Z'' - A'' \right|^2 \log^2 \left| Z'' - A'' \right|} \right), \tag{26.4.22}$$

and, if one uses again the definition (26.4.17), one obtains

$$DV = \theta e^V - \Phi - \frac{\theta}{4} \frac{d\Omega''}{d\omega}. \tag{26.4.23}$$

This scheme suggests changing the definition of U, according to

$$U \equiv u + \log \frac{d\Omega}{d\omega} + h, \tag{26.4.24}$$

where h is a single-valued analytic function that remains finite at all points of the Klein surface, but at the points which correspond to vertices of second and third kind of R_0.

At a vertex of second kind, the difference

$$h - \frac{(2n - 2)}{n} \log \left| Z'' - A'' \right|$$

must remain finite; at a vertex of third kind, it is instead the difference

$$h - 2 \log \left| Z'' - A'' \right| - 2 \log \log \left| Z'' - A'' \right|$$

which remains finite. One can always choose the function h in such a way that these conditions are satisfied, and also in such a way that the function h out of the vertices of second and third kind, and the functions

$$h - \frac{(2n-2)}{n} \log |Z'' - A''|, \quad h - 2 \log |Z'' - A''| - 2 \log \log |Z'' - A''|$$

in the neighbourhood of these vertices, are of class C^∞, with finite values of all their derivatives. Under these conditions, the function U defined in (26.4.24) is always finite and satisfies the functional equation

$$DU = 8e^U e^{-h} + D \log \frac{d\Omega}{d\omega} + Dh, \tag{26.4.25}$$

or, by defining

$$\theta \equiv 8e^{-h}, \quad -\Phi \equiv D \log \frac{d\Omega}{d\omega} + Dh, \tag{26.4.26}$$

the functional equation

$$DU = 8e^U - \Phi. \tag{26.4.27}$$

The functions θ and Φ are some given functions; the former is always positive and can never vanish; it can only become infinite at the vertices of second and third kind; the function Φ, instead, can only become infinite at the vertices of third kind.

At the vertices of second kind, θ diverges as

$$|Z'' - A''|^{\frac{(2-2n)}{n}}.$$

At the vertices of third kind, θ and Φ become infinite as

$$\frac{1}{(|Z'' - A''| \, \log |Z'' - A''|)^2}.$$

The ratio $\frac{\Phi}{\theta}$ remains therefore finite; the limit of this ratio is on the other hand equal to

$$\frac{1}{4} \frac{d\Omega''}{d\omega}.$$

It is therefore positive.

Thus, if there exist vertices of third kind, the function Φ can become infinite, but on the side of positive values only; it is therefore unbounded from above, but it is always bounded from below.

Appendix: Vertices in the Theory of Fuchsian Functions

Let $\mathbf{R}$ be the real line, and let a, b be complex numbers, with complex conjugates a' and b', respectively. Following Poincaré [113], we define

$$(a, b) \equiv \frac{(a - a')}{(a - b')} \frac{(b - b')}{(b - a')}. \tag{26.a.1}$$

Let us consider two arcs of circle, ab and cd, centred on **R**. If one has

$$(a, b) = (c, d), \tag{26.a.2}$$

this means that there exists a linear substitution with real coefficients that changes ab into cd. This is said to be the substitution

$$(a, b; c, d). \tag{26.a.3}$$

Consider now a curvilinear polygon located entirely below **R**, and whose sides are of two kinds: those of the first kind are *arcs of circles* centred on **R**; those of the second kind are *segments* of the real axis.

There exist $2n$ sides of the first kind; two consecutive sides of the first kind are separated in one of the following three ways:

(i) By a vertex located below **R**, called *vertex of the first category* (or species or kind).

(ii) By a vertex located on the real line, which is therefore a *vertex of second category*.

(iii) By a side of second kind, which is said to be a *vertex of third category* (or, again, species or kind).

By virtue of this convention, it is clear that one meets, by following the perimeter of the polygon, alternatively a side of first kind and a vertex of first, second or third category. The side that one meets after a given vertex is the subsequent side; the vertex that one meets afterwards is the subsequent vertex, and so on.

One assumes that the sides of first kind are distributed in pairs in arbitrary fashion, and a side is said to be *conjugate* of the side belonging to the same pair. The vertices are taken to be distributed into *cycles* in the following way. One starts from an arbitrary vertex; one considers the subsequent side, then its conjugate, then the subsequent vertex, then the following side, thereafter its conjugate, and afterwards the subsequent vertex, and so on, until one reverts to the original vertex. All vertices met in this way are said to belong to the same cycle.

Poincaré made the following assumptions:

(1) All vertices of a same cycle are of the same category.

(2) If all vertices of a cycle are of the first category, the sum of the corresponding angles of the curvilinear polygon is a portion of 2π.

(3) if $a_i b_i$ and $a_i' b_i'$ are two sides of a same pair, one has, with the notation of Eq. (26.a.1), the equality

$$(a_i, b_i) = (a_i', b_i'). \tag{26.a.4}$$

Under these three conditions, the group of substitutions

$$(a_i b_i;\ a_i' b_i')$$

turns out to be a Fuchsian group, and one obtains in this way all Fuchsian groups.

Part VII
The Riemann ζ-Function

Chapter 27
The Functional Equations of Number Theory

27.1 The Euler Theorem on Prime Numbers and the ζ-function

The nineteenth century mathematicians were familiar, among various concepts, with the Dirichlet series, defined by the formula

$$D(s) \equiv \sum_{n=1}^{\infty} n^{-s}. \tag{27.1.1}$$

If one writes $s = \sigma + i\tau$, and if $\delta > 0$, one sees that, viewed as a series of functions in the region $\sigma \geq 1 + \delta$ of the complex plane, this series is normally convergent. One has indeed

$$\left| \frac{1}{n^{\sigma+i\tau}} \right| = \frac{1}{n^{\sigma}} \leq \frac{1}{n^{1+\delta}}. \tag{27.1.2}$$

Since, for all n, the function $s \to n^{-s}$ is continuous in the half-plane $\sigma \geq 1 + \delta$, one sees that the sum, i.e. the function ζ_D, is continuous in this same half-plane; this being true $\forall \delta > 0$, the function ζ_D is continuous over the whole half-plane $\sigma > 1$.

Let us now consider the infinite product, when p runs through the set of all prime numbers,

$$G(s) \equiv \prod_{p} (1 - p^{-s})^{-1}. \tag{27.1.3}$$

A term whatsoever of this product is always different from zero. On the other hand, the denominator $1 - p^{-s}$ is always $\neq 0$, for $\sigma > 0$. Moreover, in this case the absolute value of p^{-s} is $p^{-\sigma} < 1$ for $\sigma > 0$, and hence we can apply the theorem according to which, for the infinite product

© Springer International Publishing AG 2017
G. Esposito, *From Ordinary to Partial Differential Equations*, UNITEXT 106,
DOI 10.1007/978-3-319-57544-5_27

$$\prod_{n=0}^{\infty}(1 + \nu_n), \ \nu_n \neq -1$$

to be absolutely convergent, it is necessary and sufficient that the series of absolute values

$$\sum_{n=1}^{\infty}|\nu_n|$$

is convergent.

The infinite product is absolutely convergent if and only if its inverse is convergent, i.e. if the series $\sum_p p^{-\sigma}$ is convergent. This is certainly the case for $\sigma > 1$, because this series has a sum majorized by the sum of the series

$$\sum_{n=1}^{\infty}n^{-\sigma},$$

which is itself convergent.

We shall therefore limit ourselves to the consideration of values $\sigma > 1$. Under these assumptions, the following result holds.

Theorem 27.1 *If $\sigma > 1$, one has the equality $G(s) = \zeta_D(s)$.*

Proof In order to prove this theorem, it is of course possible to assume s fixed once and for ever. Thus, if $\varepsilon > 0$ is given, one can, since the series $\sum_{n=1}^{\infty} n^{-\sigma}$ and the product G are convergent, find an integer m having the following properties.

(i) The remainder $\sum_{n>m} \frac{1}{n^{\sigma}}$ is majorized by $\frac{\varepsilon}{2}$.
(ii) If we denote by $G_m(s)$ the partial product formed by the first m factors of the infinite product, one has

$$|G_m(s) - G(s)| \leq \frac{\varepsilon}{2}. \tag{27.1.4}$$

One has then for every prime number p the expansion in a geometric series absolutely convergent:

$$(1 - p^{-s})^{-1} = 1 + \frac{1}{p^s} + \frac{1}{p^{2s}} + \frac{1}{p^{3s}} + \dots = \sum_{k=0}^{\infty}\frac{1}{p^{ks}}. \tag{27.1.5}$$

By virtue of the rule pertaining to the product of several absolutely convergent series, one can write

$$G_m(s) = \sum_{k_1,k_2,\dots,k_m}\left(\frac{1}{p_1}\right)^{k_1 s} \dots \left(\frac{1}{p_m}\right)^{k_m s}, \tag{27.1.6}$$

where

$$p_1 = 2, \; p_2 = 3, \; p_3 = 5, \; ..., \; p_m$$

are the first m prime numbers, and $k_1, k_2, ..., k_m$ some integers ≥ 0; which proves that one has

$$G_m(s) = \sum_{\nu} \nu^{-s}, \tag{27.1.7}$$

where the integer ν runs through the sequence of integers which, in their decomposition into prime factors, contain nothing but the first m prime numbers $p_1, p_2, ..., p_m$. If then we consider the difference $G_m(s) - \zeta_D(s)$, it consists of a part of the terms of the series $\sum_{n=1}^{\infty} n^{-s}$, which all correspond to some indices $n > m$. One has therefore the inequality

$$|G_m(s) - \zeta_D(s)| \leq \sum_{n>m} \frac{1}{n^{\sigma}} \leq \frac{\varepsilon}{2}, \tag{27.1.8}$$

from which one derives the inequality

$$|G(s) - \zeta_D(s)| \leq |G(s) - G_m(s)| + |G_m(s) - \zeta_D(s)| \leq \frac{\varepsilon}{2} + \frac{\varepsilon}{2} = \varepsilon. \tag{27.1.9}$$

By virtue of the arbitrariness of ε, one proves eventually the desired equality $G(s) = \zeta_D(s)$. $\square$

As a corollary, the function ζ_D is nowhere vanishing, for $\sigma > 1$.

27.2 Γ- and ζ-Function from the Jacobi Function

The formula obtained from the Euler theorem is only valid when $\text{Re}(s) > 1$, but the Riemann function

$$\zeta_R : s \in \mathbf{C} \rightarrow \zeta_R(s) = \sum_{n=1}^{\infty} n^{-s} \tag{27.2.1}$$

can be defined over the whole complex plane, the only singularity being located at $s = 1$. In order to prove this, let us recall the integral representation of the Γ-function

$$\Gamma : s \rightarrow \Gamma(s) = \int_0^{\infty} t^{s-1} e^{-t} dt, \quad \text{Re}(s) > 1, \tag{27.2.2}$$

which admits a meromorphic extension to the whole complex-s plane, such that $\frac{1}{\Gamma(s)}$ has simple zeros at negative integer values of s. If we set, in Eq. (27.2.2), $s = \frac{\sigma}{2}$ and

$t = n^2 \pi x$, we find

$$\pi^{-\frac{\sigma}{2}} \Gamma\left(\frac{\sigma}{2}\right) n^{-\sigma} = \int_0^\infty x^{\frac{\sigma}{2}-1} e^{-n^2 \pi x} dx. \qquad (27.2.3)$$

At this stage, we can sum on both sides of Eq. (27.2.3) over all values of n from 1 to ∞, finding that

$$\pi^{-\frac{\sigma}{2}} \Gamma\left(\frac{\sigma}{2}\right) \zeta_R(\sigma) = \int_0^\infty x^{\frac{\sigma}{2}-1} \left(\sum_{n=1}^\infty e^{-n^2 \pi x}\right) dx, \qquad (27.2.4)$$

with the understanding that $\Gamma\left(\frac{\sigma}{2}\right)$ is the analytic extension of the Γ-function originally given by (27.2.2), while

$$\psi : x \in \mathbf{R}^+ \cup \{0\} \to \psi(x) = \sum_{n=1}^\infty e^{-n^2 \pi x} \qquad (27.2.5)$$

is the Jacobi function, which satisfies the simple but non-trivial functional equation

$$\frac{2\psi(x)+1}{2\psi\left(\frac{1}{x}\right)+1} = \frac{1}{\sqrt{x}}. \qquad (27.2.6)$$

The formula (27.2.4) is the desired integral representation of the analytic extension of the Dirichlet ζ-function, i.e. the Riemann ζ-function (27.2.1) with domain the whole of $\mathbf{C}$ and a pole at $s = 1$. Such a formula can be expressed in a more convenient form by a clever use of the identity (27.2.6). For this purpose, we first exploit (27.2.6) to write

$$\psi(x) = \frac{1}{\sqrt{x}} \psi\left(\frac{1}{x}\right) + \frac{1}{2\sqrt{x}} - \frac{1}{2}. \qquad (27.2.7)$$

By virtue of (27.2.4) and (27.2.7), one finds first

$$\pi^{-\frac{\sigma}{2}} \Gamma\left(\frac{\sigma}{2}\right) \zeta_R(\sigma) = \int_0^\infty \psi\left(\frac{1}{x}\right) x^{\frac{(\sigma-3)}{2}} dx + \frac{1}{2} \int_0^\infty x^{\frac{(\sigma-3)}{2}} dx - \frac{1}{2} \int_0^\infty x^{\frac{\sigma}{2}-1} dx. \qquad (27.2.8)$$

At this stage, we split all integrals into an integral from 0 to 1 plus an integral from 1 to ∞, hence finding

$$\pi^{-\frac{\sigma}{2}} \Gamma\left(\frac{\sigma}{2}\right) \zeta_R(\sigma) = \int_1^\infty \psi\left(\frac{1}{x}\right) x^{\frac{(\sigma-3)}{2}} dx + \int_0^1 \psi\left(\frac{1}{x}\right) x^{\frac{(\sigma-3)}{2}} dx$$

$$+ \frac{1}{2} \int_0^1 \left(x^{\frac{(\sigma-3)}{2}} - x^{\frac{\sigma}{2}-1} \right) dx$$

$$+ \frac{1}{2} \int_1^\infty \left(x^{\frac{(\sigma-3)}{2}} - x^{\frac{\sigma}{2}-1} \right) dx. \tag{27.2.9}$$

Now we re-express the first integral only, exploiting again Eq. (27.2.6) to write

$$\psi \left(\frac{1}{x} \right) = \sqrt{x} \psi(x) + \frac{1}{2} \sqrt{x} - \frac{1}{2}. \tag{27.2.10}$$

This might seem an undoing of the original operations, but it is not quite so, because then

$$\int_1^\infty \psi \left(\frac{1}{x} \right) x^{\frac{(\sigma-3)}{2}} dx = \int_1^\infty \psi(x) x^{\frac{\sigma}{2}-1} dx + \frac{1}{2} \int_1^\infty x^{\frac{\sigma}{2}-1} dx - \frac{1}{2} \int_1^\infty x^{\frac{(\sigma-3)}{2}} dx. \tag{27.2.11}$$

The second and third integral on the right-hand side of (27.2.11) cancel exactly the fourth integral on the right-hand side of (27.2.9), and hence only integrable terms survive, so that one finds

$$\pi^{-\frac{\sigma}{2}} \Gamma \left(\frac{\sigma}{2} \right) \zeta_R(\sigma) = \int_1^\infty \psi(x) x^{\frac{\sigma}{2}-1} dx + \int_0^1 \psi \left(\frac{1}{x} \right) x^{\frac{(\sigma-3)}{2}} dx$$

$$+ \frac{1}{2} \int_0^1 \left(x^{\frac{(\sigma-3)}{2}} - x^{\frac{\sigma}{2}-1} \right) dx. \tag{27.2.12}$$

This formula suggests changing variable according to $\tau \equiv \frac{1}{x}$ in the second integral on the right-hand side, so that such an integral is turned into an integral from 1 to ∞ as for the first integral on the right-hand side of (27.2.12), i.e.

$$\int_0^1 \psi \left(\frac{1}{x} \right) x^{\frac{(\sigma-3)}{2}} dx = \int_1^\infty \psi(\tau) \tau^{-\frac{(\sigma+1)}{2}} d\tau, \tag{27.2.13}$$

and eventually Eq. (27.2.12) reads as

$$\pi^{-\frac{\sigma}{2}} \Gamma \left(\frac{\sigma}{2} \right) \zeta_R(\sigma) = \frac{1}{\sigma(\sigma-1)} + \int_1^\infty \psi(x) \left(x^{\frac{\sigma}{2}-1} + x^{-\frac{(\sigma+1)}{2}} \right) dx, \tag{27.2.14}$$

by virtue of the identity

$$\int_0^1 \left(x^{\frac{(\sigma-3)}{2}} - x^{\frac{\sigma}{2}-1} \right) dx = 2 \left(\frac{1}{(\sigma-1)} - \frac{1}{\sigma} \right) = \frac{2}{\sigma(\sigma-1)}. \tag{27.2.15}$$

27.3   The ξ-function: Its Functional Equation and Its Integral Representation

By inspection, we see at once that the right-hand side of (27.2.14) is invariant under the replacement $\sigma \rightarrow (1 - \sigma)$. Hence we can say that the ξ-function, defined by (now we revert to the standard notation s for our complex variable)

$$\xi : s \in \mathbf{C} \rightarrow \xi(s) \equiv \frac{1}{2}s(s-1)\pi^{-\frac{s}{2}}\Gamma\left(\frac{s}{2}\right)\zeta_R(s), \qquad (27.3.1)$$

satisfies the functional equation

$$\xi(s) = \xi(1 - s). \qquad (27.3.2)$$

Riemann pointed out that in particular, upon setting

$$s = \frac{1}{2} + it, \qquad (27.3.3)$$

one finds

$$\begin{aligned}
\xi(s) = \xi(t) &= \frac{1}{2} - \frac{1}{2}\left(t^2 + \frac{1}{4}\right)\int_1^\infty \psi(x)x^{-\frac{3}{4}}\left(x^{\frac{it}{2}} + x^{-\frac{it}{2}}\right)dx \\
&= \frac{1}{2} - \left(t^2 + \frac{1}{4}\right)\int_1^\infty \psi(x)x^{-\frac{3}{4}}\cos\left(\frac{t}{2}\log x\right)dx \\
&= 4\int_1^\infty x^{-\frac{1}{4}}\cos\left(\frac{t}{2}\log x\right)\frac{d}{dx}\left(x^{\frac{3}{2}}\psi'(x)\right)dx. \qquad (27.3.4)
\end{aligned}$$

This function is finite for all finite values of t, and can be expanded in powers of t^2 as a rapidly converging series.

Note now that the Euler product formula proved in Sect. 27.1, which states the equality of Dirichlet series (27.1.1) with the infinite product (27.1.3), shows at once that $\zeta_R(s)$ has no zeros in the half-plane $\mathrm{Re}(s) > 1$, since a convergent infinite product can vanish only if one of its factors vanishes. By virtue of the definition (27.3.1), the factors other than $\zeta_R(s)$ having only the simple zero at $s = 1$, none of the roots ρ of $\xi(\rho) = 0$ lie in the halfplane $\mathrm{Re}(s) > 1$. Moreover, we know from (27.3.2) that $1 - \rho$ is a root if and only if ρ is, hence none of the roots of ξ lie in the halfplane $\mathrm{Re}(s) < 0$ either. One can therefore find roots ρ of ξ only in the closed strip $\mathrm{Re}(\rho) \in [0, 1]$.

As a next step, Riemann stated without proof that the number of roots ρ whose imaginary parts lie between 0 and T is approximately

$$\frac{T}{2\pi}\log\frac{T}{2\pi} - \frac{T}{2\pi},$$

and that the relative error in this approximation is of the order of magnitude $\frac{1}{T}$. He provided only a heuristic argument, pointing out that the number of roots in this region is equal to the integral of $\frac{\xi'(s)ds}{2\pi i \xi(s)}$ around the boundary of the rectangle $\mathrm{Re}(s) \in [0, 1]$, $\mathrm{Im}(s) \in [0, T]$, and that this integral is equal to the above expression with a relative error $\frac{1}{T}$. The calculation was not easy, and it took almost half a century to prove that the Riemann estimate was correct [137].

27.4 Logarithm of the ζ-Function

Following [122], we now let $F(x)$, when x is not exactly equal to a prime, be equal to the number $\pi(x)$ of primes less than x, but when x is a prime let it be greater by $\frac{1}{2}$, so that, for an x where $F(x)$ jumps, one can write

$$F(x) = \frac{F(x+0) + F(x-0)}{2}.$$

In the expansion of $\log \zeta(s)$ obtainable from (27.1.3), i.e.

$$\log \zeta(s) = -\sum_p \log(1 - p^{-s}) = \sum_p p^{-s} + \frac{1}{2} \sum_p p^{-2s} + \frac{1}{3} \sum_p p^{-3s} + \dots,$$

$$(27.4.1)$$

Riemann had the clever idea to re-express the powers p^{-s}, p^{-2s}, ... by means of the identities

$$p^{-s} = s \int_p^\infty x^{-s-1} dx, \quad p^{-2s} = s \int_{p^2}^\infty x^{-s-1} dx, \dots,$$

$$(27.4.2)$$

hence obtaining

$$\frac{\log \zeta(s)}{s} = \sum_p \sum_{k=1}^\infty \frac{1}{k} \int_{p^k}^\infty x^{-s-1} dx = \int_1^\infty f(x) x^{-s-1} dx = \int_0^\infty f(x) x^{-s} d\log x,$$

$$(27.4.3)$$

with the understanding that

$$f(x) \equiv \sum_{k=1}^\infty \frac{1}{k} F(x^{\frac{1}{k}}).$$

$$(27.4.4)$$

This formula expresses the property that the number of prime squares less than x is equal to the number of primes less than $\sqrt{x}$, and, in analogous way, the number of prime nth powers less than x is equal to $F(x^{\frac{1}{n}})$.

By construction, this formula holds for all s having $\mathrm{Re}(s) > 1$. At this stage, another original idea of Riemann was to point out that, for a generic functional equation of the form

$$g(s) = \int_0^\infty h(x)x^{-s}\mathrm{d}\log x, \qquad (27.4.5)$$

one can exploit the Fourier theorem to express the function h in terms of the function g. Indeed, if h is real-valued, one has the split

$$g(s) = g(a + ib) = g_1(b) + ig_2(b), \qquad (27.4.6)$$

where

$$g_1(b) = \int_0^\infty h(x)x^{-a}\cos(b\log x)\mathrm{d}\log x, \qquad (27.4.7)$$

$$ig_2(b) = -i\int_0^\infty h(x)x^{-a}\sin(b\log x)\mathrm{d}\log x. \qquad (27.4.8)$$

This suggests multiplying Eqs. (27.4.7) and (27.4.8) by

$$(\cos(b\log y) + i\sin(b\log y))\,\mathrm{d}b$$

and integrating them from $-\infty$ to $+\infty$. One then obtains $\pi h(y)y^{-a}$ on the right-hand side in both equations, by virtue of Fourier theory. Addition of both equations and their multiplication by iy^a leads eventually to

$$2\pi ih(y) = \int_{a-i\infty}^{a+i\infty} g(s)y^s\mathrm{d}s, \qquad (27.4.9)$$

where the integration contour is such that $\mathrm{Re}(s)$ remains constant. For values of y at which a jump occurs in the value $h(y)$, the integral takes the mean of the values of h on either side of the jump. For our original problem, from the way in which the function f was defined, one can see that it has the same property, and hence we derive from (27.4.9) the general formula

$$f(x) = \frac{1}{2\pi i}\int_{a-i\infty}^{a+i\infty} \frac{\log \zeta_R(s)}{s}x^s\mathrm{d}s. \qquad (27.4.10)$$

By virtue of (27.3.1) and of the expansion of ξ in terms of its zeros, i.e.

$$\xi(s) = \xi(0) \prod_{\rho} \left(1 - \frac{s}{\rho} \right), \tag{27.4.11}$$

one finds, on denoting by $E(s) = \Pi(s) = \Gamma(s+1) = s\Gamma(s)$ the Euler function, the identity

$$\log \zeta_R(s) = \log \xi(0) + \sum_{\rho} \log \left(1 - \frac{s}{\rho} \right) - \log E \left(\frac{s}{2} \right) + \frac{s}{2} \log(\pi) - \log(s-1). \tag{27.4.12}$$

At this stage, the insertion of (27.4.12) into the integral representation (27.4.10) yields $f(x)$ as a sum of five terms. However, direct substitution would lead to divergent integrals, hence Riemann integrated first by parts to obtain the equivalent (but more manageable) formula

$$f(x) = -\frac{1}{2\pi i} \frac{1}{\log x} \int_{a-i\infty}^{a+i\infty} x^s \frac{d}{ds} \left[\frac{\log \zeta_R(s)}{s} \right] ds, \ a > 1, \tag{27.4.13}$$

which holds because the additional term obtained from integration by parts vanishes, i.e.

$$\lim_{T \to \infty} \frac{\log \zeta_R(a \pm iT)}{(a \pm iT)} x^{a \pm iT} = 0. \tag{27.4.14}$$

In order to prove the limit (27.4.14), the idea is to prove that the numerator is bounded by a constant, while the denominator tends to infinity [43]. For this purpose, one takes the absolute value of the logarithmic term, and one finds indeed that

$$|\log \zeta_R(a \pm iT)| = \left| \sum_n \sum_p \frac{1}{n} p^{-n(a \pm iT)} \right| \le \sum_n \sum_p \frac{1}{n} p^{-na} = \log \zeta_R(a), \tag{27.4.15}$$

which is the desired constant.

Now we revert to the integral representation (27.4.13) and, following closely the work in [43], we aim at studying the principal term in the formula for $f(x)$, obtained from the term $-\log(s-1)$ in (27.4.12). Such a term is

$$f_1(x) \equiv \frac{1}{2\pi i} \frac{1}{\log x} \int_{a-i\infty}^{a+i\infty} x^s \frac{d}{ds} \left[\frac{\log(s-1)}{s} \right] ds, \ a > 1. \tag{27.4.16}$$

Now fix $x > 1$ and consider the function of β defined by

$$F(\beta) \equiv \frac{1}{2\pi i} \frac{1}{\log x} \int_{a-i\infty}^{a+i\infty} x^s \frac{d}{ds} \left[\frac{\log\left(\frac{s}{\beta} - 1\right)}{s} \right] ds, \ a > 1. \qquad (27.4.17)$$

so that the desired value $f_1(x) = F(1)$. The definition of $F(\beta)$ can be extended to all real or complex numbers β other than real numbers $\beta \leq 0$ by taking $a > \text{Re}(\beta)$ and defining

$$\log\left[\frac{s}{\beta} - 1\right] \equiv \log(s - \beta) - \log(\beta), \qquad (27.4.18)$$

with the understanding that $\log z$ is defined for all z other than real $z \leq 0$ by the condition that it must be real-valued for real $z > 0$. The integral defining $F(\beta)$ is absolutely convergent by virtue of the inequality

$$\left| \frac{d}{ds} \left[\frac{\log\left(\frac{s}{\beta} - 1\right)}{s} \right] \right| \leq \frac{\left|\log\left[\frac{s}{\beta} - 1\right]\right|}{|s|^2} + \frac{1}{|s(s - \beta)|}, \qquad (27.4.19)$$

which ensures integrability of the first term in the integrand, jointly with the oscillatory behaviour of $x^s = e^{s \log x}$ on the integration line.

So far we needed only derivatives with respect to s, but now we are going to study the first derivative $F'(\beta)$. Indeed, by virtue of the identity

$$\frac{d}{d\beta} \left[\frac{\log\left(\frac{s}{\beta} - 1\right)}{s} \right] = \frac{1}{(\beta - s)\beta}, \qquad (27.4.20)$$

the tools of differentiation under the integral sign and integration by parts lead to

$$F'(\beta) = \frac{1}{2\pi i} \frac{1}{\log x} \int_{a-i\infty}^{a+i\infty} x^s \frac{d}{ds} \left[\frac{1}{(\beta - s)\beta} \right] ds$$

$$= -\frac{1}{2\pi i} \int_{a-i\infty}^{a+i\infty} \frac{x^s}{(\beta - s)\beta} ds. \qquad (27.4.21)$$

The integral on the second line of (27.4.21) can be evaluated by applying Fourier inversion to the formula

$$\frac{1}{(s - \beta)} = \int_1^\infty x^{\beta-s-1} dx, \ \text{Re}(s - \beta) > 0, \qquad (27.4.22a)$$

$$\frac{1}{(a + i\mu - \beta)} = \int_0^\infty e^{-i\lambda\mu} e^{\lambda(\beta-a)} d\lambda, \ a > \text{Re}(\beta), \qquad (27.4.22b)$$

to obtain

$$\int_{-\infty}^{\infty} \frac{1}{(a + i\mu - \beta)} e^{i\mu x} d\mu = 2\pi e^{x(\beta - a)} \text{ if } x > 0, \ 0 \text{ if } x < 0, \qquad (27.4.23)$$

from which one finds

$$\frac{1}{2\pi i} \int_{a-i\infty}^{a+i\infty} \frac{1}{(s - \beta)} y^s ds = y^\beta \text{ if } y > 1, \ 0 \text{ if } y < 1, \qquad (27.4.24)$$

provided that $a > \text{Re}(\beta)$. Since $x > 1$ by assumption, one therefore obtains

$$F'(\beta) = \frac{x^\beta}{\beta}. \qquad (27.4.25)$$

Let us now denote by C^+ the contour in the complex t-plane given by the line segment from 0 to $1 - \varepsilon$ (ε being a small positive number), followed by the semicircle in the upper half-plane $\text{Im}(t) \geq 0$ from $1 - \varepsilon$ to $1 + \varepsilon$, followed by the line segment from $1 + \varepsilon$ to x, and let us consider

$$G(\beta) \equiv \int_{C^+} \frac{t^{\beta-1}}{\log t} dt. \qquad (27.4.26)$$

The first derivatives of the functions G and F are equal, because

$$G'(\beta) = \int_{C^+} t^{\beta-1} dt = \left. \frac{t^\beta}{\beta} \right|_0^x = F'(\beta). \qquad (27.4.27)$$

The function $G : \beta \to G(\beta)$ is defined and analytic for $\text{Re}(\beta) > 0$ as is the function $F : \beta \to F(\beta)$. Hence such functions differ by a constant (possibly dependent on x) for all complex β having positive real part. This constant can be evaluated by holding $\text{Re}(\beta)$ fixed and letting $\text{Im}(\beta) \to +\infty$ in both $F(\beta)$ and $G(\beta)$. In order to evaluate the limit of $G(\beta)$, set $\beta = \sigma + i\tau$, where σ is fixed and $\tau \to \infty$. The change of variable $t \equiv e^u$ turns $G(\beta)$ into the form [43]

$$G(\beta) = \int_{i\delta-\infty}^{i\delta+\log x} \frac{e^{\beta u}}{u} du + \int_{i\delta+\log x}^{\log x} \frac{e^{\beta u}}{u} du, \qquad (27.4.28)$$

where a slight alteration of the integration path has been performed by using the Cauchy theorem. On the right-hand side of (27.4.28), we write $u = v + i\delta$ in the first integral and $u = \log x + iw$ in the second integral, finding therefore

$$G(\beta) = e^{i\delta\sigma} e^{-\delta\tau} \int_{-\infty}^{\log x} \frac{e^{\sigma v}}{(v + i\delta)} dv - i x^\beta \int_0^\delta \frac{e^{-\tau w} e^{\sigma i w}}{(\log x + iw)} dw. \qquad (27.4.29)$$

In this formula both integrals approach 0 as $\tau \to \infty$, the first because the exponential $e^{-\delta \tau}$ tends to 0, and the second because the exponential $e^{-\tau w}$ tends to 0, except at $w = 0$. Thus,

$$\lim_{\tau \to \infty} G(\beta) = 0. \tag{27.4.30}$$

To evaluate the same limit for $F(\beta)$, it is helpful to consider the function (cf. (27.4.17))

$$H : \beta \to H(\beta) \equiv \frac{1}{2\pi i} \frac{1}{\log x} \int_{a-i\infty}^{a+i\infty} x^s \frac{d}{ds} \left[\frac{\log\left(1 - \frac{s}{\beta}\right)}{s} \right] ds, \tag{27.4.31}$$

where $a > \mathrm{Re}(\beta)$, and

$$\log\left[1 - \frac{s}{\beta}\right] \equiv \log(s - \beta) - \log(-\beta) \ \forall \beta \in \mathbf{C} - \left(\mathbf{R}_+ \cup \{0\}\right). \tag{27.4.32}$$

The difference $H(\beta) - F(\beta)$ is defined for all complex numbers β not lying on the real axis, and in the upper half-plane $\mathrm{Im}(\beta) > 0$ it is equal to [43]

$$\begin{aligned}
H(\beta) - F(\beta) &= \frac{1}{2\pi i} \frac{1}{\log x} \int_{a-i\infty}^{a+i\infty} x^s \frac{d}{ds} \left[\frac{\log(\beta) - \log(-\beta)}{s} \right] ds \\
&= \frac{1}{2\pi i} \frac{1}{\log x} \int_{a-i\infty}^{a+i\infty} x^s \frac{d}{ds} \left[\frac{i\pi}{s} \right] \\
&= -\frac{1}{2\pi i} \int_{a-i\infty}^{a+i\infty} \frac{i\pi}{s} x^s ds = -i\pi, \tag{27.4.33}
\end{aligned}$$

by virtue of (27.4.24). Hence $F(\beta) = H(\beta) + i\pi$ in the upper half-plane. To obtain the limit of $H(\beta)$ as $\tau \to \infty$, one has to carry out first the differentiation in the integrand given in (27.4.31), i.e.

$$\begin{aligned}
\frac{d}{ds} \left[\frac{\log\left(1 - \frac{s}{\beta}\right)}{s} \right] &= -\frac{\log\left(1 - \frac{s}{\beta}\right)}{s^2} + \frac{1}{s(s - \beta)} \\
&= -\frac{\log\left(1 - \frac{s}{\beta}\right)}{s^2} + \frac{1}{\beta(s - \beta)} - \frac{1}{\beta s}, \tag{27.4.34}
\end{aligned}$$

then multiplication by $x^s \frac{ds}{2\pi i}$, and eventually the limit as $T \to \infty$ of the integral from $a - iT$ to $a + iT$. By virtue of the last line of (27.4.34) and of the Lebesgue bounded convergence theorem, the limit as $\tau \to \infty$ of the resulting integral where the integrand has s^2 in the denominator is equal to the integral of the limit and hence vanishes. The remaining two integrals can be evaluated by means of (27.4.24), which implies that their sum equals $\frac{(x^\beta - 1)}{\beta}$. In this sum, the numerator is bounded while the absolute value of the denominator is $|\beta| \to \infty$. Hence $H(\beta)$ approaches 0 while $F(\beta)$ approaches $i\pi$. One can thus write

$$F(\beta) = G(\beta) + i\pi \text{ if } \text{Re}(\beta) > 0, \tag{27.4.35}$$

which in turn implies, with the notation in (27.4.16),

$$f_1(x) = F(\beta = 1) = \int_0^{1-\varepsilon} \frac{dt}{\log t} + \int_{1-\varepsilon}^{1+\varepsilon} \frac{(t-1)}{\log t} \frac{dt}{(t-1)} + \int_{1+\varepsilon}^x \frac{dt}{\log t} + i\pi, \tag{27.4.36}$$

where the second integral is taken over the semicircle in the upper half-plane. In this term, as $\varepsilon \to 0$, the ratio $\frac{(t-1)}{\log t}$ approaches 1 along the semicircle, so that the integral approaches

$$\int_{1-\varepsilon}^{1+\varepsilon} \frac{dt}{(t-1)} = -i\pi. \tag{27.4.37}$$

Hence one finds eventually

$$f_1(x) = F(\beta = 1) = \text{Li}(x) = \lim_{\varepsilon \to 0} \left[\int_0^{1-\varepsilon} \frac{dt}{\log t} + \int_{1+\varepsilon}^x \frac{dt}{\log t} \right]. \tag{27.4.38}$$

One has now to consider the other four terms in the formula (27.4.12). We cannot borrow too heavily from the beautiful work in [43], but we simply quote the final result obtained by [122] when $x > 1$, i.e.

$$f(x) = \text{Li}(x) - \sum_{\text{Im}(\rho) > 0} \left[\text{Li}(x^\rho) + \text{Li}(x^{1-\rho}) \right] + \int_x^\infty \frac{dt}{t(t^2 - 1)\log t} + \log \xi(0), \tag{27.4.39}$$

where $\log \xi(0) = -\log(2)$. At this stage, Riemann exploited Eq. (27.4.4) to invert it by means of the Möbius inversion formula to find that the number $\pi(x)$ of primes less than any given magnitude x is given by

$$\pi(x) = f(x) - \frac{1}{2} f(x^{1/2}) - \frac{1}{3} f(x^{1/3}) - \frac{1}{5} f(x^{1/5}) + \frac{1}{6} f(x^{1/6}) + \cdots + \frac{\mu(n)}{n} f(x^{1/n}) + \cdots, \tag{27.4.40}$$

where $\mu(n)$ is 0 if n is divisible by a prime square, 1 if n is a product of an even number of distinct primes, -1 if n is a product of an odd number of distinct primes. This algorithm leads to the asymptotic expansion of $\pi(x)$ at large x in the next section, Eq. (27.5.6) therein.

27.5 The Riemann Hypothesis on Non-trivial Zeros of the ζ-Function

As number theory gradually emerged as one of the most important branches of mathematics and human knowledge, the Riemann hypothesis attracted attention and dedicated efforts of some of the best mathematicians of all time. With the language of previous sections, it states that the non-trivial zeros of the analytic extension of the Dirichlet series to the whole complex plane minus the point $s = 1$ lie on the line $\text{Re}(s) = \frac{1}{2}$.

To figure out what this implies, it may be better to start by considering the Liouville function [13]

$$\lambda : n \in N \rightarrow \lambda(n) \equiv (-1)^{\omega(n)}, \qquad (27.5.1)$$

where $\omega(n)$ is the number of, not necessarily distinct, prime factors in the natural number n, with multiple factors counted multiply. For example, one has

$$\lambda(2) = \lambda(3) = \lambda(5) = \lambda(7) = \lambda(8) = -1, \ \lambda(1) = \lambda(4) = \lambda(6) = \lambda(9) = \lambda(10) = 1. \qquad (27.5.2)$$

Theorem 27.2 *The Riemann hypothesis is equivalent to the statement that, for all fixed $\varepsilon > 0$,*

$$\lim_{n \to \infty} \frac{\sum_{k=1}^{n} \lambda(k)}{n^{\frac{1}{2}+\varepsilon}} = 0. \qquad (27.5.3)$$

This is a rigorous expression of the colloquial statement according to which an integer has equal probability of having an odd number or an even number of distinct prime factors.

What is on firm ground is instead another result, known under the name of prime number theorem:

Theorem 27.3

$$\lim_{n \to \infty} \frac{\pi(n)}{\frac{n}{\log(n)}} = 1, \qquad (27.5.4)$$

where $\pi(n)$ is the more standard notation for the counting function:

$$\pi(x) \equiv \{\text{number of } y's \ \in N : y \text{ is prime}, y < x.\}. \qquad (27.5.5)$$

The result (27.5.4) implies the asymptotic estimate

$$\pi(x) \sim \int_2^x \frac{dt}{\log t},$$

(27.5.6)

which provides a very good example of asymptotic expansion containing one term only [41].

Another remarkable theorem holds, according to which

$$\pi(x) \sim \mathrm{Li}(x) + \mathrm{O}(x^\theta \log x) \Longrightarrow \zeta(\sigma + it) \neq 0,$$

(27.5.7)

for $\sigma > \theta$, where $\theta \in \left[\frac{1}{2}, 1\right]$. The Riemann hypothesis states therefore that $\theta = \frac{1}{2}$ when the domain of the right-hand side of (27.5.7) is specified, and would be proved if it were possible to prove that [13, 136]

$$\pi(x) \sim \mathrm{Li}(x) + \mathrm{O}(\sqrt{x} \log x).$$

(27.5.8)

Part VIII
A Window on Modern Theory

Chapter 28
The Symbol of Pseudo-Differential Operators

28.1 From Differential to Pseudo-Differential Operators

We have come to appreciate, in Parts II and IV, the role of the leading symbol in defining the concepts of ellipticity and hyperbolicity. However, many relevant operators in mathematics and physics are not differential but, rather, pseudo-differential. They can all be characterized by their symbol, and we begin our last chapter by introducing this concept in a pedagogical way.

Suppose that $u : x \in \mathbf{R} \to u(x)$ is a differentiable function admitting Fourier transform

$$\varphi(k) \equiv \frac{1}{\sqrt{2\pi}} \int_{-\infty}^{\infty} u(y) e^{-iky} \, dy. \tag{28.1.1}$$

We can therefore re-express $u(x)$ in the integral form

$$u(x) = \frac{1}{\sqrt{2\pi}} \int_{-\infty}^{\infty} \varphi(k) e^{ikx} \, dk = \frac{1}{2\pi} \int_{-\infty}^{\infty} dk \int_{-\infty}^{\infty} dy \, e^{ik(x-y)} u(y), \tag{28.1.2}$$

with the understanding that the integration with respect to the y variable on the real line is performed first. Hence we get the action of the one-dimensional Laplacian in the integral form

$$P u(x) = -\frac{d^2}{dx^2} u(x) = \frac{1}{2\pi} \int_{-\infty}^{\infty} dk \int_{-\infty}^{\infty} dy \, k^2 e^{ik(x-y)} u(y)$$

$$= \frac{1}{2\pi} \int_{-\infty}^{\infty} dk \int_{-\infty}^{\infty} dy \, e^{ik(x-y)} p(k) u(y), \tag{28.1.3}$$

where $p : k \to p(k) = k^2$ is the symbol of the one-dimensional Laplacian.

As far as differential operators are concerned, we can consider the n-dimensional Euclidean space $\mathbf{R}^n$ rather than just the real line, and their coefficients may

© Springer International Publishing AG 2017
G. Esposito, *From Ordinary to Partial Differential Equations*, UNITEXT 106,
DOI 10.1007/978-3-319-57544-5_28

be functions rather than constants. Thus, on using the multi-index notation already introduced in Part II, we may express differential operators of order m on $\mathbf{R}^n$ in the form

$$A(x, D_x) = \sum_{|\alpha| \leq m} a_\alpha(x) D^\alpha, \qquad (28.1.4)$$

and their action, with the help of Fourier transform, reads as

$$A(x, D_x)u(x) = (2\pi)^{-n} \int_{\mathbf{R}^n} e^{\mathrm{i}x \cdot \xi} a(x, \xi) \hat{u}(\xi) \, d\xi, \qquad (28.1.5)$$

where

$$\int_{\mathbf{R}^n} dw = \int_{-\infty}^{\infty} dw_1 \ldots \int_{-\infty}^{\infty} dw_n, \ \ w = y, \xi,$$

and the function $a : (x, \xi) \to a(x, \xi)$ is called *symbol* or characteristic polynomial, whose action is defined by

$$a(x, \xi) \equiv \sum_{|\alpha| \leq m} a_\alpha(x) \xi^\alpha. \qquad (28.1.6)$$

Pseudo-differential operators are obtained by considering, instead of a symbol of polynomial nature as in (28.1.6), a more general function. More precisely, a *pseudo-differential operator* $\mathcal{P}$ with symbol $\sigma(x, \xi)$ is the operator defined by

$$(\mathcal{P}u)(x) \equiv (2\pi)^{-n} \int_{\mathbf{R}^{2n}} e^{\mathrm{i}(x-y) \cdot \xi} \sigma(x, \xi) u(y) \, dy \, d\xi. \qquad (28.1.7)$$

Strictly, the definition (28.1.7) holds for $u \in \mathcal{S}(\mathbf{R}^n)$, but can be extended to a more general class of functions, provided that the symbol $\sigma(x, \xi)$ satisfies suitable conditions. In particular, it is important to consider symbols which, for some $\delta \in \mathbf{R}$, are C^∞ functions satisfying the inequality

$$\left| D_x^\beta D_\xi^\alpha \sigma(x, \xi) \right| \leq C_{\alpha,\beta}(x) \left(1 + |\xi|^2\right)^{\frac{\delta}{2} - \frac{|\alpha|}{2}} \ \forall \alpha, \beta, \qquad (28.1.8)$$

where $C_{\alpha,\beta}$ is a continuous function. In several applications, one needs also the asymptotic expansion of the symbol. For this purpose, one can assume that $\sigma(x, \xi)$ is *polyhomogeneous*, i.e. it satisfies (28.1.8) *and* has an asymptotic expansion

$$\sigma(x, \xi) \sim \sum_{l \in \mathcal{N}} \sigma_{\delta-l}(x, \xi), \qquad (28.1.9)$$

where each $\sigma_{\delta-l}$ is a C^∞ function homogeneous of degree $\delta - l$ in ξ for $|\xi| > 1$, and

$$\sigma - \sum_{l < j} \sigma_{\delta - l}$$

is a C^∞ function satisfying the inequality (28.1.8) with δ replaced by $\delta - j$, for all $j \in \mathcal{N}$.

28.2 The Symbol of Pseudo-Differential Operators on Manifolds

The previous formulae show that assigning an operator is equivalent to assigning its symbol, and this is of course a geometric object, since the (x, ξ) coordinates pertain to $\mathbf{R}^n$ and to its cotangent space, respectively. Moreover, for a compactly supported pseudo-differential operator A on $C^\infty(\mathbf{R}^n)$, Sect. 28.1 implies that its symbol σ_A is given by [143]

$$\sigma_A(x, \xi) = e^{-ix \cdot \xi} A e^{ix \cdot \xi}. \tag{28.2.1}$$

This formula is equivalent to writing that

$$\sigma_A(x_0, \xi) = A e^{i(x - x_0) \cdot \xi}\big|_{x = x_0}, \tag{28.2.2}$$

and hence a formal application of Taylor's formula leads to (k being a multi-index $k_1 + \cdots + k_n$)

$$Af(x) e^{i\xi \cdot (x - x_0)}\big|_{x = x_0} = \sum_k \frac{i^{-k}}{k!} f^{(k)}(x_0) \frac{\partial^k \sigma_A(x_0, \xi)}{\partial \xi^k}, \tag{28.2.3}$$

from which in turn

$$e^{-i\xi \cdot x} Af(x) e^{i\xi \cdot x} = \sum_k \frac{i^{-k}}{k!} \frac{\partial^k \sigma}{\partial \xi^k} \frac{\partial^k f}{\partial x^k}. \tag{28.2.4}$$

This formula makes it possible to evaluate the symbol of a product, i.e.

$$\sigma_{AB}(x, \xi) \equiv e^{-i\xi \cdot x} AB \, e^{i\xi \cdot x} = e^{-i\xi \cdot x} A \, e^{i\xi \cdot x} \sigma_B = \sum_k \frac{i^{-k}}{k!} \frac{\partial^k \sigma_A}{\partial \xi^k} \frac{\partial^k \sigma_B}{\partial x^k}. \tag{28.2.5}$$

In this construction, an important role is played by the function l such that

$$l(x_0, \xi, x) = \xi \cdot (x - x_0), \tag{28.2.6}$$

which is linear in both x and ξ for each x_0. Its derivative with respect to x is ξ, and its derivative with respect to ξ is $x - x_0$. The desired generalization to manifolds is a real-valued function

$$l : x \in M, \ v \in T^*(M) \to l(x, v). \tag{28.2.7}$$

Linearity in ξ is mapped into linearity in v on each fibre of the cotangent bundle $T^*(M)$, but linearity in x, without introducing additional structure, has hardly a counterpart. However, *if there exists a covariant derivative operator ∇ on $T^*(M)$*, linearity at x_0 means that, for all $k \geq 2$, the symmetrized k-th covariant derivative, here denoted by $\nabla^{(k)}$, vanishes at x_0.

With the help of this definition there exists such a linear function, i.e. a real-valued function

$$l \in C^\infty(T^*(M) \times M),$$

such that $l(x, v)$ is, for fixed x, linear in each fibre of $T^*(M)$ and such that, for each $v \in T^*(M)$,

$$\nabla^{(k)} l(x, v)\big|_{x=\pi(v)} = \delta^{k1} v, \tag{28.2.8}$$

where π denotes the projection map $\pi : T^*(M) \to M$.

Once a covariant derivative ∇ is given that satisfies the condition (28.2.8), the symbol of a pseudo-differential operator A on $C^\infty(M)$ can be defined by the formula [143]

$$\sigma_A(v) \equiv A e^{il(x,v)}\big|_{x=\pi(v)}. \tag{28.2.9}$$

The uniqueness of the linear function l is lacking, and different forms of l give rise to different functional forms of the symbol, but any two of them differ by an element of a certain space as we are going to see, so that the symbol of A is an equivalence class.

28.3   Geometry Underlying the Symbol Map

The first result we need is the actual proof that the desired linear function l of Sect. 28.2 does exist, i.e.

Theorem 28.1 *There exists a real-valued C^∞ function l on the product space $T^*(M) \times M$ such that $l(x, \cdot)$ is, for each $x \in M$, linear on the fibres of the cotangent bundle of M, and such that, for each cotangent vector $v \in T^*(M)$, the condition (28.2.8) holds.*

Proof Following Widom [143], the local construction of l is first given. In a coordinate neighbourhood U with local coordinates x^i, given the m-th order covariant tensor τ, its covariant derivative reads as (see (4.4.13))

$$\tau_{i_1\ldots i_m;i} = \frac{\partial \tau_{i_1\ldots i_m}}{\partial x^i} - \sum_{\nu} \Gamma\{j,[i,i_\nu]\}\tau_{i_1\ldots i_{\nu-1}\,j\,i_{\nu+1}\ldots i_m}. \tag{28.3.1}$$

Hence we deduce by induction that, for any scalar function f,

$$f_{i_1;\ldots;i_k} = \frac{\partial^k f}{\partial x^{i_1}\ldots\partial x^{i_k}} + \sum_{|j|<k} \gamma_{i_1\ldots i_k j}\frac{\partial^j f}{\partial x^j}, \tag{28.3.2}$$

where the $\gamma_{i_1\ldots i_k j}$ are polynomials in the derivatives of Christoffel symbols of second kind. Thus, for all $k > 1$, the condition (28.2.8) is equivalent to the requirement that each derivative

$$\frac{\partial^k l}{\partial x^{i_1}\ldots\partial x^{i_k}}$$

should be equal, at $\pi(v)$, to some linear combination of lower-order derivatives, with coefficients that are C^∞ functions of $\pi(v)$. Thus, starting with the requirements

$$l(v,\pi(v)) = 0, \quad \frac{\partial l(x,v)}{\partial x^i}\bigg|_{x=\pi(v)} = v\left(\frac{\partial}{\partial x^i}\right), \tag{28.3.3}$$

this determines the value of the symmetrized k-th covariant derivative of l, for all multi-indices k, which is needed in order to satisfy (28.2.8). A theorem of Borel ensures that there exists a C^∞ function having partial derivatives arbitrarily prescribed, and the proof [104] shows that the function l may be chosen to be linear and C^∞ in v as well.

Having established local existence, we can say that, for each coordinate neighbourhood U_i in M, there exists a function $l_i \in C^\infty(T^*(U_i) \times U_i)$ with the desired properties. One then takes finitely many sets U_i covering M, a family of functions $\varphi_i \in C_0^\infty(U_i)$ providing a partition of unity, and other functions $\psi_i \in C_0^\infty(U_i)$ equal to 1 on a neighbourhood of the support of φ_i. The desired function l can then be expressed in the form

$$l(x,v) = \sum_{i=1}^N \varphi_i(\pi(v))l_i(x,v)\psi_i(x), \tag{28.3.4}$$

which is globally defined and satisfies all the requirements. $\square$

Now the analogue of Taylor's theorem for a smooth function $f \in C^\infty(M)$ can be derived. For this purpose, the derivative $f^{(k)}(x_0)$ is replaced by $\nabla^k f(x_0)$. As far as the analogue of $x - x_0$ is concerned, one can point out that since, for fixed $x_0, x \in M$, the function $l(x, v)$ is linear for $v \in T^*_{x_0}$, it can be viewed as an element of T_{x_0}. One can instead regard $l(x, \cdot)$ as a vector field on M. With this understanding, if we write

$$l(x, \cdot)^k \equiv l(x, \cdot) \otimes \cdots \otimes l(x, \cdot), \tag{28.3.5}$$

we mean a symmetric k-th order contravariant tensor field, and hence

$$\nabla^k f(x_0) \cdot l(x, x_0)^k$$

is a meaningful operation, and it coincides with

$$\nabla^{(k)} f(x_0) \cdot l(x, x_0)^k,$$

since $l(x, x_0)^k$ is symmetric. The following useful theorem on the analogue of Taylor's expansion holds [143]:

Theorem 28.2 *For each point x_0 of the manifold M and each integer N, the function*

$$f(x) - \sum_{n=0}^{N} \frac{1}{n!} \nabla^n f(x_0) \cdot l(x, x_0)^n$$

vanishes to order $(N + 1)$ at x_0.

Proof The identity (28.2.8) for $k \neq 1$ becomes now

$$\nabla^{(k)} l(x, x_0)\big|_{x=x_0} = 0, \tag{28.3.6}$$

while the identity for $k = 1$, i.e. $d_x l(x, v)|_{x=\pi(v)} = v$ has the following counterpart:

$$d_x l(x, x_0)|_{x=x_0} = \text{identity on } T_{x_0}. \tag{28.3.7}$$

Thus, $\nabla^{(k)} l(x, x_0)^n$ vanishes at x_0 for $n \neq k$, and for $n = k$ it equals $k!$ times the symmetrization operator on $\otimes_k T_{x_0}$. Thus, for all values of the index $k \leq N$, one finds

$$\nabla^{(k)} \sum_{n=0}^{N} \frac{1}{n!} \nabla^n f(x_0) \cdot l(x, x_0)^n \Bigg|_{x=x_0} = \nabla^{(k)} f(x_0), \tag{28.3.8}$$

and hence the desired proof is obtained. □

Interestingly, if all symmetrized covariant derivatives vanish up to a given order, the same holds for all partial derivatives up to the same order, as follows by applying induction to Eq. (28.3.2). The next step in symbolic calculus is the evaluation of unsymmetrized covariant derivatives

$$\nabla^k l(v) \equiv \nabla^k l(x, v)\big|_{x=\pi(v)} . \qquad (28.3.9)$$

These turn out to be polynomials in the torsion and curvature tensors of the given connection, and their covariant derivatives. This is clearly seen by recalling the application of Ricci identity to a covariant tensor of order m, i.e.

$$T_{i_1 \ldots i_m; j; k} = T_{i_1 \ldots i_m; k; j} + \sum_{\nu} \sum_{p=1}^{n} T_{i_1 \ldots i_{\nu-1} \, p \, i_{\nu+1} \ldots i_m} R^p_{i_\nu j k}$$

$$- \sum_{p=1}^{n} T_{i_1 \ldots i_m; p} T^p_{jk}, \qquad (28.3.10)$$

where T and R are the torsion and curvature tensors, tensor fields of type $(1, 2)$ and $(1, 3)$, respectively.

Thus, by applying induction, for every permutation α of the indices $1, \ldots, k$, the difference

$$f_{i_1; \ldots; i_k} - f_{i_{\alpha(1)}; \ldots; i_{\alpha(k)}}$$

is a sum of terms, each of which is a product of $(k-1)$-st or lower order covariant derivatives of f and covariant derivatives of torsion and curvature, followed by contraction on certain indices [143]. In particular, when f is the function l, one finds at $x = \pi(v)$ that

$$\sum_{\alpha} l_{i_{\alpha(1)}; \ldots; i_{\alpha(k)}} = 0, \qquad (28.3.11)$$

which leads, by induction, to the evaluation of $\nabla^k l$. From the second covariant derivatives of l one finds [142]

$$\left(\nabla^2 l\right)_{ij} = \frac{1}{2} \sum_{p=1}^{n} v_p T^p_{ij}, \qquad (28.3.12)$$

whereas, from the third covariant derivatives with vanishing torsion, one obtains

$$\left(\nabla^3 l\right)_{ijk} = \frac{1}{3} \sum_{p=1}^{n} v_p \left(R^p_{ijk} + R^p_{jik}\right). \qquad (28.3.13)$$

Operators That Generalize $\frac{\partial^k}{\partial \xi^k}$ and $\frac{\partial^k}{\partial x^k}$.

If σ is a C^∞ function on the cotangent bundle of M, the k-th directional derivative of σ along the fibres of $T^*(M)$ is denoted $D^k \sigma$. If $x_0 = \pi(v)$, one thinks of σ as a function on the cotangent space at x_0, and its k-th derivative $D^k \sigma$ is evaluated at v. This is a k-linear function on $T^*_{x_0}$, and hence may be identified with an element of the space $\otimes_k T_{x_0}$. Thus, $D^k\sigma$ is a contravariant k-tensor, and provides the analogue of $\frac{\partial^k\sigma}{\partial\xi^k}$, but of course k is a single integer in D^k and a multi-index in

$$\frac{\partial^k}{\partial\xi^k} = \frac{\partial^{k_1}}{\partial\xi_1^{k_1}} \cdots \frac{\partial^{k_n}}{\partial\xi_n^{k_n}}.$$

The covariant derivatives ∇^k act on C^∞ functions on M rather than C^∞ functions on the cotangent bundle of M, and hence the meaning of $\nabla^k\sigma$ is unclear if we require that σ should belong to the space $C^\infty(T^*(M))$. Thus, one resorts to the definition [143]

$$\nabla^k\sigma(v) \equiv \nabla^k\sigma\Big(d_x l(x, v)\Big)\Big|_{x=\pi(v)}. \tag{28.3.14}$$

The crucial feature is that, although the function $l(\cdot, v)$ is not unique, all its derivatives are determined at $\pi(v)$, and therefore there are no residual ambiguities in the definition (28.3.14) of $\nabla^k\sigma$ as a covariant k-tensor. Furthermore, the mixed derivative $\nabla^k D^j\sigma$ is defined in analogous way by

$$\nabla^k D^j\sigma(v) \equiv \nabla_x^k\, D_v^j\,\sigma\Big(d_x l(x, v)\Big)\Big|_{x=\pi(v)}. \tag{28.3.15}$$

Examples If

$$\sigma(v) = \sum_{i,j=1}^n v_i v_j (g^{-1})^{ij} \tag{28.3.16}$$

is the leading symbol of the Laplacian on a n-dimensional Riemannian manifold with Levi-Civita connection, one finds the directional derivatives

$$(D\sigma)^i = 2\sum_{j=1}^n v_j(g^{-1})^{ij}, \quad (D^2\sigma)^{ij} = 2(g^{-1})^{ij}. \tag{28.3.17}$$

Moreover, since

$$\sigma(dl) = \sum_{i,j=1}^n (g^{-1})^{ij}, \tag{28.3.18}$$

and both l_i and the metric g have vanishing covariant derivatives at $\pi(v)$, the definition (28.3.14) yields

$$\nabla\,\sigma = 0. \tag{28.3.19}$$

However, by virtue of (28.3.13), $\nabla^3 l$ does not vanish and hence one finds [143]

$$\left(\nabla^2 \sigma\right)_{kl} = \frac{2}{3} \sum_{p,q=1}^{n} v_p v_q \left(R^{pq}_{\ kl} + R^{p\ q}_{k\ l}\right), \tag{28.3.20}$$

while an example of mixed derivative (28.3.15) is provided by

$$\left(\nabla^2 D\sigma\right)^p_{\ kl} = \frac{4}{3} \sum_{q=1}^{n} v_q \left(R^{pq}_{\ kl} + R^{p\ q}_{k\ l}\right), \tag{28.3.21}$$

28.4   Symbol and Leading Symbol as Equivalence Classes

A space of fundamental importance is the space of functions σ of class C^∞ on the Cartesian product of $\mathbf{R}^n$ with $\mathbf{R}^m$ and satisfying, for all multi-indices j and k, the condition

$$\frac{\partial^j}{\partial x^j} \frac{\partial^k}{\partial \xi^k} \sigma(x, \xi) = O\left((1 + |\xi|)^{\omega - \rho|k| + (1-\rho)|j|}\right), \tag{28.4.1}$$

in a uniform way on compact x-sets. Such a space is denoted by $S^\omega_\rho\left(\mathbf{R}^n \times \mathbf{R}^m\right)$.

On replacing Euclidean space with a manifold M, one arrives at the spaces $S^\omega_\rho(M \times \mathbf{R}^m)$ and $S^\omega_\rho(T^*(M))$ which consist of functions satisfying (28.4.1) in terms of local coordinates. If it is sufficiently clear, from the general context, what the underlying space is, one writes simple S^ω_ρ, and one defines the spaces

$$S^\infty_\rho \equiv \cup_\omega S^\omega_\rho, \quad S^{-\infty} \equiv \cap_\omega S^\omega_\rho, \tag{28.4.2}$$

where in $S^{-\infty}$ the subscript ρ is missing since such a space is independent of ρ, unlike all its constituents.

Starting from these spaces of symbol maps one can define the space $L^\omega_\rho(M)$, given by those operators on $C^\infty(M)$ which, locally, are pseudo-differential operators whose symbols lie in S^ω_ρ. The manifold M is often taken to be compact, although this requirement is not mandatory.

If we revert for a moment to operators A on the Euclidean space $\mathbf{R}^n$, it is useful to summarize how one obtains the asymptotic expansion of $Af(x)e^{it\varphi(x)}$ as $t \to \infty$. Indeed, on defining, for a particular x_0, the function

$$\psi(x, x_0) \equiv \varphi(x) - \varphi(x_0) - \varphi'(x_0) \cdot (x - x_0), \tag{28.4.3}$$

one can write for a particular x_0

$$e^{-it\varphi(x_0)} Af(x)e^{it\varphi(x)} = Af(x)e^{it\psi(x,x_0)}e^{it\varphi'(x_0)\cdot(x-x_0)}. \tag{28.4.4}$$

At this stage, the application of Eq. (28.2.4) makes it possible to obtain

$$\sum_m \frac{i^{-m}}{m!} \frac{\partial^m}{\partial x^m} f(x) e^{i\psi(x,x_0)} \frac{\partial^m}{\partial \xi^m} \sigma(x_0, t\varphi'(x_0)). \tag{28.4.5}$$

In this formula, the derivatives of order m of $\psi(x, x_0)$ are equal to $\varphi^{(m)}(x_0)$ if $|m| > 1$, and vanish otherwise. Thus, the power-series expansion of $e^{i\psi(x,x_0)}$ and the Leibniz formula lead to

$$\frac{\partial^m}{\partial x^m} f(x) e^{i\psi(x,x_0)} = \sum_{k=0}^{\infty} \frac{i^k}{k!} \frac{m!}{m_0! \dots m_k!} f^{(m_0)}(x_0) \varphi^{(m_1)}(x_0) \dots \varphi^{(m_k)}(x_0), \tag{28.4.6}$$

where $m_0 \geq 0, m_1, \dots, m_k \geq 2$. This formula is now inserted into (28.4.5), and replacement of x_0 with x in (28.4.4) yields eventually

$$e^{-it\varphi(x)} A f(x) e^{it\varphi(x)} = \sum_k \frac{i^{k-\sum_i m_i}}{k! m_0! \dots m_k!} f^{(m_0)}(x) \varphi^{(m_1)}(x)$$

$$\times \dots \varphi^{(m_k)}(x) \frac{\partial^{\sum_i m_i}}{\partial \xi^{\sum_i m_i}} \sigma(x, t\varphi'(x)). \tag{28.4.7}$$

This formula means that, for any N, one subtracts from the left-hand side of (28.4.7) all terms on the right which do not belong to the space S_ρ^{-N}. The resulting difference belongs therefore to the space S_ρ^{-N}.

In the case of a manifold M with connection, one considers the associated function $l(x, v)$ built as we have shown before. Since $dl(\pi(v), v) = v$, one can find a function ψ of class C^∞ on the product $M \times M$, equal to 1 on a neighbourhood of the diagonal and such that

$$\psi(x, x_0) \neq 0 \text{ and } 0 \neq v \in T_{x_0}^* \implies d_x l(x, v) \neq 0. \tag{28.4.8}$$

Now for any operator A in the space L_ρ^ω one defines first the symbol

$$\sigma_A(v) \equiv A\psi\left(x, \pi(v)\right) e^{il(x,v)}\Big|_{x=\pi(v)}. \tag{28.4.9}$$

By virtue of (28.4.7), this is a function belonging to the space S_ρ^ω. If different forms of the functions ψ and l are used, they give rise to symbol maps (28.4.9) that differ by an element of the space $S^{-\infty}$ defined in (28.4.2). This suggests defining *the symbol of A as the corresponding equivalence class in the quotient space* $\frac{S_\rho^\omega}{S^{-\infty}}$. Such an equivalence class is again denoted by σ_A.

Now we are going to prove a property that makes it possible to obtain a clear conceptual framework also for the leading symbol of a pseudo-differential operator. This is our main goal since we have learned so far that invariant concepts are entirely ruled by the leading symbol (which is why some authors do not want to define the mere symbol, as is stressed by Bleecker and Booss-Bavnbek [12]).

Theorem 28.3 *For any function f of class C^∞ and any operator A in the space L_ρ^∞, one has*

$$A\psi(x, \pi(v))f(x)\, e^{il(x,v)}\Big|_{x=\pi(v)} = \sum_{k=0}^{\infty} \frac{i^{-k}}{k!}\nabla^k f(\pi(v))D^k\sigma_A(v), \quad (28.4.10)$$

with the understanding that the difference between the left-hand side and the sum of those terms on the right not belonging to S_ρ^{-N} does indeed belong to the space S_ρ^{-N}, and that all estimates hold uniformly if the function f belongs to any bounded set in C^∞.

Proof Since $Dl(x, v) = l(x, \pi(v))$, one finds from the definition (28.4.9)

$$\nabla^k f(\pi(v))D^k\sigma_A(v)$$
$$= i^k A\psi(x, \pi(v))\nabla^k f(\pi(v))l(x, \pi(v))^k\, e^{il(x,v)}\Big|_{x=\pi(v)}. \quad (28.4.11)$$

Thus, the left-hand side of (28.4.10) minus the sum of terms on the right corresponding to $k < N$ equals

$$A\psi(x, \pi(v))\left\{ f(x) - \sum_{k=0}^{N-1} \frac{1}{k!}\nabla^k f(\pi(v))l(x, \pi(v))^k \right\} e^{il(x,v)}\Big|_{x=\pi(v)}. \quad (28.4.12)$$

By virtue of Theorem 28.2, the expression in curly brackets in (28.4.12) vanishes to order N at $x = \pi(v)$, and hence in light of (28.4.7) the whole expression belongs to the space [143]

$$\cup_{m_0 \geq N} S^{\omega - \rho \sum_{i=1}^{k} m_i + k} \subset S^{\omega - \rho N},$$

because $\rho \sum_{i=1}^{k} m_i \geq k$. $\square$

Theorem 28.4 *For any function f of class C^∞, $\varphi \in C^\infty(M, R)$ such that $d\varphi \neq 0$ on the support of f, and any operator $A \in L_\rho^\infty$, one has*

$$e^{-it\varphi(x)}Af(x)e^{it\varphi(x)}$$
$$= \sum_k \frac{i^{k-\sum_i m_i}}{k!m_0!\ldots m_k!}t^k\left(\nabla^{m_0}f\right)\left(\nabla^{m_1}\varphi\right)\ldots\left(\nabla^{m_k}\varphi\right)$$
$$D^{\sum_i m_i}\sigma(t\, d\varphi), \quad (28.4.13)$$

with the understanding that the difference between the left-hand side and the sum of those terms on the right not belonging to $S_\rho^{-N}(M \times \mathbf{R})$ does indeed belong to the space $S_\rho^{-N}(M \times \mathbf{R})$. *Furthermore, the estimates pertaining to these statements hold uniformly for* f *belonging to a bounded set in* $C^\infty(M)$ *and* φ *belonging to a compact set.*

Proof Theorem 28.4 is the analogue of Theorem 28.2 and follows from Theorem 28.3 as Eq. (28.4.7) follows from Eq. (28.2.4) [142]. Note that, in (28.4.13), the tensor product

$$\left(\nabla^{m_0} f\right) \ldots \left(\nabla^{m_k} \varphi\right)$$

represents a covariant tensor of order $\sum_i m_i$ based at x. The last factor $D^{\sum_i m_i} \sigma$ is instead a contravariant tensor of order $\sum_i m_i$ based at x. $\square$

The next question of interest is how the symbol changes under a change a connection. The answer is as follows [143].

Theorem 28.5 *Suppose that the pseudo-differential operator A has symbols* σ_A *and* $\sigma_{A'}$ *with respect to the connections* ∇ *and* ∇', *respectively. If* λ *is the linear function associated with* ∇' *one then finds*

$$\sigma'_A = \sum_{m_1 \ldots m_k \geq 2} \frac{i^{k - \sum_i m_i}}{k! m_1! \ldots m_k!} \left(\nabla^{m_1} \lambda\right) \ldots \left(\nabla^{m_k} \lambda\right) D^{\sum_i m_i} \sigma_A. \qquad (28.4.14)$$

Proof One has to apply Theorem 28.4 with

$$f(x) = \psi(x, \pi(v)), \quad \varphi(x) = \lambda(x, v), \qquad (28.4.15)$$

where v belongs to a compact subset of $T^*(M)/0$ containing points of all fibres of $T^*(M)$. $\square$

Note that, if the operator $A \in L_\rho^\omega$, the symbols σ_A and σ'_A belong to the space S_ρ^ω, and a term of the sum corresponding to any $k > 0$ belongs to the space

$$S_\rho^{\omega - \rho \sum_i m_i + k} \subset S_\rho^{\omega - (2\rho - 1)k},$$

because each $m_i \geq 2$. Thus, the difference

$$\sigma_A - \sigma'_A \in S_\rho^{\omega - 2\rho + 1}, \qquad (28.4.16)$$

and this property suggests defining the leading symbol as an element of the quotient space

$$S_\rho^\omega / S_\rho^{\omega - 2\rho + 1},$$

which is independent of the connection. *The leading symbol is any representative of this equivalence class.*

28.5   A Smooth Linear Equation Without Solution

The question of solvability of pseudo-differential equations is another important branch of the theory of functional equations studied in our monograph, and we here begin by presenting in detail an example from the theory of partial differential equations.

The work by Lewy [94] proved that there exist linear partial differential equations with coefficients of class C^∞ which possess not a single smooth solution in any neighbourhood. The example studied therein was an equation of first order in three independent variables with complex-valued coefficients and unknown function. Lewy began his analysis by proving the following result.

Theorem 28.6 *Let x_1, x_2, y_1 be independent real variables, u a dependent complex-valued variable and $\psi(y_1)$ a real function of class C^1. Consider the linear equation*

$$\left[-\frac{\partial}{\partial x_1} - i\frac{\partial}{\partial x_2} + 2i(x_1 + ix_2)\frac{\partial}{\partial y_1} \right] u = \psi'(y_1). \qquad (28.5.1)$$

Assume that there exists a solution $u(x_1, x_2, y_1)$ of Eq. (28.5.1) in a neighbourhood $N = N(0, 0, y_1^0)$ of the point $(0, 0, y_1^0)$, with u of class C^1. The function ψ is then holomorphic at $y_1 = y_1^0$.

Proof Following Lewy [94], we integrate Eq. (28.5.1) over a circle in the neighbourhood N described by the equations

$$(x_1)^2 + (x_2)^2 = \text{const.} = y_2, \quad y_1 = \text{const.} \qquad (28.5.2)$$

The introduction of the angle θ by

$$x_1 + ix_2 = \sqrt{y_2}\, e^{i\theta} \qquad (28.5.3)$$

yields

$$\frac{\partial}{\partial x_1} + i\frac{\partial}{\partial x_2} = \alpha\left(\frac{\partial}{\partial \log \sqrt{y_2}} + i\frac{\partial}{\partial \theta} \right), \qquad (28.5.4)$$

where α is determined by applying Eq. (28.5.4) on $\log \sqrt{y_2}$. Hence one finds

$$\alpha = \frac{(x_1 + ix_2)}{y_2}, \qquad (28.5.5)$$

and

$$\int_0^{2\pi} \left(\frac{\partial}{\partial x_1} + i \frac{\partial}{\partial x_2} \right) u \, d\theta = \int_0^{2\pi} \frac{e^{i\theta}}{\sqrt{y_2}} \left[\frac{\partial}{\partial \log \sqrt{y_2}} + i \frac{\partial}{\partial \theta} \right] u \, d\theta$$

$$= \int_0^{2\pi} \frac{e^{i\theta}}{\sqrt{y_2}} \left[\frac{\partial}{\partial \log \sqrt{y_2}} + 1 \right] u \, d\theta$$

$$= 2 \frac{\partial}{\partial y_2} \int_0^{2\pi} e^{i\theta} \sqrt{y_2} \, u \, d\theta. \tag{28.5.6}$$

This formula suggests defining the function

$$U : (y_1, y_2) \to U(y_1, y_2) \equiv i \int_0^{2\pi} e^{i\theta} \sqrt{y_2} \, u \, d\theta, \tag{28.5.7}$$

for which we obtain, from Eq. (28.5.1), the linear equation

$$\left(\frac{\partial}{\partial y_1} + i \frac{\partial}{\partial y_2} \right) U = \pi \, \psi'(y_1), \tag{28.5.8}$$

subject to the condition $U(y_1, 0) = 0$ from the definition (28.5.7). Moreover, the function

$$V(y_1, y_2) = V(y) \equiv U(y_1, y_2) - \pi \, \psi(y_1) \tag{28.5.9}$$

is of class C^1 and satisfies

$$\left(\frac{\partial}{\partial y_1} + i \frac{\partial}{\partial y_2} \right) V = 0. \tag{28.5.10}$$

This means that V is a holomorphic (and hence analytic) function whose domain of existence includes all those points (y_1, y_2) for which $y_2 > 0$ is sufficiently small and y_1 is such that the triplet (x_1, x_2, y_1) lies in N if

$$(x_1)^2 + (x_2)^2 \le y_2.$$

Since $V = -\pi \, \psi$ on $y_2 = 0$ as we pointed out before, with ψ real by hypothesis, V can be continued across $y_2 = 0$ as analytic function of y. Thus, V is analytic on $y_2 = 0$, which implies analyticity of ψ at and near $y_1 = y_1^0$. $\square$

Lewy applied Theorem 28.6 in a negative sense. He considered Eq. (28.5.1) in which $\psi(y_1)$ is real and of class C^∞, but not analytic at $y_1 = y_1^0$. But then Eq. (28.5.1) can have no solution u which is of class C^1 in any neighbourhood $N = N(0, 0, y_1^0)$.

The Lewy Example of an Equation Without Solution

With the help of a periodic real function $\psi(y_1)$ of class C^∞, which is nowhere analytic, Lewy constructed a function $F(x_1, x_2, y_1)$ of class C^∞ such that the linear equation

$$\left[-\frac{\partial}{\partial x_1} - i\frac{\partial}{\partial x_2} + 2i(x_1 + ix_2)\frac{\partial}{\partial y_1} \right] u = F(x_1, x_2, y_1) \qquad (28.5.11)$$

has no solution of type H^1, no matter what open set in the variables (x_1, x_2, y_1) is taken as domain of existence. By definition, a function is said to be of type H^1 if its first partial derivatives satisfy a Hölder condition with positive exponent, provided the distance of the points involved does not exceed 1.

Lewy considered a countable set of points $P_1, P_2, \ldots$, which is dense in the (x_1, x_2, y_1)-space, and positive radii $\rho_1, \rho_2, \ldots$ such that $\lim_{n\to\infty} \rho_n = 0$, and denoted by N_j the sphere of radius ρ_j about P_j. An arbitrary open set always contains some sphere N_j. The x_1 and x_2 coordinates of the point P_j can be denoted by p_j and q_j (not related to the (p, q) variables of classical mechanics), and it is useful to define

$$c_j \equiv \max\left[j, |p_j|, |q_j| \right]. \qquad (28.5.12)$$

Lewy considered the sequences of real numbers $\varepsilon_1, \varepsilon_2 \ldots$ such that

$$\text{least upper bound}_{j=1,2,\ldots} |\varepsilon_j| < \infty, \qquad (28.5.13)$$

and he set

$$F_\varepsilon(x_1, x_2, y_1) \equiv \sum_{j=1}^\infty \varepsilon_j c_j^{-c_j} \, \psi'(y_1 + 2q_j x_1 - 2p_j x_2). \qquad (28.5.14)$$

The function F_ε is itself of class C^∞ because, if the νth derivative D^ν is constructed by termwise differentiation, one finds

$$D^\nu F_\varepsilon(x_1, x_2, y_1) = \sum_{j=1}^\infty \varepsilon_j c_j^{-c_j} \psi^{(\nu+1)}(y_1 + 2q_j x_1 - 2p_j x_2)(q_j)^{\nu_1}(-p_j)^{\nu_2} 2^{\nu_1+\nu_2}, \qquad (28.5.15)$$

where $\nu_1 + \nu_2 \leq \nu$. The series on the right-hand side of (28.5.15) converges absolutely and uniformly, since the derivative $\psi^{(\nu+1)}$ is bounded by virtue of the periodicity of ψ, and

$$\sum_{j=1}^\infty c_j^{-c_j+\nu} < \infty. \qquad (28.5.16)$$

The sequences ε with norm (28.5.13) form a complete metric space, and hence such a space is not exhausted by a countable sum of non-dense sets. This property makes it possible to prove the existence of a sequence ε^* in this space such that, for $F = F_{\varepsilon^*}$, there exists no open set of points (x_1, x_2, y_1) on which Eq. (28.5.11) has a solution of type H^1 as defined before.

If a function has the property of having first partial derivatives satisfying a Hölder condition of exponent $\frac{1}{n}$ and constant m, with $n, m = 1, 2, 3, \ldots$, let us denote by H_{nm}^1 such a feature. The property H^1 is then given by

$$H^1 = \sum_{n,m=1}^{\infty} H_{nm}^1. \tag{28.5.17}$$

All functions that vanish at P_j and satisfy H_{nm}^1 are compact, which implies that the functions of type H^1 that vanish at P_j and exist in N_j are all contained in a countable sum of compact sets. Lewy denoted by E_{jnm} the set of sequences ε such that Eq. (28.5.11) with $F = F_{\varepsilon}$ has a solution existing in N_j and of type H_{nm}^1 therein. One can always assume that the solutions vanish at P_j and, out of any sequence of solutions that belong to various ε of the set E_{jnm} and tend to a limit sequence, one can select a suitable sub-sequence of sequences ε and corresponding solutions u of Eq. (28.5.11) with $F = F_{\varepsilon}$ such that the solutions converge in N_j together with their first partial derivatives. The limit function satisfies then the limit equation in N_j and is of type H_{nm}^1, hence the set E_{jnm} is closed.

But the set E_{jnm} is non-dense. In light of its closure, it suffices to verify that every sequence of E_{jnm} is limit of sequences of the complete metric space that the sequences ε belong to, none of which lies in E_{jnm}. Note that, if α and β are elements of the set E_{jnm}, the same holds for $\frac{(\alpha - \beta)}{2}$. Consider the particular sequence consisting of Kronecker symbols

$$\delta^j \equiv \left(\delta_1^j, \delta_2^j, \ldots, \right). \tag{28.5.18}$$

Equation (28.5.11) has no solution of class C^1 in N_j for $F = F_{\delta^j}$ (see below). Thus, by virtue of the linearity of (28.5.11), none of the sequences

$$\varepsilon + \lambda \delta^j, \ \lambda \neq 0, \ \varepsilon \in E_{jnm} \tag{28.5.19}$$

can belong to E_{jnm}. Upon taking $|\lambda|$ arbitrarily small, the sequence ε of E_{jnm} is obtained as limit of sequences not in E_{jnm}. Thus, the operation

$$\sum_{j,n,m=1}^{\infty} E_{jnm}$$

does not exhaust the space of all sequences ε with (28.5.13). If ε^* is a sequence not belonging to $\sum_{j,n,m=1}^{\infty} E_{jnm}$, then Eq. (28.5.11) with $F = F_{\varepsilon^*}$ has no solution of type

H^1 in any sphere N_j, $j = 1, 2, \ldots$ or any other open set. It now remains to prove what follows.

Theorem 28.7 *The equation*

$$Lu \equiv \left[-\frac{\partial}{\partial x_1} - i\frac{\partial}{\partial x_2} + 2i(x_1 + ix_2)\frac{\partial}{\partial y_1} \right] u$$
$$= \psi'(y_1 + 2q_j x_1 - 2p_j x_2) \qquad (28.5.20)$$

has no solution of class C^1 in the sphere N_j of radius ρ_j about the point P_j.

Proof Following once more Lewy [94], we consider the transformation

$$X_1 \equiv x_1 - p_j, \quad X_2 \equiv x_2 - q_j, \quad Y_1 \equiv y_1 + 2q_j x_1 - 2p_j x_2, \qquad (28.5.21)$$

whose inverse reads as

$$x_1 = X_1 + p_j, \quad x_2 = X_2 + q_j, \quad y_1 = Y_1 - 2q_j X_1 + 2p_j X_2, \qquad (28.5.22)$$

which leads to the useful formulae for first derivatives

$$\frac{\partial}{\partial x_1} = \frac{\partial}{\partial X_1} + 2q_j \frac{\partial}{\partial Y_1}, \qquad (28.5.23a)$$

$$\frac{\partial}{\partial x_2} = \frac{\partial}{\partial X_2} - 2p_j \frac{\partial}{\partial Y_1}, \qquad (28.5.23b)$$

$$\frac{\partial}{\partial y_1} = \frac{\partial}{\partial Y_1}. \qquad (28.5.23c)$$

Thus, Eq. (28.5.20) is re-expressed in the form

$$Lu \equiv \left[-\frac{\partial}{\partial X_1} - i\frac{\partial}{\partial X_2} + 2i(X_1 + iX_2)\frac{\partial}{\partial Y_1} \right] u = \psi'(Y_1). \qquad (28.5.24)$$

If Eq. (28.5.20) had a solution of class C^1 in N_j, Eq. (28.5.24) would have a solution of class C^1 in the neighbourhood of the transform of the centre of the sphere N_j, whose new coordinates are

$$X_1 = X_2 = 0, \quad Y_1 = Y_1^0 = y_1(P_j). \qquad (28.5.25)$$

But here we can exploit Theorem 28.6 according to which, for the solution to exist, the function $\psi(Y_1)$ would have to be analytic at this point Y_1^0, which contradicts the hypothesis about ψ. $\square$

28.6   Solving Pseudo-Differential Equations

Since Lewy published his famous result, the emphasis has been gradually put on solving more complicated equations, involving pseudo-differential operators rather than linear operators [90] and the many references therein). In particular, the outstanding work by Dencker ([40] has studied the question of local solvability of classical pseudo-differential operators P on a C^∞ manifold M. For these operators, the symbol has an asymptotic expansion given by a sum of homogeneous terms. Dencker has considered, among classical pseudo-differential operators, those whose homogeneous leading symbol $p = \sigma(P)$ is of principal type, which means that

$$d \, p \neq 0 \text{ when } p = 0. \tag{28.6.1}$$

The concept of local solvability of P on a compact set $K \subset M$ means that the equation

$$Pu = v \tag{28.6.2}$$

has a local solution of distributional nature $u \in \mathcal{D}'(M)$ in a neighbourhood of K for any v of class C^∞ on M. Dencker has succeeded in proving the Nirenberg–Treves conjecture, according to which the local solvability of principal-type pseudo-differential operators is equivalent to the so-called condition (Ψ). This condition rules out sign changes, from negative to positive, of the imaginary part of the leading symbol along the oriented bicharacteristics of the real part.

We have preferred not to provide a long list of results without proof, but rather stimulate the intellectual curiosity of the general reader by describing in detail the pioneering Lewy analysis in the previous section. Excellent references on partial and pseudo-differential operators are, for example, Hörmander [79], Smirnov [127], Hörmander [80], Palais [107], Seeley [126], Hörmander [81, 82], Gilkey [70], Grubb [73], Hörmander [83–86], Evans [55], while for the ordinary differential equations of part I we still recommend the work of Coddington and Levinson [27] as a bridge between classical and modern theory.

For the readers interested in physical applications, we mention our joint work with our teacher and colleague, G. Marmo, on symbol maps in quantum mechanics [52, 53], and our attempt to describe the quantum birth of the universe by means of a pseudo-differential boundary-value problem [49, 50].

References

1. S. Agmon, *Lectures on Elliptic Boundary Value Problems* (Van Nostrand, New York, 1965)
2. L. Ahlfors, *Complex Analysis* (McGraw-Hill, New York, 1966)
3. E. Almansi, Sull'integrazione dell'equazione differenziale $\triangle^{2n} u = 0$. Ann. Matematica Ser. III(2), 1–51 (1898)
4. L. Ambrosio, *Lecture Notes on Elliptic Partial Differential Equations* (Scuola Normale Superiore, Pisa, 2013)
5. I.G. Avramidi, G. Esposito, Gauge theories on manifolds with boundary. Commun. Math. Phys. **200**, 495–543 (1999)
6. N. Aronszajn, *Polyharmonic Functions* (Clarendon Press, Oxford, 1984)
7. J.K. Beem, P.E. Ehrlich, *Global Lorentzian Geometry* (Dekker, New York, 1981)
8. J.J. Benavides Navarro, E. Minguzzi, Global hyperbolicity is stable in the interval topology. J. Math. Phys. **51**, 112504 (2011)
9. S. Bergman, M. Schiffer, *Kernel Functions and Elliptic Differential Equations in Mathematical Physics* (Academic Press, New York, 1953)
10. L. Bers, An outline of the theory of pseudoanalytic functions. Bull. Am. Math. Soc. **62**, 291–331 (1956)
11. L. Bianchi, *Lezioni di Geometria Differenziale*, vol. 1 (Enrico Spoerri, Pisa, 1902)
12. D.D. Bleecker, B. Booss-Bavnbek, *Index Theory with Applications to Mathematics and Physics* (International Press, Boston, 2013)
13. P. Borwein, S. Choi, B. Rooney, A. Weirathmuller, *The Riemann Hypothesis* (Springer, Berlin, 2006)
14. R. Caccioppoli, Limitazioni integrali per le soluzioni di un'equazione lineare ellittica a derivate parziali. Giornale Mat. Battaglini Ser. IV **80**, 186–212 (1950)
15. R. Caccioppoli, *Lezioni di Analisi Matematica, Parte 2* (Libreria Internazionale Treves, Napoli, 1951a)
16. R. Caccioppoli, Elementi di una teoria generale dell'integrazione k-dimensionale in uno spazio n-dimensionale. Atti Congresso U.M.I. **2**, 41–49 (1951b)
17. R. Caccioppoli, Misura e integrazione sugli insiemi dimensionalmente orientati. Rend. Acc. Naz. Lincei Ser. 8 **12**, 3–11 (1952a); ibid. 137–146 (1952b)
18. R. Caccioppoli, Fondamenti per una teoria generale delle funzioni pseudo-analitiche di una variabile complessa. Rend. Acc. Naz. Lincei Ser. 8 **13**, 197–204 (1952c); ibid. 321–329 (1952d)
19. R. Caccioppoli, Funzioni pseudo-analitiche e rappresentazioni pseudo-conformi delle superfici riemanniane. Ricerche Mat. **2**, 104–127 (1953)
20. S. Campanato, Proprietà di una famiglia di spazi funzionali. Ann. Scuola Normale Superiore Classe di Scienze, Terza Serie **18**, 137–160 (1964)

© Springer International Publishing AG 2017
G. Esposito, *From Ordinary to Partial Differential Equations*, UNITEXT 106,
DOI 10.1007/978-3-319-57544-5

21. S. Campanato, *Sistemi Ellittici in Forma Divergenza e Regolarità all'Interno* (Scuola Normale Superiore, Pisa, 1980)
22. J.F. Carinena, A. Ibort, G. Marmo, G. Morandi, *Geometry from Dynamics, Classical and Quantum* (Springer, Berlin, 2015)
23. S.S. Chern, *Complex Manifolds Without Potential Theory* (Springer, Berlin, 1979)
24. Y. Choquet-Bruhat, Hyperbolic partial differential equations on a manifold, *Battelle Rencontres*, ed. by C. DeWitt-Morette, J.A. Wheeler (Benjamin, New York, 1968), pp. 84–106
25. S. Christianovich, Le problème de Cauchy pour les équations non linéaires hyperboliques. Rec. Math. Moscow, N. s. 2 (1937)
26. C.H. Chu, A.T. Lau, *Harmonic Functions on Groups and Fourier Algebras*, vol. 1782. Lecture Notes in Mathematics (Springer, New York, 2002)
27. E.A. Coddington, N. Levinson, *Theory of Ordinary Differential Equations* (McGraw-Hill, New York, 1955)
28. R. Courant, D. Hilbert, *Methods of Mathematical Physics. II. Partial Differential Equations* (Wiley, New York, 1962)
29. G. Darboux, Mémoire sur les solutions singulières des équations aux dérivées partielles du premier ordre. Mémoires des Savants étrangers **27** (1880)
30. G. Darboux, *Lecons sur la Théorie Générale des Surfaces et les Applications Géometriques du Calcul Infinitésimal Première Partie. Généralités. Coordonnées Curvilignes, Surfaces Minima* (Gauthier-Villars, Paris, 1887)
31. G. Darboux, *Lecons sur la Théorie Générale des Surfaces et les Applications Géometriques du Calcul Infinitésimal. Deuxième Partie. Les Congruences et les Équations Linéaires aux Dérivées Partielles des Lignes Tracées sur les Surfaces* (Gauthier-Villars, Paris, 1889)
32. G. Darboux, *Lecons sur la Théorie Générale des Surfaces et les Applications Géometriques du Calcul Infinitésimal. Troisième Partie. Lignes Géodésiques et Courbure Géodésique. Paramètres Différentiels, Déformation des Surfaces* (Gauthier-Villars, Paris, 1894)
33. E. De Giorgi, Osservazioni relative ai teoremi di unicità per le equazioni differenziali a derivate parziali di tipo ellittico con condizioni al contorno di tipo misto. Ricerche Mat. **2**, 183–191 (1953)
34. E. De Giorgi, Su una teoria generale della misura $(r - 1)$-dimensionale in uno spazio ad r dimensioni. Ann. Mat. Pur. Appl. **36**, 191–213 (1954)
35. E. De Giorgi, Sulla differenziabilità e l'analiticità delle estremali degli integrali multipli regolari. Mem. Accad. Sci. Torino Cl. Sci. Fis. Mat. Nat. Ser. 3 **3**, 25–43 (1957)
36. E. De Giorgi, *Appunti di Analisi Matematica* (Scuola Normale Superiore, Pisa, 1968); published as *Semicontinuity Theorems in the Calculus of Variations* (Quaderno 56 Accademia Pontaniana, Napoli, 2009)
37. E. De Giorgi, F. Colombini, L. Piccinini, *Frontiere Orientate di Misura Minima e Questioni Collegate* (Editrice Tecnico Scientifica, Pisa, 1972)
38. E. Delassus, Sur les équations linéaires aux derivées partielles à caracteristiques réelles, Ann. Scient. E.N.S. 3e Serie **12**, 53–123 (1895)
39. E. Delassus, Sur les équations linéaires aux derivées partielles, Ann. Sci. E.N.S. 3e Serie **13**, 339–365 (1896)
40. N. Dencker, The resolution of the Nirenberg–Treves conjecture. Ann. Math. **163**, 405–444 (2006)
41. J. Dieudonné, *Calcul Infinitesimal* (Hermann, Paris, 1980)
42. F.G. Dressel, The fundamental solution of the parabolic equation. *Duke Math. J.* **7**, 186–203 (1940); **13**, 61–70 (1946)
43. H.M. Edwards, *Riemann's Zeta Function* (Academic Press, New York, 1974; Dover, New York, 2001)
44. S.D. Eidelman, On fundamental solutions of parabolic systems. Math. Sbornik **38**, 51–92 (1956)
45. A. Enneper, Analytisch-geometrische untersuchungen. Z. Math. Phys. **9**, 96–125 (1864)
46. A. Erdélyi, The Fuchsian equation of second order with four singularities. Duke Math. J. **9**, 48–58 (1942)

47. V.P. Ermakov, *Univ. Izv. Kiev, Ser. III* **9**, 1 (1880)
48. G. Esposito, Mathematical structures of space-time. Fortschr. Phys. **40**, 1–30 (1992)
49. G. Esposito, Non-local boundary conditions in Euclidean quantum gravity. Class. Quantum Grav. **16**, 1113–1126 (1999a)
50. G. Esposito, New kernels in quantum gravity. Class. Quantum Grav. **16**, 3999–4010 (1999b)
51. G. Esposito, C. Stornaiolo, A new family of gauges in linearized general relativity. Class. Quantum Grav. **17**, 1989–2005 (2000)
52. G. Esposito, G. Marmo, G. Sudarshan, *From Classical to Quantum Mechanics* (Cambridge University Press, Cambridge, 2004)
53. G. Esposito, G. Marmo, G. Miele, G. Sudarshan, *Advanced Concepts in Quantum Mechanics* (Cambridge University Press, Cambridge, 2015)
54. G. Esposito, E. Battista, E. Di Grezia, Bicharacteristics and Fourier integral operators in Kasner space-time. Int. J. Geom. Methods Mod. Phys. **12**, 1550060 (2015)
55. L.C. Evans, *Partial Differential Equations* (American Mathematical Society, Providence, 2010)
56. H. Federer, *Geometric Measure Theory* (Springer, Berlin, 1969)
57. G. Fichera, Analisi esistenziale per le soluzioni dei problemi al contorno misti, relativi all'equazione ed ai sistemi di equazioni del secondo ordine di tipo ellittico, autoaggiunti. Ann. Scuola Norm. Pisa, Ser. 3 **1**, 75–100 (1949)
58. A.R. Forsyth, *Theory of Differential Equations, Vol. 1: Exact Equations and Pfaff's Problem* (Cambridge University Press, Cambridge, 1890)
59. A.R. Forsyth, *Theory of Differential Equations, Vol. 2: Ordinary Equations, Not Linear* (Cambridge University Press, Cambridge, 1900)
60. A.R. Forsyth, *Theory of Differential Equations, Vol. 3: Ordinary Equations, Not Linear* (Cambridge University Press, Cambridge, 1900)
61. A.R. Forsyth, *Theory of Differential Equations, Vol. 4: Ordinary Linear Equations* (Cambridge University Press, Cambridge, 1902)
62. A.R. Forsyth, *Theory of Differential Equations, Vol. 5: Partial Differential Equations* (Cambridge University Press, Cambridge, 1906a)
63. A.R. Forsyth, *Theory of Differential Equations, Vol. 6: Partial Differential Equations* (Cambridge University Press, Cambridge, 1906b)
64. Y. Fourès-Bruhat, Théorème d'existence pour certains systèmes d'équations aux dérivées partielles non linéaires, Acta Math. **88**, 141–225 (1952). English translation, by G. Esposito, published in *Max Planck Institute for the History of Science, Preprint Series*, number 480 (2016)
65. F.G. Friedlander, *The Wave Equation on a Curved Space-Time* (Cambridge University Press, Cambridge, 1975)
66. M. Gaczkowski, P. Gorka, Harmonic functions on metric measure spaces: convergence and compactness. Potential Anal. **31**, 203–214 (2009)
67. P.R. Garabedian, *Partial Differential Equations* (Chelsea, New York, 1964)
68. R.P. Geroch, Singularities, in *Relativity*, ed. by M. Carmeli, S.I. Fickler, L. Witten (Plenum Press, New York, 1970), pp. 259–291
69. R.P. Geroch, Domain of dependence. J. Math. Phys. **11**, 437–449 (1970b)
70. P.B. Gilkey, *Invariance Theory, the Heat Equation and the Atiyah-Singer Index Theorem* (CRC Press, Boca Raton, 1995)
71. E. Goursat, Démonstration du theorème de Cauchy. Acta Math. **4**, 197–200 (1884)
72. E. Goursat, *A Course in Mathematical Analysis, vol. 2, Part 2. Differential Equations* (Dover, New York, 1917)
73. G. Grubb, *Functional Calculus of Pseudo-Differential Boundary Problems* (Birkhauser, Boston, 1996)
74. J. Hadamard, *Lectures on Cauchy's Problem in Linear Partial Differential Equations* (Yale University Press, New Haven, 1923; Dover, New York, 1952)
75. S.W. Hawking, Stable and generic properties in general relativity. Gen. Rel. Grav. **1**, 393–401 (1971)

76. S.W. Hawking, G.F.R. Ellis, *The Large-Scale Structure of Space-Time* (Cambridge University Press, Cambridge, 1973)

77. E.R. Hedrick, Non-analytic functions of a complex variable. Bull. Am. Math. Soc. **39**, 75–96 (1933)

78. K. Heun, Zur theorie der Riemann's sehen functionen zweiter ordnung mit vier verzweigungspunkten. Math. Ann. **33**, 61 (1889)

79. L. Hörmander, On the theory of general partial differential operators. Acta Math. **94**, 161–248 (1955)

80. L. Hörmander, Pseudo-differential operators. Commun. Pure Appl. Math. **18**, 501–517 (1965)

81. L. Hörmander, Fourier Integral Operators. I. Acta Math. **127**, 79–183 (1971)

82. L. Hörmander, The Weyl calculus of pseudo-differential operators. Commun. Pure Appl. Math. **32**, 359–443 (1979)

83. L. Hörmander, *The Analysis of Linear Partial Differential Operators. I. Distribution Theory and Fourier Analysis* (Springer, Berlin, 2003)

84. L. Hörmander, *The Analysis of Linear Partial Differential Operators. II. Differential Operators with Constant Coefficients* (Springer, Berlin, 2005)

85. L. Hörmander, *The Analysis of Linear Partial Differential Operators. III. Pseudo-Differential Operators* (Springer, Berlin, 2007)

86. L. Hörmander, *The Analysis of Linear Partial Differential Operators. IV. Fourier Integral Operators* (Springer, Berlin, 2009)

87. J.K. Hunter, *Notes on Partial Differential Equations* (University of California at Davis, Davis, 2014)

88. O.D. Kellogg, *Foundations of Potential Theory* (Dover, New York, 1953)

89. J. Leray, *Hyperbolic Differential Equations* (Institute for Advanced Study, Princeton, 1953)

90. N. Lerner, Solving pseudo-differential equations, in *Proceedings of The International Congress of Mathematicians*, vol. II, pp. 711–720, Beijing, arXiv:math/0304335 (2002)

91. J. Le Roux, Sur les intégrales des équations linéaires aux dérivés partielles du second ordre a deux variables indépendantes. Ann. Scient. E.N.S.3ᵉ Serie **12**, 227–316 (1895)

92. T. Levi-Civita, *Lezioni di Calcolo Differenziale Assoluto* (Spoerri, Pisa, 1925); *The Absolute Differential Calculus* (Blackie & Son, London, 1926; Dover, New York, 1977)

93. T. Levi-Civita, *Caratteristiche dei Sistemi Differenziali e Propagazione Ondosa* (Zanichelli, Bologna, 1931)

94. H. Lewy, An example of a smooth linear partial differential equation without solution. Ann. Math. **66**, 155–158 (1957)

95. E.H. Lieb, M. Loss, *Analysis, Graduate Studies in Mathematics 14* (American Mathematical Society, Providence, 1997)

96. G. Lysik, On the mean-value property for polyharmonic functions. Acta Math. Hung. **133**, 133–139 (2011)

97. G. Lysik, A characterization of polyharmonic functions. Acta Math. Hung. **147**, 386–395 (2015)

98. E. Magenes, G. Stampacchia, I problemi al contorno per le equazioni differenziali di tipo ellittico. Ann. Scuola Normale Superiore Ser. **12**(3), 247–358 (1958)

99. F. Maggi, *Finite Perimeter Sets and Geometric Variational Problems. An Introduction to Geometric Measure Theory* (Cambridge University Press, Cambridge, 2012)

100. C. Miranda, *Partial Differential Equations of Elliptic Type* (Springer, Berlin, 1970)

101. C. Miranda, *Istituzioni di Analisi Funzionale Lineare* (Unione Matematica Italiana, Bologna, 1972)

102. C.B. Morrey, Second order elliptic system of differential equations. Ann. Math. Stud. **33**, 101–159 (1954)

103. J. Moser, A new proof of De Giorgi's theorem concerning the regularity problem for elliptic differential equations. Commun. Pure Appl. Math. **13**, 457–468 (1960)

104. R. Narasimhan, *Analysis on Real and Complex Manifolds* (North Holland, Amsterdam, 1968)

105. J. Nash, Continuity of solutions of parabolic and elliptic equations. Am. J. Math. **80**, 931–954 (1958)

106. M. Nicolesco, *Les Fonctions Polyharmoniques* (Hermann, Paris, 1936)
107. R.S. Palais, *Foundations of Global Non-Linear Analysis* (Benjamin, New York, 1968)
108. R. Penrose, *Techniques of Differential Topology in Relativity* (Society for Industrial & Applied Mathematics, Bristol, 1983)
109. I.G. Petrovsky, *Lectures on Partial Differential Equations* (Interscience, New York, 1954; Dover, New York, 1991)
110. E. Pinney, The nonlinear differential equation $y'' + p(x)y' + cy^{-3} = 0$. Proc. Am. Math. Soc. **1**, 681 (1950)
111. H. Poincaré, Sur les fonctions fuchsiennes. Comptes R. Acad. Sci. **92**, 333–335 (1881a)
112. H. Poincaré, Sur les fonctions fuchsiennes. Comptes R. Acad. Sci. **92**, 395–398 (1881b)
113. H. Poincaré, Sur les fonctions fuchsiennes. Comptes R. Acad. Sci. **92**, 1484–1487 (1881c)
114. H. Poincaré, Sur les intégrales irrégulières des équations linéaires. Acta Math. **8**, 295–344 (1886)
115. H. Poincaré, Les fonctions fuchsiennes et l'équation $\Delta u = e^u$. Comptes R. Acad. Sci. **136**, 627–630 (1898a)
116. H. Poincaré, Les fonctions fuchsiennes et l'équation $\Delta u = e^u$. J. Math. Cinq. Série **4**, 137–230 (1898b)
117. H. Poincaré, Fonctions modulaires et fonctions fuchsiennes. Ann. Fac. Sci. Toulouse. Ser. 3(3), 125–149 (1912)
118. C. Pucci, Discussione sul problema di Cauchy per le equazioni di tipo ellittico. Ann. Mat. Pura Appl. **46**, 131–153 (1958)
119. G. Ricci-Curbastro, Di alcune applicazioni del Calcolo differenziale assoluto alla teoria delle forme differenziali quadratiche binarie e dei sistemi a due variabili. Atti Ist. Veneto Serie VII **IV**, 1336–1364 (1893)
120. G. Ricci-Curbastro, Sulla teoria intrinseca delle superficie ed in ispecie di quelle di secondo grado. Atti Ist. Veneto Serie VII **VI**, 445–488 (1895)
121. G. Ricci-Curbastro, *Lezioni sulla Teoria delle Superficie* (Drucker, Padova, 1898)
122. B. Riemann, On the number of prime numbers less than a given quantity (1859), in *Oeuvres Mathématiques de Riemann*, ed. by L. Laugel (Gauthier-Villars, Paris, 1898; Jacques Gabay, Paris, 1990)
123. H. Ringström, Origin and development of the Cauchy problem in general relativity. Class. Quantum Grav. **32**, 124003 (2015)
124. A. Ronveaux, *Heun's Differential Equations* (Oxford University Press, Oxford, 1995)
125. L. Schwartz, *Cours d'Analyse* (Hermann, Paris, 1967)
126. R.T. Seeley, Topics in pseudo-differential operators, in *CIME Conference on Pseudo-Differential Operators*, ed. by L. Nirenberg (Edizioni Cremonese, Roma, 1969), pp. 167–305
127. V.I. Smirnov, *A Course of Higher Mathematics, vol. 4, part 2: Partial Differential Equations* (Pergamon Press, Oxford, 1964)
128. S.L. Sobolev, On some estimates of families of functions having square-integrable derivatives. Dokl. Akad. Nauk SSSR **1**, 267–270 (1936a)
129. S.L. Sobolev, Méthode nouvelle a résoudre le problème de Cauchy pour les équations linéaires hyperboliques normales. Rec. Math. Moscou, N. s. 1 (1936b)
130. S.L. Sobolev, On a theorem of functional analysis. Am. Math. Soc. Transl. Ser. 2(34), 39–68 (1963a)
131. S.L. Sobolev, *Applications of Functional Analysis in Mathematical Physics*, vol. 7. Translations of Monographs Series (American Mathematical Society, Providence, 1963b)
132. L.W. Thomé, Zur theorie der linearen differentialgleichungen. J. Reine Angew. Math. **75**, 265–291 (1873)
133. F. Treves, *Introduction to Pseudo-differential and Fourier Integral Operators*, vol. 2 (Plenum Press, New York, 1980)
134. G. Valiron, *Équations Fonctionnelles. Applications* (Masson, Paris, 1945)
135. I. Vekua, *Generalized Analytic Functions* (Pergamon Press, Oxford, 1962)
136. H. von Koch, Sur la distribution des nombres premiers. Acta Math. **24**, 159–182 (1901)

137. H. von Mangoldt, Zur Verteilung der Nullstellen der Riemannschen Funktion $\xi(s)$. Math. Ann. **60**, 1–19 (1905)

138. R.M. Wald, *General Relativity* (Chicago University Press, Chicago, 1984)

139. K. Weierstrass, Untersuchungen über die flächen deren miettlere krümmung überall gleich null ist. Monats Berl. Akad. 612, 855 (1866)

140. K. Weierstrass, Über eine besondere gattung von minimalflächen. Monats Berl. Akad., p. 511 (1867)

141. E.T. Whittaker, G.N. Watson, *Modern Analysis* (Cambridge University Press, Cambridge, 1927)

142. H. Widom, Families of pseudo-differential operators, in *Topics in Functional Analysis*, ed. by I. Gohberg and M. Kac (Academic Press, New York, 1978), pp. 345–395

143. H. Widom, A complete symbolic calculus for pseudo-differential operators. Bull. Sci. Math. Ser. **104**(2), 19–63 (1980)

144. F. Zucca, The mean value property for harmonic functions on graphs and trees. Ann. Mat. Pura Appl. **181**(4), 105–130 (2002)

Index

© Springer International Publishing AG 2017
G. Esposito, *From Ordinary to Partial Differential Equations*, UNITEXT 106,
DOI 10.1007/978-3-319-57544-5

Printed in the United States
By Bookmasters